**McGRAW-HILL
BOOK COMPANY**

New York
St. Louis
San Francisco
Düsseldorf
Johannesburg
Kuala Lumpur
London
Mexico
Montreal
New Delhi
Panama
Paris
São Paulo
Singapore
Sydney
Tokyo
Toronto

HORACE ROBERT BYERS, Sc.D.

*Professor of Meteorology
Texas A & M University*

General
Meteorology

FOURTH EDITION

This book was set in Times New Roman.
The editors were Robert H. Summersgill, Brete C. Harrison, and Susan Gamer;
the cover was designed by Joseph Gillians;
and the production supervisor was Thomas J. LoPinto.
New drawings were done by John Cordes, J & R Technical Services, Inc.
The Maple Press Company was printer and binder.

Library of Congress Cataloging in Publication Data

Byers, Horace Robert, date
    General meteorology.

    First ed. published in 1937 under title: Synoptic
and aeronautical meteorology.
    1. Meteorology. I. Title.
QC861.2.B9 1974   551.5   73-19886
ISBN 0-07-009500-0

**GENERAL
METEOROLOGY**

CONTENTS

PREFACE TO THE FOURTH EDITION

New meteorological knowledge continues to unfold at a rapid rate, offering a challenge to the teacher, the professional, and the textbook writer to keep pace with the developments. Since 1959, when the third edition was published, new areas of interest have developed and our understanding of old concepts has improved. Better quantitative treatments have become possible. At the same time, students entering the field are better equipped to grasp its principles and those of many other fields of science and technology.

This book is intended to present the broad picture of atmospheric processes to the serious student of meteorology before he takes up more complex concepts of fluid mechanics and meteorological analysis and before he branches out into the various meteorological specialties.

Major changes made in this edition include the following: (1) The former chapters on instruments and observations, which are usually treated in a separate course, have been placed in appendixes in order to provide better continuity in the presentation. (2) Exercises have been added at the end of most chapters. (3) An appendix covering the system of units and giving some useful constants has been added. (4) Some of the previous chapters have been combined into one chapter instead of two. (5) An Introduction has been added to lay some groundwork for the course.

(6) Introductory remarks have been added for various chapters and topics to provide a more connected presentation.

Again, the author is indebted to his colleagues, and others, for suggestions and assistance in putting together the material for this book and, in some cases, critically reading the manuscript.

HORACE ROBERT BYERS

EXCERPT FROM THE PREFACE TO THE THIRD EDITION

This book, written at about the level of some of the textbooks of general physics, is intended for the serious student of physical sciences and technology. It presupposes that the reader has taken elementary courses in mechanics and heat and that he has some familiarity with or is concurrently learning calculus. It can form the basis of a course in meteorology to be given as early as the sophomore year for those technical students who have seriously prepared for it, or it may be given as late as the senior year.

This third edition differs from the earlier ones in several aspects. Primarily, of course, it brings the material up to date in this rapidly changing field. Also, the book is now more fundamental in nature than previously. The emphasis on basic material is necessary in order to place the student in a position to branch out into areas of meteorology in which quantitative methods have only recently been used. Furthermore, the applications have become so specialized that a separate book is needed to treat each one, and such specialized books have been appearing in increasing numbers. The author of a book such as this has the satisfaction of knowing that the applications become outdated more rapidly than do the fundamentals.

HORACE ROBERT BYERS

1

INTRODUCTION

When astronauts describe their experiences and sensations of space travel, they invariably speak of the majestic beauty of the planet Earth. The most prominent features they see are the great cloud masses marking the major storm areas, covering a large part of the continents and oceans in delicate white veils or dense bands. The clouds are embedded in the faintly tinted blue atmosphere, which deepens in hue around the rim of the planet where the line of sight goes tangentially through a longer atmospheric path.

If time-lapse pictures are taken of the earth, it is noted that the cloud patterns are continually changing. There are swirling masses of clouds in the higher latitudes which can be recognized as tracing the air motion along fronts, and spiraling around the low-pressure disturbances. From the equatorial regions dense white masses may be sending off prominent streams of clouds into middle latitudes.

It is apparent that the earth has a restless atmosphere. That is why we must consider its properties not only in their spatial distributions but also in terms of changes with time. With all its waves and whirls the atmosphere nevertheless rotates with the earth as an integral part of the system. Its internal motions serve to balance the unequal heating and cooling of different parts of the earth.

In this book we shall examine how the heat from the sun is distributed, concentrated in low latitudes, to provide the thermal drive for the atmospheric motions; how the rotation of the earth affects these motions; and how storminess becomes an essential part of the picture.

SOME CHARACTERISTICS OF THE STUDY

We shall place our emphasis on the atmosphere up to 35 km, which is that part in which the activity related to weather occurs.   This spherical shell, having a thickness with respect to the size of the earth that is comparable with the skin of an apple, is the seat of the most active natural phenomena that man can observe.   The atmosphere may be regarded as a huge thermodynamic engine operating between an equatorial heat source and a sink at the poles.   Its circulation is complicated by special effects produced by the rotation of the earth and by reactions to various local drives.

The weather map appearing on television or printed in the newspapers is a greatly simplified version of the complicated, interrelated, three-dimensional structures that evolve in time, which we shall be discussing—the substratospheric jet streams, the wave-like perturbations in the free atmosphere, the great air exchanges that make up the general circulation, as well as the details of modifications caused by phase changes of the water substance, frictional effects, and local heat sources and sinks.   All these phenomena can be described in exact mathematical-physical terms. Meteorologists, aided by various forms of automation and advanced numerical computers, are approaching a state of their science where complete and accurate predictions might be possible.

A glance at pages of this book selected at random will indicate the kinds of mathematical-physical relations, diagrams, and map representations used to describe properties and processes.   If the student becomes confused, it is not surprising, because he may be accustomed to studying science through controlled experiment rather than in an open system that is running without control.   If he finds the drama of the atmosphere exciting and the problems of meteorology challenging, then he has experienced some of the fascination of the subject.

NAMING THE SCIENCE

The Greeks had the word *meteōrologia*, from *meteōros*, meaning things up above, plus *logos*, translated as discourse; but today the study of the atmosphere has become so divided into specialties that the all-inclusive word *meteorology* which we have perpetuated from the ancient Greeks does not satisfy everybody.   Thus we have the word *aeronomy* for the study of that part of the atmosphere where ionization and dissociation are important, generally above about 35 km, while *atmospheric science(s)* is favored by some as the inclusive designation. The former International Association of Meteorology has *and Atmospheric Physics* added to its name, and some university meteorology departments have dropped the broad term meteorology in favor of words that mean the same but suggest something different.

Part of the motive in adopting new names, freely admitted by those who have suggested them, is to get rid of a designation that is misunderstood by the general public and even by some scientists. A meteorologist is regarded by too many people as a person who displays a crude map and gives a weather forecast on television or, at best, as a forecaster in a government weather office. His efforts are seen as unscientific or, in some cases, downright amusing. Even the sophisticated citizen is likely to view meteorology as a largely descriptive subject bearing on the study of geography. Then there are those who think meteorologists are a breed of astronomers who study meteors.

The misunderstanding is not helped by the treatment of the subject in some high school and junior college courses. It is often included in a teaching unit with other descriptive material described as "earth sciences." Indeed, geographers, because of their interest in weather and climate as elements in shaping the physical and cultural landscape, are often called upon to teach it.

The student, and perhaps teachers, may be surprised to find that meteorology is a specialized branch of advanced physics, using sophisticated mathematical tools and leaning heavily on all the physical sciences. Chiefly it involves electromagnetic radiation theory, thermodynamics, classical mechanics, physics of fluids, physical chemistry, and boundary-layer theory. If the upper atmosphere is included, then solar physics, spectroscopy, plasma physics, ionization, particle physics, x-ray phenomena, optics, cosmic-ray physics, excitation phenomena, electrodynamics, magnetohydrodynamics, radio propagation, and other related processes must be understood.

CHARTING OUR COURSE

The chapters of this book present topics in a sequence that leads from basic forces to the complicated interactions of the atmosphere. First, in Chap. 2, we consider the source of the energy that drives the atmosphere—the sun—and how its energy is received by the earth. To understand its effects and the response of the earth to the resulting uneven heating, the student learns some fundamental concepts of radiation physics, then applies these to an accounting of the heat budget of the earth and its atmosphere (Chap. 3). With this background, we then take a somewhat detailed look, in Chap. 4, at the distribution of temperature in the global atmosphere.

Air behaves as would any of the ordinary gases under heating and cooling, expansion and compression; so in Chap. 5 we explore gas thermodynamics and learn how and why the pressure changes with height and under what conditions the atmosphere can become thermally unstable, as in thunderstorms, or strongly stable, as in fogs and air-pollution incidents. The air carries a most important substance, water, usually in the invisible-vapor form, but through condensation in clouds and occasional falling as rain or snow, the water not only moistens the land but also produces formidable thermodynamic changes in the atmosphere (Chap. 6).

Having understood the thermal or thermodynamic machinery that drives the atmosphere, we next study the resulting air motions (Chap. 7). Here we encounter something entirely new to most students of physics—a noninertial or apparent force. Because of the rotation of the earth and the need to represent motions in this rotating frame of reference, an apparent deflecting force arises. We show how this force relates to other forces and what the motions (the winds) look like when represented on a weather map. This leads us to more sophisticated concepts of fluid motions applicable to weather prediction and to an understanding of global circulations (Chap. 8).

With our knowledge of thermodynamics and mechanics of the atmosphere, we are ready to examine and explain the general circulation of the air (Chap. 9), which serves to keep various parts of the earth in temperature equilibrium over the long term. We discover the great importance of perturbations (storms) in exchanging air between equatorial and polar regions. The alternating bursts of cold or warm air accompanying these disturbances constitute the mechanism of the weather observed in all latitudes except the tropics (Chap. 10).

We are now looking at day-to-day weather, the tropical and polar air masses and their contrasting properties (Chap. 11). Along with this view of weather, we learn something about the analysis of weather charts for the various levels and in vertical cross sections (Chap. 12). Then, to complete our picture of the weather, we turn our attention to the tropics, where we see a different set of weather problems and the most dramatic phenomenon of all, the hurricane or typhoon (Chap. 13).

We come next to what all the preceding has been leading to—weather prediction (Chap. 14). Although hints at various elements of prediction have been brought up before, we now concentrate on some of the modern numerical methods, such as in the big U.S. National Weather Service prediction center.

In our final two chapters (Chaps. 15 and 16) we examine important smaller-scale phenomena. We consider the physics of clouds and precipitation. (It is on this subject that weather modification is based.) And finally we summarize what is known about thunderstorms and other severe weather manifestations.

The descriptive and quantitative details revealed in the text are more meaningful if accompanied by practical work in observing, measuring, analyzing, and predicting the weather. If the student is not concurrently getting such practice in another course, he should give some attention to displays of weather charts and whatever other materials are at hand. In Appendixes A and B the essentials of common instruments and observations are presented. These can be covered in a separate course, but the student who is lacking such exposure would do well to refer to the appendixes for information on the sources of meteorological data. Appendix C is helpful in providing essential information for quantitative applications and problem solving.

EXERCISES

1 Review the system of units summarized in Appendix C, and

(a) Construct a graph relating Fahrenheit temperature to Celsius temperature from -60 to $40°C$.

(b) Change 600 cal g^{-1} (the heat of vaporization of water near 0°C) to joules per kilogram and 2 g-cal cm^{-2} min^{-1} (the approximate solar heating outside the atmosphere) to watts per square meter.

(c) Convert 8.5×10^5 dynes cm^{-2} to newtons per square meter; to millibars; to the equivalent millimeters of mercury; to the equivalent inches of mercury.

(d) Convert grams per cubic centimeter to kilograms per cubic meter.

(e) Convert miles per hour to meters per second.

(f) Compare the volume of an average cloud droplet 5 μm in radius with a raindrop 2 mm in radius and a marble-size hailstone 1 cm in radius; then think about how many cloud droplets have to come together to make a raindrop or a hailstone.

2

THE SUN, THE EARTH, AND RADIATION

That great thermonuclear reactor, the sun, is the original source of the energy of the atmosphere. When we burn coal in our homes and industrial plants, we are using stored-up energy from the sun. Coal represents the fossilized remains of forests that grew thousands of years ago and, like the forests of today, required sunlight for their growth. Other forms of heat and mechanical energy also are derived from the sun indirectly: oil and its refined products, natural gases, and many chemicals were produced by the sun acting on what once were living organisms. Waterfalls owe their energy to the elevation of water in the evaporation-precipitation cycle driven by the atmospheric energy derived from the sun. In fact, even our own existence depends on the sun, for without it there would be no plants for food, no fish in the sea, and no animals or other living creatures on earth.

Except as man has learned to duplicate on a relatively minute scale the approximate thermonuclear reactions of the sun, we owe to this one relatively small star essentially all forms of energy available on this planet. The energy furnished us by other stars and celestial bodies is negligible by comparison, as is also the heat coming out from the hot interior of the earth. Only the lunar tides represent an appreciable source of energy on the earth not derived from the sun, but they are several orders of magnitude smaller than the solar radiant energy. The sun is of great importance to meteorology; all natural phenomena can be traced to the manner in which the energy from the sun is received and utilized over different parts of the earth.

Most of the sun's energy is wasted, as far as we are concerned, by passing out into endless space in all directions; only an infinitesimally small portion of the whole output is intercepted by the earth. Yet to us this relatively small amount of heat which we receive represents a huge store of energy.

Because of the importance of the sun every student of meteorology should be familiar with it, particularly in its relationship to the earth. For the purposes of this book only a few outstanding facts will be considered. Elementary textbooks of astronomy usually contain a fairly complete chapter on the sun, and the student is referred to them.

FEATURES OF THE SUN

The sun, having a mass some 330,000 times that of the earth and a radius about 110 times the earth's, is at a mean distance of very nearly 1.5×10^8 km—about 93 million miles—with the earth and other planets in orbit around it. Only the solar atmosphere, representing a tiny fraction of its mass and volume, is accessible to observation, yet there are distinct layers in this atmosphere. The *photosphere*, having a depth of about 0.0005 solar radius, covers the normally visible disk and is the direct source of practically all the observed solar radiation. Surrounding it to an additional thickness of about 0.02 solar radius is the *chromosphere* consisting of relatively transparent gases in a more or less homogeneous layer from which emerge spikes or spicules. Its existence was first noted in the eclipse of 1870. Beyond it and extending outward without fixed limits is the *corona*, a pearly veil of extremely hot gas, mostly in highly ionized, atomic form, seen only at times of eclipse or with a coronagraph, which is an instrument especially designed to display the corona without the aid of an eclipse.

While the temperature at the center of the sun is estimated to be around 15 million Kelvin, the photosphere, from which most of the detectable radiation is emitted, has a temperature ranging from 7300 K at the bottom to 4500 K at the top. The effective blackbody temperature, which is the temperature a perfect radiator would have in order to produce the measured luminance, is about 5800 K.

The sun emits radiation through the entire electromagnetic spectrum from x-rays and cosmic rays to radio waves up to wavelengths of 15 m or more. The radio emissions originate in regions over large sunspots and are especially intense at times of flares. Our eyes are constructed to make maximum use of the sun; we see in the part of the spectrum where the sun emits its maximum energy. Its infrared rays provide us with heat.

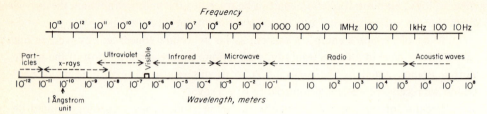

FIGURE 2-1
The electromagnetic spectrum.

A layout of the electromagnetic spectrum, on a logarithmic scale, is displayed in Fig. 2-1. Frequency v is related to the wavelength λ by the obvious relation $c = v\lambda$, where c is the speed of propagation of the waves, which is at the speed of light—approximately 3×10^8 m per sec. Use is made also of the wave number, which is the number of waves in a unit distance, or $1/\lambda$. The unit of frequency ($\sec^{-1}$) is expressed as the hertz, abbreviated Hz.

Some of the radiation is in what is called the visible range; that is, it can be detected by the human eye. That of a wavelength too short to be observed by eye may be classed as *ultraviolet* radiation, and the long-wave type as *infrared* radiation. Just as in the case of light radiation, all radiant energy travels in straight paths through space and, at all wavelengths, with the speed of light. The wavelengths of solar and terrestrial radiation are of the order of 10^{-6} m, called microns (abbreviated μm), but sometimes given in 10^{-9} m (nanometers, nm) or in a special radiation unit of 10^{-10} m called the angstrom (Å). The visible range lies between about 0.4 and 0.7 μm (400 to 700 nm, 4000 to 7000 Å). In the main body of the atmosphere the wavelengths of about 0.1 to 30 μm are the practically important ones, but in the high atmosphere photons in the ultraviolet, x-rays, and particle radiations produce important reactions.

MOTIONS OF THE EARTH

The movements of the earth with which we are most familiar are its rotation about an axis through the poles and its revolution in an orbit around the sun. Both these motions are in a counterclockwise direction if viewed from over the North Pole; that is, the rotation is from west to east, as is also the revolution. Our 24-hr day and timepieces are based on the time required for the earth to make one rotation with respect to the sun. This unit of time is called the *solar day*. We adjust our calendars to the period of the earth's orbit around the sun—365.242 solar days.

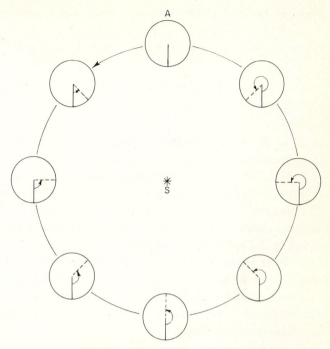

FIGURE 2-2
The sidereal day in relation to motions of the earth.

The true speed of the earth's rotation is not determined with reference to the sun, because as the earth proceeds in its orbit, it turns a little more than one rotation to "look back" at the sun. After the earth in 1 year has gone through its complete orbit, the additional turning has totaled a full rotation. The effect is shown in Fig. 2-2, where the earth is viewed from above the North Pole. As the earth moves along in its orbit in the direction indicated by the arrow, a selected meridian represented by the heavy radial line is shown at *A* to be facing the bottom of the page and also facing the sun. At a later stage in its orbit, when that same meridian faces the sun, the earth has gone beyond the 360° of rotation that would be determined by the bottom of the page (a fixed reference point). The angle represented by the small internal arrow is the additional angle of rotation required to line up the meridian with the sun. It is seen that as the earth goes through its orbit, the additional fractions of a rotation add up to 180° when the earth is opposite *A*; and when in 1 year it returns to *A*, the added angle has amounted to the full rotation. Considering the year of 365.242 solar days, it is seen that the earth has turned in each solar day 1/365.242 of a rotation beyond one. The true rotation time is called the *sidereal day*, and if we designate by S the number of hours, minutes, and seconds in a sidereal day, then

$$S + \frac{1}{365.242} S = 24$$

$$S = \frac{24}{1 + (1/365.242)}$$

$$= 23.93447 \text{ hr}$$

or 23 hr 56′4.09″.   Thus the actual rate of rotation, once per sidereal day, is slightly faster than our timepieces indicate.

When we come to considering air motions over the earth—the winds—we shall find that the rotation of the earth has a profound effect, and we will have to use the true speed of 2π rad per sidereal day.   Incidentally, when we speak precisely of the year, we should say it has 365.242 *mean* solar days, because the orbit is not circular as in Fig. 2-2 but is slightly elliptical.

There are two movements of the earth that are of little consequence in meteorology.   One is the precessional motion, which is a slow conical movement of the axis, once around in 26,000 years.   The other is the solar motion, in which the whole solar system, including the earth, is flying through space at 19 km per sec in the general direction of the star Vega.

The rotation and revolution of the earth are the motions of most significance in meteorology.   The rotation indirectly accounts for the diurnal changes in the weather, such as the warming up during the daytime and the cooling off at night.   Furthermore the rotation imposes on the earth and in the atmosphere an acceleration second in importance only to the gravitational acceleration, namely, the centripetal acceleration.

THE SEASONS

The revolution of the earth is associated with the seasonal changes.   If the plane of the orbit were in the plane of the earth's equator, there would be very little seasonal change.   At perihelion, when the earth describing its ellipse in space reaches the major axis at the end nearest the sun, the greatest intensity of solar radiation would be received; and when the earth is at aphelion, which is the farthest end of the major axis, a minimum of solar heating would be experienced.   This difference in amount of solar radiation received over the earth as a whole is extremely small compared with the seasonal variations known to exist from a different cause which we now investigate.

A study of Fig. 2-3 reveals the real explanation of the seasons.   The plane of the equator is inclined at an angle of $23\frac{1}{2}°$ from the plane of the earth's orbit.   This means that the axis is inclined at an angle of $23\frac{1}{2}°$ from the perpendicular to the plane of the orbit.   The direction toward which the axis is inclined is very nearly in the major

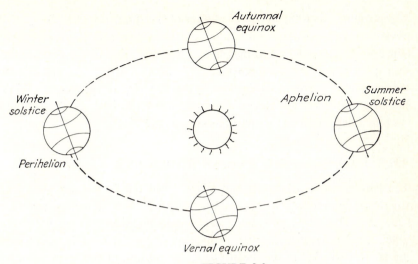

FIGURE 2-3
The revolution of the earth and the seasons.

axis of the ellipse.   Therefore, the *solstices*, the places where this inclination is toward
the sun, are very near the points of perihelion and aphelion.   The winter solstice,
when the Southern Hemisphere has its maximum exposure to the sun, occurs just a few
days before perihelion.   (Note that the earth is nearest the sun during the winter of
the Northern Hemisphere; the difference between perihelion and aphelion in the solar
energy received by the earth is less than 7 percent.)   At the winter solstice the sun is
directly overhead at noon in lat $23\frac{1}{2}°$S.   The summer solstice, when the Northern
Hemisphere has its maximum exposure to the sun, occurs just a few days before
aphelion.   At that time the sun is directly overhead at noon in lat $23\frac{1}{2}°$N.   At two
points midway between the solstices, a line drawn from the sun to the earth is perpendicular to the plane of inclination of the earth's axis; so the sun shines equally in
the Northern and Southern Hemispheres.   These are the *equinoxes*, the vernal equinox
occurring in the spring and the autumnal equinox in the fall of the Northern Hemisphere.   The approximate dates of these significant points or events are vernal equinox,
March 21; summer solstice, June 22; autumnal equinox, September 22; and winter
solstice, December 22.   The dates vary slightly on account of our system of leap
years.   The time from the vernal equinox to the summer solstice is sometimes denoted
as spring; from then until the autumnal equinox as summer; autumn from the autumnal
equinox to the winter solstice; then winter until the next vernal equinox.   An examination of weather records shows that these dates have only a very general meaning as far
as meteorology is concerned.

For purposes of reference and computations it is convenient to assume that the
earth is fixed in space and to speak of the "apparent motion" of the sun and stars.

The plane in which the apparent motion of the sun is observed is called the *ecliptic plane*. It is obvious that this is the same as the plane of the earth's orbit. It is inclined at an angle of $23\frac{1}{2}°$ to the plane of the *celestial equator*, which is the extension into space of the earth's equatorial plane. The equinoxes are at the intersection of the two planes. The vernal equinox is found where the sun in its apparent motion in the ecliptic crosses the celestial equator going northward, and the autumnal equinox is at the intersection as the sun is going southward.

THE TROPICS AND POLAR CIRCLES

The latitude circles of $23\frac{1}{2}°$N and S are called the *Tropic of Cancer* and *Tropic of Capricorn*, respectively. They are the highest latitudes from the equator where the sun can be observed directly overhead at noon, and then only one day each during the year. As a consequence of the inclination of the earth's axis, when the sun shines directly on lat $23\frac{1}{2}°$ in one hemisphere, the portion of the earth poleward from $90 - 23\frac{1}{2} = 66\frac{1}{2}°$ in the other hemisphere is without sunlight. For the Northern and Southern Hemispheres these latitude circles are called the *Arctic* and *Antarctic Circles*. Every point poleward from these circles has at least 24 hr of continuous darkness once during the year. At the poles there are 6 months without sun and 6 months with continuous sunlight between the equinoxes.

In Fig. 2-3 the polar circles and the two tropics are shown, the equator being omitted. If the rays from the sun are considered as parallel lines, the darkening of the arctic regions during winter is apparent, as is also the preponderance of daylight in these latitudes during summer. It is to be noted that these conditions occur in opposite phase in the antarctic regions.

DURATION AND INTENSITY OF SUNSHINE

The lengthening or shortening of the period of daylight at a given latitude follows the increase or decrease of the angle of the sun at noon above the southern horizon in the Northern Hemisphere and above the northern horizon in the Southern Hemisphere. Thus, during summer the temperatures are higher in our latitudes not only because the sun shines more directly and therefore more intensely on the surface of the earth, but also because it shines for a greater number of hours.

In considering the direct effect of angle of the sun above the horizon, we introduce the concept of the rate of flux of energy F (joules per second or watts) and the flux per unit area (watts per square meter). Consider an area[1] A normal to rays of parallel

---

[1] In illustrating the effect we represent area by a line segment because the other dimension is not in the plane of the incoming rays and therefore is of no concern in the discussion.

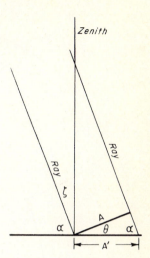

FIGURE 2-4

radiation, as in Fig. 2-4.   The same flux spreads over an area A' on the surface, which lies at an angle θ to the area A.   From the figure it is seen that $A = A' \cos \theta$.   The flux per unit area on the surface $E = F/A' = F \cos \theta/A$.   Thus the flux is a fraction of that which would reach the surface at normal incidence when $\theta = 0$, $\cos \theta = 1$, and $E = F/A$.   The angle ζ seen in the figure is denoted as the zenith angle of the ray, and since the zenith is, by definition, perpendicular to A' while A is perpendicular to the ray, $90 - \zeta = 90 - \theta$ and $\zeta = \theta$.   The point of this discussion is that more energy comes in per unit area the smaller the zenith angle.

The duration of sunlight can be computed precisely as a function of latitude from spherical trigonometry for a smooth and perfect sphere.   Since the earth is not a perfect sphere, the published tables are computed with more complex formulas.   Local corrections have to be applied for terrain features.   The gross picture can be seen in Fig. 2-5, which is a view of the earth from the direction of the sun at the summer solstice and precisely along the sun's rays.   It is noon on the facing meridian, and the other meridians, which are 45° apart, have the times as indicated.   At $23\frac{1}{2}$°N the sun is rising at 6 A.M. and setting at 6 P.M., while farther north the sun rises earlier and sets later, until at the Arctic Circle the sun's rays are tangential at midnight.   Southward into the Southern Hemisphere fewer meridians are illuminated, until all areas are in darkness beyond the Antarctic Circle.

Table 2-1 gives the values of the declination of the sun—its angle above the celestial equator—during the year.   The noon altitude angle α above the horizon as the sun crosses the observer's meridian is related to the declination δ and the latitude φ by the relation $\alpha = 90 - \varphi + \delta$, as shown in Fig. 2-6.   Thus, for any given latitude in the Northern Hemisphere, a constant term $90 - \varphi$ is added to the values of Table 2-1 to

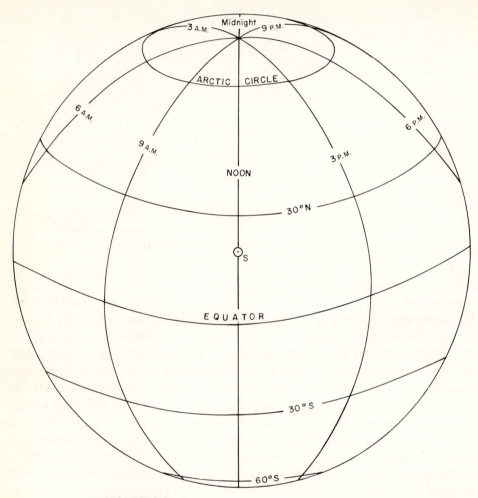

FIGURE 2-5
Exposure of the earth to the sun at the summer solstice, with subsolar point at S marking noon. Times on other meridians at 45° intervals of longitude are shown.

Table 2-1 DECLINATION OF THE SUN ON VARIOUS DATES THROUGH THE YEAR

Jan. 21	−20°54′	July 21	20°30′
Feb. 21	−10°50′	Aug. 21	12°23′
Mar. 21	0°	Sep. 21	1°1′
Apr. 21	11°35′	Oct. 21	−10°25′
May 21	20°2′	Nov. 21	−19°45′
June 21	23°27′	Dec. 21	−23°26′

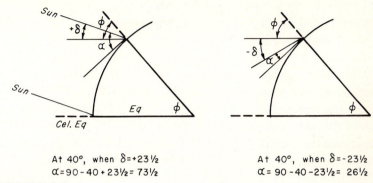

At 40°, when δ=+23½
α=90-40+23½= 73½

At 40°, when δ=-23½
α= 90-40-23½= 26½

FIGURE 2-6
Illustrating relationship between declination δ and altitude angle α at latitude
$\varphi = 40°N$.

obtain the sun's altitude on the meridian. In the Southern Hemisphere, δ is sub-tracted.   The values are useful in showing the different rates at which the sun's altitude changes through the year.   Near the time of the vernal equinox the sun is rapidly assuming a higher and higher noon position in the Northern Hemisphere, but as sum-mer approaches, the sun's altitude does not change so much from day to day.   The same characteristic is noted in a reverse sense in the fall.   In midsummer and mid-winter the changes are relatively slow, and the sun hovers near its highest or lowest position before starting its apparent ascent or descent.   This makes the spring and autumn appear as transition seasons between the two more constant seasons, winter and summer, as the vernal warming and autumnal cooling are favored by the year's greatest change of noon altitude of the sun.

THE LAG OF THE SEASONS

If the temperature depended solely on the amount of radiation received from the sun at a given time, it would be highest in June and lowest in December; or to be more specific, May, June, and July would be the three warmest months and November, December, and January the three coldest.   Actually, June, July, and August are the warmest months in most Northern Hemisphere locations and December, January, and February the coldest.

The lag is accounted for on the basis of the time required for heating and cooling. One may have a roaring fire in a stove to heat a room in the early morning on a cold day, but the temperature in the room will not reach its highest point until later on, even though the fire may have died down by that time.   Conditions in the heating of the earth by the sun are somewhat analogous. The same lag is also noticed in the diurnal period of solar heating. The highest temperatures occur not at noon when the sun is most intense but a few hours later.

Any object, including the earth, can give off heat as well as receive it. If the heat coming in equals that going out, there will be no temperature change. If they do not balance, the temperature will increase or decrease. For the earth as a whole there is no net gain or loss of heat, but there are gains and losses through the year at a given latitude. For example, in the Northern Hemisphere, outside the tropics, the heat received exceeds the heat lost until some time in August, when the heat lost begins to exceed that received. Cooling then predominates until some time in February, when the heat gained begins to exceed the heat lost. The process is complicated somewhat by the transport of heat and cold from various regions by the winds.

Another factor contributing to the lag of the seasons in high latitudes and high elevations is the freezing and thawing of ice. Heat added to ice causes it to melt while its temperature remains at 0°C. Until the required amount of heat is added to the frozen mass in the spring, its temperature will not rise above the melting point, and similarly in the fall there is a lag in the cooling while the freezing proceeds.

SOLAR RADIATION

Not all the solar radiation measured in the atmosphere comes directly from the sun. If the measuring instrument is pointed at a portion of the sky away from the sun, an appreciable amount of the incoming energy recognizable as of solar origin is detected. This is called *sky radiation*. It is the downward-directed component of solar radiation that is scattered in all directions by the air molecules and by fine, dust-like particles that are suspended in the atmosphere. Scattering in the atmosphere is treated in greater detail later in this chapter.

Most measurements of solar radiation are concerned with the total radiation—direct solar plus sky radiation. To establish standards and to study solar-terrestrial relationships, direct solar-radiation measurements are necessary. Most of the standards and other data used in this work were originated by the Smithsonian Institution during several years of pioneering effort at its solar observatories at Mount Wilson and Table Mountain, California. In particular the Smithsonian measurements were aimed at obtaining the value of the solar constant. This is defined as the amount of solar radiant energy received per minute outside the atmosphere on a surface of 1 cm^2 normal to the incident radiation at the earth's mean distance from the sun. Direct measurements from space platforms essentially confirm the Smithsonian standard value of the solar constant, originally computed in 1913, as 1.94 g-cal cm^{-2} min^{-1}.

Meteorologists are more concerned with what is usually called the *insolation*, which is the rate at which the total solar energy—direct plus sky radiation—is received on a horizontal surface.   This would be the same as the solar constant only if the surface were exposed at the outside of the atmosphere at normal solar incidence at the earth's mean distance from the sun.   The insolation received at the surface of the earth depends upon the solar constant, the distance from the sun, the inclination of the sun's rays, and the amount depleted while passing through the atmosphere.   The last two are the important variable factors.   From common everyday experience it is obvious that the greatest depletion results from a cloud cover.

The solar radiation is measured by instruments called *pyrheliometers*.   They operate on the principle of an indirect measurement based upon temperature effects of the radiation falling on an absorbing element, such as a blackened silver disk.   For obtaining records of total solar and sky radiation on a horizontal surface, the instrument is usually referred to as a *pyranometer*.   In a commonly used system equal areas of black and of white surfaces are exposed in a horizontally placed disk.   Various thermoelectric arrangements are used to measure and record the difference in temperatures of the two areas, which is proportional to the amount of radiation received. The record can be made on a recording potentiometer, or the data may be fed into a computer for automatic processing.

The measurements may be expressed in watts per square meter, in ergs per square centimeter per second, or in g-cal per square centimeter per minute.   Frequently the *langley* is used as the unit, equivalent to 1 g-cal cm^{-2}, and the insolation is then stated in langleys per minute.   The unit is named in honor of S. P. Langley (1834–1906), noted physicist and astronomer of the Smithsonian Institution.

TYPES OF HEAT TRANSFER

Three types of heat transfer are recognized by physicists and meteorologists: conduction, convection (horizontal as well as vertical), and radiation.   Heat conduction is a form of heat transfer with which everyone is familiar through ordinary observation, such as when one holds a metal rod with one end in a hot flame and experiences a heating of the other end by conduction from the hot part.   Most metals are good conductors of heat, whereas fluids and particularly gases are poor conductors.   In meteorological problems heat conduction need not be taken into account, because in the atmosphere it is of quite small magnitude compared with the other processes of heat transfer.   It is important in considering the small details of heat transfer, especially under some circumstances within centimeters of the surface.

Convection involves the transfer of heat by means of mass motions of the medium in which the heat is transferred.   This is possible only in fluids and gases, because they alone have internal mass motions.   In rigid bodies, of course, this type of transfer cannot occur.   In convection, the moving masses carry with them heat acquired by conduction in their previous positions.   The motion itself is referred to as convection, but convection may be stated as a measure of the rate of heat transfer by the mass motions.   Since the atmosphere is a medium in which mass motions are easily started, convection is found to be one of the chief ways in which heat is transferred there.   This transfer may be accomplished either by vertical or by horizontal motions, or by a combination of both.   In ordinary parlance, however, the vertical motions are usually meant when speaking of convection.   Horizontal convection is on a much larger scale, involving the slow heat transfer over great distances such as from the equatorial to the polar regions, whereas vertical convection may cover an area about the size of a single cloud, and proceeds at a fairly rapid rate.

Horizontal convective transport is called *advection*.   For the earth as a whole it is the most important means of atmospheric heat transport.   The pronounced day-to-day changes in weather observed in most middle-latitude regions are due to the interplay of great advective currents.   The horizontal components of air motions are more sustained than the vertical components.

Part of the convective system of heat transfer is in the form of heat exchanged in evaporation and condensation of water.   When water is evaporated from the surface of the earth, heat is removed that is carried latent in the water vapor to be released again when the vapor condenses to form clouds or surface condensation products such as dew or frost.   The heat transported to the clouds and released there by this process is quite appreciable.

Also included in the convective transport as we are now considering it is that type of vertical convection which is forced by wind action.   This is usually called *mechanical turbulence*.   It accounts for important heat exchanges next to the ground. For example, by transporting heat downward, it partly compensates for the loss of heat by radiation during a windy night.

Both conduction and convection depend on the existence of a material medium. Radiation is the only means by which heat can be transferred through space without the aid of a material medium.   In our everyday experience we meet with examples of convection and radiation.   The ordinary hot-water radiator used in homes and buildings is really not a radiator in its principal action, because it is so constructed that the heating of the room is mainly accomplished through convection.   If we were to depend upon heating a room by radiation only, a fireplace would be more in order. In this case the convection currents go mainly up the chimney, and much of the heat is lost in this way.   We can warm our bodies by standing in front of the fireplace; but considering the amount of fuel used, the temperature of the room is not raised much.   The heating of the air in the room itself depends on the small amount of convective transport that may occur either from the fire directly into the room or from walls and objects in the room that may themselves have been heated by the radiation from the fire.   The air itself absorbs directly only a very small amount of this radiation.

In the space between the sun and the earth, where a relatively minute quantity of matter exists, radiation is the only important form of heat transfer.   A material medium through which radiant energy can pass is said to be *transparent* to the radiation.

RADIANT ENERGY AND LIGHT

Radiant energy or radiated heat cannot be easily distinguished from light.   As a matter of fact, the differentiation between the radiation of heat and the radiation of light is based only on the scope of our visual perceptions.   Energy that comes to us as visible light represents only a certain portion of the radiation from a hot body such as the sun.   A large part of it is invisible.   All bodies, at whatever temperature, emit radiation, but unless they are very hot, the radiation is not visible.   From common experience we know that objects have to be quite hot even to glow in the dark.

Although it is difficult to define the nature of all forms of radiation, it can be said that, with respect to many of its most important properties, radiant energy can be considered to be transmitted in the form of waves.   Certain aspects of it are better considered as particle emanations, especially in radioactivity, cosmic rays, etc., which, in the wave concept, correspond to extremely small wavelengths (10^{-10} to 10^{-8} cm). In meteorology the radiation is usually of such a form that the wave theory can be applied without ambiguity.

BLACKBODY RADIATION

The amount of energy radiated from a body depends largely on the temperature of the body.   It has been shown by experiment that, at a given temperature, there is an upper limit to the amount of energy that can be emitted in a given time by a unit surface of a body.   The maximum amount of radiation for a given temperature is called the *blackbody radiation*.   A body that radiates for every wavelength the maximum intensity of radiation possible at a given temperature is known as a *blackbody*. At a given temperature this maximum is the same for every blackbody regardless of its structure or composition.   A blackbody can also be described in terms of absorption by considering the fact that all the radiant energy reaching the surface of a blackbody is absorbed by the body.

The term *blackbody* is misleading in that it implies the concept of color. As a matter of fact, objects that do not under ordinary conditions appear black may radiate as blackbodies. The sun itself is an example, and even the energy radiated from a bright snow surface is very nearly that of a blackbody. In dealing with laboratory measurements of great accuracy, one is best able to produce blackbody conditions in the form of radiation that passes out through a small cavity in a solid body at a uniform temperature. Surfaces especially prepared for producing blackbody radiation usually consist of a screen or grid made up of a myriad of needlepoints, or some form of honeycomb structure that provides many cavities.

All blackbodies emit a continuous spectrum. Gases, however, have a discontinuous spectrum, showing emission and absorption in various parts of the spectrum called *lines*. These lines are characteristic for each gaseous substance and serve as a means of identification.

DEFINITIONS

The terminology used in discussing the physics of radiation is not completely standardized. In some cases several different words are used to describe the same process or measurement. The definitions which follow and the words applied to them are presented in a form to produce a minimum of confusion.

The amount of radiant energy in unit time (the radiant power) passing through any surface, whether it be the surface of an emitter, an absorber, or an imaginary surface in the space between, is generally referred to as the radiation *flux* through that surface. It may be expressed in several ways:

> The total flux F, watts, from the entire surface of an emitter or through a surface, real or imaginary, that completely encloses it. The total flux from an emitting body is often called its *luminosity*.
>
> The flux per unit solid angle, watts per steradian, usually called the *intensity I*.
>
> The flux per unit area, watts per square meter, which may be called the *specific flux*, the *flux density*, or the *irradiance E*. The area may be normal to the beam or at some other angle. From an emitting surface, the quantity is called the *emittance*. When expressed as a function of wavelength, it is called *monochromatic* emittance.
>
> The flux per unit solid angle per unit normal area, or the intensity per unit normal area; designated by different names—*radiance, luminance, brightness B*.

An understanding of the different expressions for flux may be helped by showing their interrelationships. Consider a radiation field that is spherically symmetric around a point or spherical source. Since there are 4π steradians in a complete sphere, the total flux through any concentric spherical surface at any distance in space from the emitter would be $F = 4\pi I$. In the absence of an absorbing medium the intensity is independent of the distance, but the flux density E must be smaller the greater the distance r, because of the fanning out of the radial beams. Considering an area on a concentric spherical surface, one notes that $F = 4\pi r^2 E$ and that $I = Er^2$. The radiance may be expressed as $B = I/A \cos \theta$, where θ is the angle between the area A and the normal to the beam. The intensity may therefore be written as $I = B \cdot A \cos \theta$, and for a spherical surface at distance r around a source, $F = 4\pi I = 4\pi \cdot B4\pi r^2 = 16\pi^2 r^2 B = 4\pi r^2 E$, and $B = E/4\pi$.

The total flux, or luminosity, of the sun can be determined from a radiation measurement taken outside the atmosphere—the solar constant—from the relation $F = 4\pi r^2 E$, where r is the distance of the earth from the sun. If E is taken approximately as 2 langleys min^{-1} or 2 g-cal cm^{-2} min^{-1}, at mean solar distance ($r = 1.5 \times 10^8$ km $= 1.5 \times 10^{13}$ cm), the luminosity of the sun is $F = 4\pi \cdot 2(1.5 \times 10^{13})^2 = 5.65 \times 10^{27}$ g-cal min^{-1} $= 3.94 \times 10^{26}$ W $= 3.94 \times 10^{20}$ MW, approximately.

Several dimensionless ratios are defined as follows:

Emissivity The ratio of the observed flux emitted by a body or surface to that for a blackbody under the same conditions. It varies with wavelength and temperature, and may be expressed for a single wavelength, a band of wavelengths, or the entire spectrum.

Absorptivity The fractional part of the incident radiation that is absorbed by the surface in question. It also varies with the wavelength of the radiation and the temperature of the body. Like the emissivity, it is a dimensionless number between 0 and 1. It can also be applied to a given thickness of a gas or liquid.

Reflectivity The fractional part of the incident radiation that is reflected by a surface.

Transmissivity In a partly transparent medium such as the atmosphere or the oceans, the transmissivity is the fractional part of the radiation transmitted through the medium per unit of distance or mass along the path of the radiant beam. The amount of penetration is often called the *optical thickness* or *optical path*, expressed in units of mass or number of molecules of the absorbing and back-scattering medium.

Although absorptivity and reflectivity are most often defined in terms of a surface, they are real effects, too, in the atmosphere. If the molecules of a gas, because of their temperature, are in a state of oscillation compatible with the wavelength of the radiation, they will absorb energy through resonance. Thus the gases can absorb a portion of the wavelengths or lines of the spectrum. Reflectivity occurs in a gas through that component of the scattering which is directed against the oncoming radiation. Thus we can speak of absorptivity and reflectivity in the atmosphere. Since absorptivity, reflectivity, and transmissivity represent different fractions of the same radiation, their sum must be unity, or $A + R + T = 1$.

RADIATION LAWS

In order to understand meteorological radiation processes, it is necessary to know some of the classic radiation laws that are based on experimental and theoretical work of physicists of the late nineteenth and early twentieth centuries. Some of these radiation principles are easy to understand from common experience. For example, one can see that if two bodies of about the same emissivity but at different temperatures are suspended near each other, the warmer one will lose energy by radiation, part of which will be absorbed by the cooler one, so that the temperatures of the two bodies will approach some intermediate value. If the surface of the cooler body is initially at the same temperature as other surfaces in its surroundings, it will not lose energy through its own radiation until it is otherwise warmed to a higher temperature.

A demonstration of the effects of different values of absorptivity and emissivity can be performed by taking a thin metal sheet or aluminum foil as the cool body and trying two types of surface. The sheet can be coated with lampblack on one side to make the absorptivity and emissivity on that side close to that of a blackbody, and the other side can be left in its more or less polished condition. If the blackened side is exposed to the radiation from the hot body, the temperature of the sheet will rise very fast because the absorptivity on the black side is high and the emissivity on the unblackened side is low. It should be noted that energy is radiated from the blackened side but that this can be at most half of the blackbody emission from the sheet, since it represents only one of the two surfaces of the sheet. The surface area on the other side emits something less than half; so the total emission is considerably less than that of an equivalent blackbody while the absorption is at approximately the blackbody rate. If the sheet is exposed with the unblackened side toward the hot body, its warming will be very slow because the sheet will emit its energy toward the surroundings on the blackened side approximately as a blackbody and absorb poorly on the other side. Actually, for the same temperature, the same amount of energy would be emitted as in the first case, but the absorption rate would be less. Intermediate rates of warming would be experienced if both sides were the same, with an all-black sheet warming up faster than an all-polished one.

Kirchhoff's Radiation Law

This law relates the emission of radiation of a given wavelength at a given temperature to the absorption of that radiation. Some consequences of the law have already been anticipated by the preceding discussions and definitions.

To visualize the fundamental form of the law, let us consider a closed cavity, the walls of which are kept at a constant high temperature and are radiating as a blackbody. If a nonblackbody is suspended by nonconducting threads in the cavity, a balance between it and the walls will be reached. This balance will involve the blackbody energy emitted by the walls F_B, the radiant energy emitted by the nonblackbody F_λ, and the unabsorbed energy reflected from the nonblackbody. If the absorptivity of this body is a_λ, then the unabsorbed energy is $(1 - a_\lambda)F_B$. The balance is

$$F_B - F_\lambda - (1 - a_\lambda)F_B = 0 \qquad (2\text{-}1)$$

If we divide by F_B, we obtain

$$\frac{F_\lambda}{F_B} = a_\lambda \qquad (2\text{-}2)$$

By definition, the emissivity is

$$e_\lambda = \frac{F_\lambda}{F_B}$$

Therefore

$$a_\lambda = e_\lambda \qquad (2\text{-}3)$$

Kirchhoff's radiation law requires that the absorptivity of a body for radiant energy of a given wavelength at a given temperature is equal to its emissivity at that wavelength and temperature. The subscript λ which we have introduced is to indicate that a_λ, F_λ, and e_λ have different values for the different wavelengths at a given temperature.

Put another way, Kirchhoff's law states that the emissivity at a given wavelength and temperature is equal to the absorptivity for radiation of the same wavelength from a blackbody at the same temperature. The law also stipulates that the ratio of the emitted flux to the absorptivity is equal to the emitted flux of a blackbody at the same wavelength and temperature. This is shown by taking the relationship given in Eq. (2-2):

$$\frac{F_\lambda}{a_\lambda} = F_B \qquad (2\text{-}4)$$

For a given temperature, the rules hold for the total emission spectrum. They are also applicable to gases, with proper consideration given to the selective nature of absorption and emission in the characteristic *lines* of the spectrum. At the same temperature the emission spectrum and the absorption spectrum of a body match exactly. This is also true of a gas, even to the details of the spectral lines.

Planck's Law

This, one of the great fundamental laws of physics, gives the relationship of the energy emitted by a blackbody to the wavelength or frequency and the temperature. It is based on the postulate of Max Planck put forth in 1900 to the effect that an oscillator can acquire energy only in discrete units, called *quanta*. Atoms and molecules can occupy several energy states. Left to themselves without excitation, they will spontaneously emit radiation and fall to lower energy levels. The frequency of the radiation emitted is proportional to the difference in energy ε between the energy states before and after the emission. The relation is $\varepsilon = h\nu$, where h is Planck's constant, having the value 6.626×10^{-27} erg sec (6.626×10^{-34} J sec). In other words, $\varepsilon = h\nu$ is the magnitude of the quantum of energy. Einstein later gave the name *photon* to this quantum.

In an earlier development, Boltzmann had shown that if n_0 is the number of molecules in any given energy state, the number n in a state whose energy is higher by an amount ε is

$$n = n_0 e^{-\varepsilon/kT} \qquad (2\text{-}5)$$

where k is Boltzmann's constant (1.3806×10^{-16} erg K^{-1}). But with Planck's quantum rule, the possible values of ε must be 0, $h\nu$, $2h\nu$, $3h\nu$, etc.; so there is a distribution

$$n_1 = n_0 e^{-h\nu/kT}$$
$$n_2 = n_0 e^{-2h\nu/kT}$$
$$n_3 = n_0 e^{-3h\nu/kT}$$

and so on, or for all states

$$n = n_0 + n_0 e^{-h\nu/kT} + n_0 e^{-2h\nu/kT} + n_0 e^{-3h\nu/kT} + \cdots \text{ etc.}$$

The derivation of Planck's law is beyond the intended scope of this book. From a physical-mathematical treatment of the problem an expression is obtained for the energy emitted in photons from a finite volume as detected at a distance on a plane surface. The flux of energy per unit area measured through an infinitesimal frequency range $d\nu$ as a function of temperature and frequency is given by

$$E_\nu(T) = \frac{2\pi h}{c^2} \frac{\nu^3 d\nu}{e^{h\nu/kT} - 1} \qquad (2\text{-}6)$$

where c is the speed of light. In the following discussions wavelength will be referred to in place of frequency. Since $\nu = c/\lambda$ and $d\nu = -(c/\lambda^2)d\lambda$ (the minus sign showing that the frequency and wavelength change inversely), the expression becomes

$$E_\lambda(T) = \frac{2\pi h}{c^2} \frac{c^3 \lambda^{-3} c\lambda^{-2}(-d\lambda)}{e^{hc/\lambda kT} - 1} = \frac{2\pi hc^2(-d\lambda)}{\lambda^5(e^{hc/\lambda kT} - 1)} \qquad (2\text{-}7)$$

It is convenient to compute the flux for unit intervals of wavelength ($d\lambda = 1$) to obtain

$$E_\lambda(T) = \frac{2\pi hc^2}{\lambda^5(e^{hc/\lambda kT} - 1)} \qquad (2\text{-}8)$$

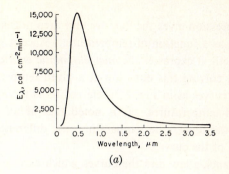

(a)

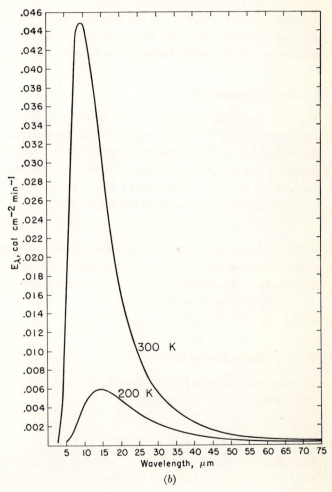

(b)

FIGURE 2-7
Planck curves for blackbody emittance at *a*, 6000 K, and *b*, 300 K (upper curve)
and 200 K (lower curve).

This expression gives the flux per unit area normal to the rays at any specified wavelength λ as a function of temperature. Since it gives the value for each unit step of wavelength, it expresses the *monochromatic emittance* as defined on page 20.

By solving this equation at a given temperature for various wavelengths, one obtains curves as in Fig. 2-7, referred to as the *Planck curves* for a blackbody at the specified temperatures. It is noted that at all wavelengths the energy emitted is greater the higher the temperature, the difference being most pronounced near the maxima of the curves.

Planck's law and the curves which can be plotted from his equation underlie the physical relations of radiation applied to the atmosphere. The characteristics of the emission from a blackbody are fundamental to the study of radiation exchanges in many branches of physics as well as in meteorology. We shall now proceed to examine two relationships derived from Planck's equation.

Stefan-Boltzmann Law

The total blackbody radiation over all wavelengths at a given temperature is measured by the area under the Planck curve for that temperature. The total is obtained by integrating Eq. (2-7) at constant temperature through the wavelengths from zero to infinity. The integration is not easy to perform from the point of view of elementary calculus; so only the result is given here. In terms of the energy detected per unit area (specific flux, or irradiance) the expression is

$$E = \sigma T^4 \qquad (2\text{-}9)$$

where σ is referred to as the Stefan-Boltzmann constant and the relationship itself as the Stefan-Boltzmann law. This important and useful law shows that the energy radiated by a blackbody is proportional to the fourth power of its Kelvin temperature. The constant $\sigma = 5.670 \times 10^{-5}$ erg sec^{-1} cm^{-2} K^{-4}, or 5.670×10^{-8} W m^{-2} K^{-4}.

Wien's Displacement Law

The maximum of each Planck curve is obtained in the usual way by setting equal to zero the derivative at constant T, or

$$\left(\frac{\partial E}{\partial \lambda}\right)_T = 0 \qquad (2\text{-}10)$$

This problem may be solved by substituting the variable x for $hc/\lambda kT$, so that $\lambda^{-5} = h^{-5}c^{-5}k^5T^5x^5$. Then (2-8) becomes

$$E = \frac{2\pi k^5 T^5}{h^4 c^3} \frac{x^5}{e^x - 1} \qquad (2\text{-}11)$$

It is easier to find the maximum by taking the derivative of the natural logarithms, starting with

$$\ln E = \ln \frac{2\pi k^5 T^5}{h^4 c^3} + 5 \ln x - \ln(e^x - 1) \qquad (2\text{-}12)$$

and differentiating at constant temperature

$$\left(\frac{\partial \ln E}{\partial x}\right)_T = 0 + \frac{5}{x} - \frac{1}{e^x - 1} \frac{\partial (e^x - 1)}{\partial x} = \frac{5}{x} - \frac{e^x}{e^x - 1}$$

$$= \frac{5}{x} - (1 + e^{-x} + e^{-2x} + e^{-3x} + \cdots) \qquad (2\text{-}13)$$

Only the first two terms in the parentheses are significant; so we place the maximum of the curve where

$$\frac{5}{x} - 1 - e^{-x} = 0 \qquad \text{or} \qquad x + xe^{-x} = 5 \qquad (2\text{-}14)$$

which is satisfied when $x = 4.9651 = hc/\lambda_{max} kT$, and $\lambda_{max} T = hc/4.9651k = 0.288$ cm deg. Thus at 288 K we find $\lambda_{max} = 0.288/288 = 10^{-3}$ cm or 10 μm; at 6000 K, the approximate mean temperature of the photosphere of the sun, $\lambda_{max} = 0.288/6000 = 4.8 \times 10^{-5}$ cm $= 0.48$ μm. The relationship derived here is known as *Wien's displacement law*. It gives the wavelength at which a body of a given temperature emits its maximum energy, and we see that for the sun the peak is in the visible range but for the earth the maximum is in the infrared. However, we should keep in mind that at all wavelengths the hotter body emits more than the cooler one.

SCATTERING

Sky radiation is energy from the sun that has been *scattered* by the air molecules and fine dusts. The blueness of the sky is accounted for by this process, because in the visible range the violets and blues are most subject to scattering.

In scattering, no energy transformation occurs. Each scattering molecule or particle distributes the energy in all directions, forward, back, and all around. The back-scatter as seen in space or above a haze layer is part of the light reflected from the atmosphere; so to an orbiting observer the scattered sunlight is part of the brightness of the earth in sunlight. The part scattered toward the eye of an observer within the atmosphere from any sky angle is the blue sky radiation. Insolation measurements include both the direct solar and the sky radiation.

The first adequate theory of scattering was that developed by Lord J. W. S. Rayleigh (1842–1919). The theory is obtained from concepts of electromagnetics in which each scattering center is considered as an oscillating dipole having a frequency of oscillation the same as the impressed frequency. The complete physical-mathematical development is beyond the scope of this book. The result is a scattering coefficient which depends on the density and electrical properties of the medium, but most of all is sensitive to the sizes of the scatterers and the inverse fourth power of the wavelength λ^{-4}. The coefficient is obtained by deriving the ratio of the flux scattered by a scatterer to the initial flux density received from the source, or F_s/E_0; but the coefficient is taken for all the N scatterers contained in a cubic centimeter, or $k_s = NF_s/E_0$. The expression for scatterers of volume V each is

$$k_s = 24\pi^3 N \left(\frac{m^2 - 1}{m^2 + 1}\right)^2 \frac{V^2}{\lambda^4} \qquad (2\text{-}15)$$

where m is the *refractive index* of the material. The refractive index of a medium or substance is the ratio of speed of electromagnetic waves (light) in a vacuum to that in the substance, and is related to the *dielectric constant*.

It is seen that the coefficient of Rayleigh scattering is highly dependent on the size of the particles. Since V is proportional to the cube of the radius or diameter, k_s is proportional to the sixth power of the radius. Also it should be noted that λ^{-4} enters into the expression. The ratio of k_s for blue light at 0.425 μm to that for red light at 0.650 μm under the same conditions would be $(650/425)^4 = 5.48$. Thus the scattering of blue light is 5.48 times the scattering of red light.

Air molecules themselves contribute significantly to the scattering. In writing an expression for their scattering, we do not wish to consider their volumes. We also find that the refractive index of dry air depends on its density alone. J. J. Johnson[1] has shown that Eq. (2-15) can easily be transposed into an expression for a gas by factoring as follows:

$$N \left(\frac{m^2 - 1}{m^2 + 2}\right)^2 V^2 = \frac{1}{N} \left[N \frac{(m + 1)(m - 1)}{m^2 + 2} V\right]^2 \qquad (2\text{-}16)$$

In the atmospheric gas, m is very close to unity, and the result is essentially unchanged by taking m equal to 1 in the terms $m + 1$ and $m^2 + 2$; so we can write the right-hand side of Eq. (2-16) as

$$\frac{1}{N} \left[N \frac{2(m - 1)}{3} V\right]^2 = \frac{4}{9N} [N(m - 1)V]^2 \qquad (2\text{-}17)$$

But N and V refer to molecules, and V, the volume of a molecule, is constant for any gas. N, the number of molecules per cubic centimeter, is an expression of the gas

[1] J. J. Johnson, "Physical Meteorology," John Wiley & Sons, Inc., New York, 1954.

density, and since $m - 1$ is a function of the air density, we see that $N(m - 1)V =$ constant $\times$ density.   We define the $m - 1$ for a gas as given by this last quantity, writing it $(m - 1)_{gas} = N(m - 1)V$.   This gives, from Eq. (2-15),

$$k_s = \frac{32\pi^3}{3N\lambda^4} (m - 1)^2_{gas} \qquad (2\text{-}18)$$

Thus for the molecules of the atmospheric gases it is seen that the scattering depends only on the air density and the inverse fourth power of the wavelength.

Rayleigh scattering is applicable only when the ratio of the diameter of the scatterer to the wavelength is less than about one-tenth.   Thus the Rayleigh theory applies to the scattering of blue light, for instance, by gas molecules and the smallest dust particles.   At larger values of the ratio of particle diameter to wavelength, a theory developed by G. Mie in 1908 describes a more complex relationship.   At a ratio somewhere between 1 and 10, the phenomenon is better described as diffraction, and at ratios of the order of 10 occurs the ill-defined lower limit of validity of simple geometric optics.

Sunlight, which is basically white, often reaches the earth with a reddish tinge. This is especially noticeable at sunset when the light passes through its longest path of atmosphere, and is explained by the fact that the blue light has been scattered by the atmosphere and only the reddish portions reach us directly.   The same effect is noted when the sun shines through a smoky atmosphere.   Cloud particles do not change the color of the light because of their relatively large size.   The ability of clouds to scatter microwaves (0.1 to 10 cm), however, makes weather radar possible.

In conditions of clear skies, the amount of light sent back into space through atmospheric scattering is slightly greater than that reflected from the surface of the earth.   The light scattered to the earth as sky radiation exceeds somewhat that scattered out into space.

On the earth, scattering is the process mainly responsible for reducing the visibility or distance from which objects can barely be seen.   Under hazy or dusty conditions the light from a distant object may be completely attenuated by scattering before reaching the eye.   Direct absorption by the haze particles is of some importance, but scattering is the main effect.   The visibility is also affected by the scattered light, or "air light," coming from all directions with enough relative strength to weaken or overpower visual contrast between distant objects.

ABSORPTION IN THE ATMOSPHERE

Gas molecules, dust, haze, smoke, and cloud particles absorb part of the radiant energy passing through the atmosphere.   The gases absorb only in certain wavelengths to form an absorption spectrum characteristic of each gas.   The absorption lines are concentrated over certain bands of wavelength.   The two most important gases, nitrogen and oxygen, have appreciable absorption bands only in the ultraviolet, from about 200 to 2000 Å.   Ozone, O_3, which is concentrated at heights of 20 km or so but even there seldom consists of more than 0.2 part per million, is a strong absorber of ultraviolet radiation from about 1000 to 3000 Å.   This absorption saves us from exposures to ultraviolet radiation which over a period of time would be lethal.

Meteorologically the most important absorption by gases is in the infrared, and two variable gases, water vapor and carbon dioxide, which usually occur in concentrations of less than 1 percent, are mainly responsible.   Ozone also has an isolated and striking absorption band at 9.6 μm.   Water vapor absorbs strongly in several bands from 1 to about 6.5 μm, then is essentially transparent from 7 to 12 μm.   It absorbs effectively at wavelengths of 18 μm and longer.   Carbon dioxide exhibits bands in the vicinity of 2.7 and 4.2 μm.   Its most marked absorption is at 14 μm and higher.   Details of infrared absorption are treated in the next chapter—see, for example, Fig. 3-2.

The transparent region with wavelengths of from about 7 to 12 μm, where neither water vapor nor carbon dioxide absorbs appreciably, is referred to as the atmospheric "window," because with clear skies the energy of that wavelength band passes through the atmosphere unimpeded.   The 9.6-μm ozone band presents a slight interruption to the transparent band.   It is interesting to note that this atmospheric window is in the region of the maximum of the Planck curve for blackbody radiation at terrestrial temperatures.   Instruments on aircraft, balloons, or satellites designed to measure outgoing radiation from the surface or from cloud tops obviously operate in the wavelengths of the window.

Clouds play a dominant role in all aspects of radiation.   They are by far the most important absorbers of radiation at essentially all wavelengths.   In sunlight they reflect a high percentage of the incident solar radiation and account for most of the brightness of the earth as seen from space.   Their role in absorbing and reradiating the terrestrial radiation from the surface is most striking at night.   The importance of clouds and water vapor in the radiation balance of the earth will be treated more fully in the next chapter.

THE DEPLETION OF SOLAR RADIATION IN THE ATMOSPHERE

In passing through a finite distance in the atmosphere, the radiation from the sun is reduced by scattering and absorption.   The depletion by gases can be computed from laboratory measurements on the individual gases and from their known concentrations in the atmosphere.   The atmosphere consists of a mechanical mixture of several gases, sometimes called the *permanent* gases, that remain in fixed proportion to the total, and other gases that vary markedly with time and location.   A number of tabulations of the composition of the air have been made by various investigators, all in fairly close agreement.   Table 2-2 is taken from various sources compiled by Glueckauf.[1]

Through the appreciable atmosphere the permanent gases exist in nearly the same proportion at all heights, the remaining constituents showing generally a rapid decrease in percentage with height.

It is interesting to note that although water vapor usually comprises less than 3 percent of the gases even with moist conditions at sea level, it absorbs nearly 6 times as much solar radiant energy as do all the other gases combined.   Furthermore, it accounts for nearly all the gaseous absorption of the terrestrial radiation.

Ozone is of special interest because of its photochemical reaction to certain wavelengths of the solar radiation, particularly in the ultraviolet.   The height of maximum ozone concentration is between 20 and 30 km.   Oxygen molecules are dissociated by the solar radiation into atomic oxygen which combines with other oxygen molecules to form ozone $(O + O_2 \rightarrow O_3)$ which itself is dissociated to form O_2 and O.   The three forms O, O_2, and O_3 react to achieve an equilibrium mixture which is different at various heights.

[1] E. Glueckauf, The Composition of Atmospheric Air, *Compendium of Meteorology*, 1951, pp. 1–10.

Table 2-2   COMPOSITION OF THE ATMOSPHERE

Constituent	Percent by volume	Parts per million
Nitrogen	78.084 ± 0.004	
Oxygen	20.946 ± 0.002	
Carbon dioxide	0.033 ± 0.001	
Argon	0.934 ± 0.001	
Neon		18.18 ± 0.04
Helium		5.24 ± 0.004
Krypton		1.14 ± 0.01
Xenon		0.087 ± 0.001
Hydrogen		0.5
Methane (CH_4)		2
Nitrous oxide (N_2O)		0.5 ± 0.1

Important variable gases	
Water vapor	0–3 percent
Ozone	0–0.07 ppm (ground level)
	0.1–0.2 ppm (20–30 km)

Suspended in the atmosphere are certain particles. By far the most important are the liquid water and ice particles of clouds. Not only in the radiation balance but also in the whole problem of heat transfer in the atmosphere they play a predominant part. Other particles are present which have less direct meteorological effects. They include dust, smoke, industrial effluents, various chemical particles that may be classified separately from dust and smoke, things of an organic nature such as pollens, spores, bacteria, and fibers, and particles recognizable only as ions or radiation particles.

TRANSMISSIVITY AND EXTINCTION

Since the atmosphere is partly transparent to both the solar and terrestrial radiation, a great deal of attention must be given to the problem of atmospheric transmission and extinction of radiant energy in the various wavelengths.

If radiation flux E_0 of wavelength λ reaches a point in a gas and is reduced to E_λ after a unit path length through the gas, the transmissivity τ of that gas for the wavelength in question will be expressed by the ratio

$$\tau = \frac{E_\lambda}{E_0} \qquad (2\text{-}19)$$

An extinction coefficient k can be expressed as the fractional change in flux in passing through unit length L, where k, with the units of reciprocal length, would be the sum of the scattering coefficient k_s and an absorption coefficient k_a. In general differential form the expression would be

$$\frac{dE_\lambda}{E_\lambda} = -k\,dL \qquad (2\text{-}20)$$

where the negative sign indicates decreasing flux with path length L.

This equation may be integrated between the limits $E_{0\lambda}$ and E_λ, letting $L = 0$, where $E_\lambda = E_{0\lambda}$, to obtain

$$k = \frac{1}{L} \ln \frac{E_{0\lambda}}{E_\lambda} \qquad (2\text{-}21)$$

$$E_\lambda = E_{0\lambda} e^{-kL} \qquad (2\text{-}22)$$

When it is more convenient to use the common logarithms (base 10 instead of base e), a " decimal coefficient of extinction " α can be employed such that

$$E = E_0 \times 10^{-\alpha L} \qquad (2\text{-}23)$$

where $\alpha = 0.4343k$.

Since the number of molecules in a given path length varies with the air density, the coefficient is usually expressed in terms of unit length at normal temperature and pressure (N.T.P.) for most gases. An exception is water vapor, which cannot exist in gaseous form at such a high vapor pressure below the temperature of its boiling point. The so-called "precipitable water" in the length L, meaning the thickness of the liquid film that would be formed by converting all the vapor to liquid, that is, to a density of 1 g per cm^3, is used.

The relationships presented here apply only to parallel radiation. In the case of scattering of the magnitude observed in the atmosphere, this restriction can lead to appreciable error if not taken into account, especially in the case of solar radiation at low altitude angles of the sun. Since the nonparallel effects are hard to correct for, situations in which these errors are large are usually avoided in making measurements and computations.

REFLECTION

Everyone is familiar with the process of the reflection of light. In general, our visual impression is approximately correct in indicating the wavelengths of greatest reflectivity, for they fall mostly in the visible spectrum. This means that only the solar radiation is reflected appreciably, while the reflection of the long-wave radiation from the earth and atmosphere may be neglected.

There is some variation in the reflectivity of various natural surfaces. The general reflectivity is called the *albedo*. Although this term is defined somewhat loosely, we shall think of it as the reflectivity over the significant wavelengths for surfaces viewed from above. The albedo of the entire planet Earth is, on the average, considered to be about 0.36 (36 percent reflected into space). By far the most important part of this reflectivity comes from cloud covers. The albedos of natural surfaces may be summarized as follows:

Upper surfaces of clouds From 40 to 80 percent, depending on thickness, but with an average of about 55 percent.

Snow surfaces Over 80 percent for cold, fresh snow; as low as 50 percent for snow several days old.

Land surfaces Most common surfaces, such as forests, grassy fields, plowed fields, and rocky deserts, have an albedo averaging about 10 to 20 percent but ranging from 5 percent for dark, coniferous forests to 45 percent for dry sand.

Water In general, the lowest albedos are found over water. It depends, however, on the zenith distance (degrees) of the sun. The Smithsonian Meteorological Tables give the following relation to the sun's zenith distance:

Zenith distance	0	20	40	50	60	70	80	85
Percent reflected	2.0	2.1	2.5	3.4	6.0	13.4	34.8	58.4

When the wind speed is sufficient to raise whitecaps, the albedo is greater than that of a nonfoaming water surface.

The data indicate that, except for the effects of a snow cover and some of the deserts, there is no important geographical difference in the albedo of surfaces on the earth itself. Therefore, over the earth as a whole we find that the albedo will be determined mainly by the extent and thickness of clouds. This fact is amply brought out by satellite measurements, photographs, and observations. Extensive snow covers are also important. Under cloudless conditions, the areas of greatest reflectivity are snow-covered mountain and polar regions.

Obviously, the diminution of sunlight that we observe on an overcast day is due not so much to absorption within the cloud as to the large amount of light reflected from the top of the overcast layer. It should be noted that the albedo measured in space also includes the light scattered outward by the atmosphere.

GREENHOUSE EFFECT

The atmosphere conserves the heat energy of the earth because it absorbs radiation selectively. Most of the solar radiation in clear skies is transmitted to the earth's surface, but a large part of the outgoing terrestrial radiation is absorbed and radiated back to the surface. This is called the *greenhouse* effect. A greenhouse permits most of the short-wave solar radiation to pass through the glass roof and sides to be absorbed by the floor or ground and plants inside. These objects reradiate energy at their temperatures of about 300 K, and therefore with principal intensity around 10 μm. The glass absorbs the energy at these wavelengths and sends part of it back into the greenhouse, causing the inside of the structure to become warmer than the outside. The atmosphere acts similarly, transmitting and absorbing in somewhat the same way as the glass. Fleagle and Businger[1] show that the earth's surface has an average temperature 35°C warmer than would be the case if no long-wave back-radiation were coming down from the atmosphere.

EXERCISES

1 From the sun's declination on various dates as given in Table 2-1, make a graph of the altitude angle of the noon sun through the year at your latitude.

2 If you constructed a clock that would complete its circuit of 12 divisions twice each sidereal day instead of twice each solar day, and set it at 12 o'clock when the sun was on

[1] R. G. Fleagle and J. A. Businger, "An Introduction to Atmospheric Physics," p. 154, Academic Press, Inc., New York, 1963.

your meridian at the summer solstice, what time would it read with the sun on the meridian at the autumnal equinox? At the winter solstice? (Assume that your clock keeps perfect time and that the earth's orbit is circular.)

3 Approximately 90 percent of the mass of the atmosphere lies below 30 km. Considering that level as the effective top of the atmosphere, compare the path length of the sun's rays through the atmosphere at sunset with the path length when the sun is directly overhead.

4 Given the luminosity of the sun as 3.94×10^{20} MW, compute the solar constant for Venus, Mars, and Jupiter, taking the mean solar distances to be 1.081×10^8 km, 2.277×10^8 km, and 7.773×10^8 km, respectively.

5 Considering that a planet presents itself as a disk to the rays of the sun, show that with an albedo A and a solar constant S, the earth-atmosphere system absorbs solar energy at a rate given by $(1 - A)S/4$.

  Using this result and taking the solar constant as 1.94 langleys per minute and the albedo of the earth as 0.35, apply the Kirchhoff and Stefan-Boltzmann laws to obtain the effective radiating temperature of the earth. (Be careful of units.)

6 At a wavelength of 16 μm, 250 cm of CO_2 at normal temperature and pressure (NTP) is capable of complete absorption. What is the mass of such a column of CO_2 per square centimeter? Hint: You might find it useful to work through the gram-molecular volume.

7 Consider Planck curves for blackbodies at 253, 273, 293, and 303 K. Determine the wavelength of the maximum of each curve and the total area under each curve.

8 In a certain wavelength of the water-vapor absorption spectrum the intensity of the radiation is reduced to $1/e$ of its original value (where e is the base of the natural logarithms) as it passes through 1 cm of precipitable water. Find the extinction coefficient for absorption only.

9 A common value of the index of refraction for the gases in the lower atmosphere is $m = 1.000315$ or $m - 1 = 315 \times 10^{-6}$, and $(m - 1)^2 = 315^2 \times 10^{-12}$. From this value compute the scattering coefficient k_s for blue light of wavelength 0.5 μm. The number of molecules per cubic centimeter is the number per mole (Avogadro's number) divided by the gram-molecular volume, or 2.66×10^{19}. Give the units of k_s. The number of molecules contained in a vertical air column through the entire atmosphere is equivalent to 8 km at NTP. Multiply k_s by this height in compatible units to obtain the total scattering of blue light of this wavelength from the total atmospheric path length.

10 At a certain wavelength it is found that the decimal coefficient of extinction by water vapor is $\alpha = 0.15$. What would be the transmissivity through 2 cm of precipitable water at this wavelength?

3

THE HEAT BALANCE OF THE ATMOSPHERE

Having developed the physical background of solar and terrestrial radiation, we are now ready to look at the effects produced in the atmosphere.   This chapter and the next one will be largely descriptive as we detail the broad picture of radiation exchange between the sun, the earth, and the atmosphere, and examine the results.

In the preceding chapter it was noted that the radiant energy emitted depends on the fourth power of the Kelvin temperature of the radiating body.   This means that the more the sun heats up the earth the greater the amount of energy that will be sent back into space by terrestrial radiation.   If this rate of heat loss did not on the average exactly balance the amount of heat received from the sun, then the earth would become continuously hotter or continuously colder, as the case may be.

In geological history there have been periods when the earth or parts of it have been warmer or cooler than at the present time, and it seems at once evident that this must have been caused by changes in the incoming and outgoing radiation.   However, several theories in explanation of the climates of the past do not require that the heat balance between the sun and the earth should have changed.   If the earth is now getting hotter or colder, it is doing so at such a slow rate that for our purposes we may consider that there is, over a long period of time, a perfect heat balance of the earth and its atmosphere in relation to solar radiation.

The mechanism by means of which the heat balance is maintained on the earth and in its atmosphere is extremely complicated.   It is necessary to consider the various wavelengths of the radiation separately because of the selective absorption and emission of the atmosphere.   In this chapter the mechanism of radiation balance will be discussed, both to explain the observed conditions and to form a basis for the discussion of various radiation problems with which practical meteorologists must deal.

In the cosmic sense, radiation is the only means of maintaining a complete heat balance, because it is only by radiation that heat energy can be transferred through space.   If the outgoing terrestrial radiation were greater than the incoming solar radiation, the earth would become progressively colder.   Not only must there be a balance between the earth's surface and the incoming radiation, but also a balance must exist that includes the atmosphere.   Heat escapes through the outer atmospheric boundary.

SOLAR RADIATION AND INSOLATION

Outside the atmosphere the flux density of solar radiation at normal incidence and mean solar distance is 1.94 g-cal cm^{-2} min^{-1} (langleys per minute) according to agreed standards—the solar constant.   At the surface of the earth the insolation, defined on page 17, is very much less than the solar constant because (1) the rays are normal at only one point in the tropics at any given time, and therefore the radiation falling on every point would be less than the solar constant even if there were no atmosphere; and (2) a considerable fraction of the radiation is depleted in the atmosphere.   The depletion depends on the length of the optical path, determined by latitude, date, and time, and on the nature and amount of absorbing, reflecting, and scattering matter in the atmosphere.   Again it should be pointed out that in this depletion process the reflection from clouds is the most important factor.

A chart prepared by Landsberg[1] of the global distribution of insolation is presented in Fig. 3-1.   It is taken from solar radiation measurements at representative land stations, and interpreted over the oceans from observations of cloudiness at sea.   As satellite data are collected over the years, the evaluation of the cloudiness factor can be improved.   On the chart the maxima over inland deserts and the minima over the oceans demonstrate the effect of absence of clouds in the one area and their presence in the other.   The tropical forest areas such as the Amazon Basin, the Congo, and the Vietnam–South China area are low in insolation because of prevalent cloudiness.

[1] H. E. Landsberg, Solar Radiation at the Earth's Surface, *Solar Energy*, vol. 5, no. 3, pp. 95–98, 1961.

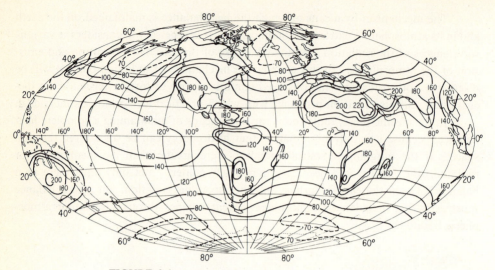

FIGURE 3-1
Generalized distribution of radiation received at the surface of the earth over the course of the year (kcal cm^{-2} year^{-1}). (*After Landsberg.*)

On the average for all locations over the earth, clouds cover slightly more than half the sky. In the clear areas about 70 percent of the incident radiation reaches the surface of the earth and is absorbed there, while under cloudy skies less than a third of the energy gets through. Reflection is the principal effect of the clouds.

Computations of the solar input to the heat balance take into consideration the average cloudiness. When these computations are made from observations within the atmosphere and at the surface of the earth, the average cloudiness is difficult to apply. In photographs or observations along the sun's rays from a distant satellite the earth appears as a disk. If clouds partially cover the earth in a uniform distribution, then the maximum cloudiness illuminated by the sun appears near the limb and the minimum at the subsolar point in the noon position. The solar beam reaches the clouds from a side angle as the earth curves away from this central point. The mean cloudiness has to be corrected to the apparent disk, giving increasing weight to clouds with increasing distance from the subsolar point. However, averaged over a period of time, the longitudinal or east-west direction from the subsolar point has no effect, because every point at a given latitude has equal exposure as the earth rotates. Only the latitudinal or north-south curvature has to be considered in the averages. The subsolar point is on the ecliptic; so the correction is a function of altitude angle from that line, or the latitude minus the sun's declination, the latter being considered negative in the opposite hemisphere.

TERRESTRIAL RADIATION

From the application of Planck's radiation formula, through Wien's displacement law, it is found that the energy emitted from a body having the temperature of the earth would be in the infrared, with the dominant wavelengths centered on approximately 10 μm of this radiation.   A large proportion is absorbed by the water vapor and carbon dioxide of the atmosphere, with water vapor playing by far the predominant role as far as gases are concerned.   Clouds absorb radiation in these wavelengths even more effectively.

From Kirchhoff's law it is apparent that the radiant energy from the earth that is absorbed in the atmosphere will be reradiated.   This emission would be in all directions; however, since the horizontal temperature differences are negligible, horizontal radiation fluxes will cancel out, each mass losing in a horizontal direction the same amount that it receives horizontally.   Only the upward and downward fluxes need be considered.   It follows that since the earth is normally warmer than the overlying atmosphere, the atmosphere will receive more radiation than it emits, and with a decreasing temperature with height in clear air, the net flux must be upward. Under certain circumstances, an appreciable layer of the atmosphere is found to be warmer than the ground, and the net flux from that layer is downward.   This flux, of course, is in addition to and at a different wavelength from that coming downward from the sun.

For a study of radiation processes in the atmosphere, the nature of the absorption by the most effective absorbing gas, water vapor, must be known.   Determinations of water-vapor absorption have been made in most parts of the spectrum associated with terrestrial radiation.   They show the absorption spectrum to be made up of a myriad of absorption lines distributed through the various wavelengths.   The computation of transmission functions in all these spectral lines from the laboratory data can be accomplished by electronic computers.   A complication arises from the fact that in each spectral line the absorption coefficient varies from a very high value in the center of the line to much smaller values on the sides.   In addition, there is an effect on this factor due to pressure and temperature, known as the "broadening" of the absorption lines such that the half-value width of each line increases with air pressure and temperature.   In the atmosphere where the decrease of pressure with height is large (half an atmosphere in the first 6000 m), the pressure broadening is an important effect.   However, the amount of water vapor decreases so rapidly with height that the correction from sea-level values is operated upon a very small quantity and thus loses importance.

The existence of a band of wavelengths (approximately 8 to 13 μm) in which the transmissivity is essentially unity, called the *transparent band* or atmospheric "window" has already been mentioned.   It lies in the vicinity of the peak of the Planck curve for

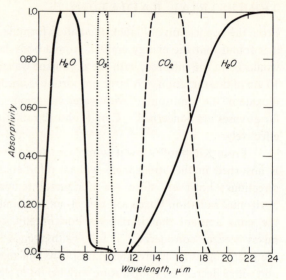

FIGURE 3-2
Absorptivity of water vapor, carbon dioxide, and ozone in a typical atmosphere, as a function of wavelength in the infrared.

the terrestrial range of temperatures. At other infrared wavelengths the atmosphere is a good absorber. It is the absorption spectrum of water vapor that gives the atmosphere its unique radiation character in the infrared. Carbon dioxide, especially in its strong absorption lines between about 13 and 17 μm, is next in importance, although its absorption effectiveness in the atmosphere is greatly outweighed by that of water vapor. At places near sea level there is enough water vapor in the total overlying atmosphere in cloudless conditions to account for almost complete absorption of terrestrial radiation in the wavelengths other than those of the atmospheric window. The window contains a narrow obstruction in the form of the 9.6-μm ozone absorption band.

Figure 3-2 depicts the absorptivity by the principal absorbing gases of a typical atmosphere, one having 1 cm of precipitable water (1 g per cm^2), 250 cm carbon dioxide, and 0.3 cm ozone, the latter two reduced to normal temperature and pressure (NTP). An examination of the Planck curves in Fig. 2-7b shows that the wavelengths of appreciable emission by a blackbody at terrestrial temperatures is in the range of 4 to about 50 μm, which is the range covered here, except that since above 24 μm the absorptivity continues as 1, it has not been considered necessary to extend the diagram beyond that wavelength. The window from about 8 to 12 μm is strikingly revealed, interrupted by the sharp 9.6-μm ozone band.   The ozone is concentrated at high levels in the atmosphere where the temperature is so low (220 to 240 K) that even with an absorptivity and emissivity of 1, the actual radiance, according to the fourth-power law, is about one-third of what it would be at 300 K.   Kirchhoff's law requires that the total absorption similarly should be small.   The CO_2 absorptivity accounts for the right-hand frame of the window, effective together with the increasing absorptivity of the water vapor.   At the crossover of the two curves both have values greater than 0.5; so their sum ensures that the absorptivity is 1 at all wavelengths beyond 14 μm.

The data in Fig. 3-2 are taken from Elsasser and Culbertson,[1] and are computed from *slab* absorptivities, that is, from absorptivity exhibited by an extensive layer for radiation from an extensive source as contrasted with the laboratory *beam* absorptivity in gases contained in a tube.   The 1-cm value of water vapor is typical of large parts of the United States in summer.   Shorter path lengths, i.e., smaller values of precipitable water, would have a smaller absorptivity, and of course the outward path length would decrease markedly with height.   Depending on the vertical distribution of the absorbing gases, especially water vapor, there would be some altitude at which the path length along the vertical upward would be the same as that downward.   The water-vapor content of the air normally decreases rapidly with height.

Most natural surfaces on the earth, including water, snow, bare ground, grass, and tree-covered areas, have emissivities in the infrared of between 0.90 and 0.95. Matching emissivity with absorptivity from Kirchhoff's law, one sees that a humid atmosphere which also contains carbon dioxide and ozone can, under extreme conditions, produce an emissivity approaching these values.   Under cloudy skies near-black emissivities in the infrared are to be expected.

RADIATION IN THE FREE ATMOSPHERE

Three factors determine the flux of radiation through the atmosphere: (1) the vertical distribution of temperature, (2) the vertical distribution of the absorbing gases, and (3) the presence or absence of clouds.   The temperature effect is seen through the application of the concept of the Stefan-Boltzmann law.   Although a layer of absorbing gas is not a blackbody, its emissivity ε may be applied to give, for extensive layers,

$$E = \varepsilon\sigma T^4$$

Since temperature normally decreases with height, the flux must decrease with height. Under these conditions the net flux is upward.   This can be seen by considering absorbing layers in the atmosphere.   The first layer absorbs the energy from the ground and

[1] W. M. Elsasser and M. F. Culbertson, Atmospheric Radiation Tables, *American Meteorological Society Monographs*, vol. 4, no. 23, Boston, 1960.

reradiates a lesser amount of energy because it has a lower temperature.   The emission from the layer is divided into two parts, that going upward and that going downward. Thus the layer next above it receives only that portion of the energy which is radiated upward.   This layer in turn radiates in both directions a lesser amount of energy because it has a still lower temperature.   Of course the quantity that is radiated downward is absorbed again by the layers underneath, but this quantity of energy is less than that radiated upward by these lower layers, because the temperature is higher near the ground.   The final layer at the "top" of the atmosphere emits energy to space at its temperature.   Thus an atmosphere that in summation is opaque to infrared radiation emits blackbody radiation proportional to the fourth power of the temperature of the top layer.   And as long as the temperature decreases with height, the net flow is upward.

The net upward flux is accentuated by the energy that escapes through the transparent band and by the decrease with height of the amount of absorbing gas, especially water vapor.

A cloud layer of considerable thickness acts essentially as a blackbody, absorbing the terrestrial radiation in all wavelengths, including that coming upward through the water-vapor window.   From the top of the cloud layer the radiation proceeds as if the surface of the earth were at that height.   Satellite infrared measurements obtain the temperatures of cloud tops treated as a blackbody as detected through the water-vapor window.   Satellite measurements in the 6-μm band obtain an "effective radiating temperature" of the atmosphere.   This temperature should approximate that of the uppermost moist layers, at some indefinite height determined by the water-vapor distribution.   Data from a radiometer at 6.7 μm give an excellent portrayal of the water vapor between 5 and 10 km altitude and the air circulation there by means of this water vapor as a tracer.

The effects of a temperature inversion, that is, a layer of air in which the temperature increases with height, are readily seen.   If the overlying warm air contains an amount of water vapor comparable with that of the cooler air beneath, then the net transfer of heat by radiation will be downward.   In many temperature inversions existing in the atmosphere, however, there is also a sharp decrease in the water-vapor content in passing upward into the warmer air.   Considering the fact that absorption depends on the quantity of water vapor present as well as on the absorption coefficient, one conceives that it is possible that in the case of a very dry inversion there will be a net flux upward, just as in the case of a normal temperature distribution.   It is a difficult problem to determine the magnitudes and respective moisture distributions in temperature inversions required for a balance between upward and downward radiation.   Recent studies of this problem in connection with the formation of fogs and low clouds along the California coast indicate that the strongest temperature inversions are often associated with the lowest quantities of water vapor in the upper air.   In some cases, however, these inversions are so strong that even a relatively minute quantity of water vapor will radiate with sufficient intensity to cause an addition of heat to the layers beneath the inversion by radiation from those above. Weaker inversions, which nevertheless show a minute quantity of water vapor above them, represent the most favorable conditions for radiation heat loss outward through the atmosphere.

RADIATION FLUX AND COOLING IN THE ATMOSPHERE

The flux of radiant energy is from warm to cold layers except in cases where the emissivities of the layers differ in a counteracting sense.   For a radiation flux to occur through layers of the atmosphere without changing the temperature of those layers, there must be for each vertical distribution of temperature a given vertical distribution of water vapor or other absorbing material.   These may be called the *steady-state distributions*.   Between two layers of the atmosphere the upward flux is given by the difference between the energy radiated upward from the lower layer and that going downward from the upper layer.   If this difference increased with height, the upward flux would increase with height, and the layers would be sending more heat upward than they were receiving.   Thus they would cool.   This change of net flux with height is called the *flux divergence*.   If the net flux is represented by F_n, positive upward in height z, the flux divergence could be expressed in relation to the heat gained or lost as

$$\frac{dF_n}{dz} > 0 \qquad \text{cooling (positive flux divergence)}$$

$$\frac{dF_n}{dz} = 0 \qquad \text{steady state}$$

$$\frac{dF_n}{dz} < 0 \qquad \text{warming (negative flux divergence)}$$

Calculations show that the atmospheric gases absorb solar radiation equivalent to a heating of as much as 1°C or more per day, depending on height and latitude, but that heat losses from infrared emission, mainly by water vapor and carbon dioxide, more than compensate for the gain from solar heating; so a net cooling is found to predominate.   Nonradiative forms of heat exchange make up for the net loss.   Solar heating of the free atmosphere is at a maximum in two distinct parts of the atmosphere: (1) at 2 to 6 km, accounting for as much as 1°C per day in the summer hemisphere, due almost entirely to absorption by water vapor; (2) at 20 km and upward, equivalent to as much as 2°C per day at 30 km, due mainly to ozone absorption.   Cooling by infrared emission from the atmospheric gases shows three maxima: (1) at 6 to 12 km,

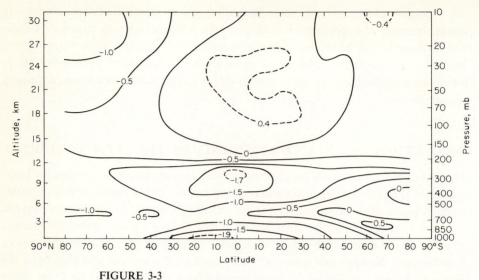

FIGURE 3-3
Rates of heating and cooling (−) of the atmosphere (°C per day), December–
February. (*After Dopplick.*)

equivalent to more than −2°C per day over the tropics due to water-vapor emission
toward the cold and dry air above; (2) near the surface, caused by a concentration of
water vapor there, especially in low latitudes, and about equal to the 6- to 12-km
cooling; (3) at 30 km and higher due to carbon dioxide emission, but almost exactly
canceled in the summer hemisphere by the ozone absorption of solar radiation. The
high-altitude CO_2 radiation is in agreement with the finding that this gas, which has
a high infrared emissivity, is well mixed through the whole atmosphere at 320 ppm.

The *net* heating and cooling by radiation is pictured in the two pole-to-pole
vertical cross sections of Figs. 3-3 and 3-4 for northern winter and summer, respectively, taken from a study by Dopplick.[1] Lines connecting points of equal heating or
cooling at intervals of 0.5°C are shown, plus dashed lines at lesser intervals to reveal
centers of maxima and minima. In both winter and summer, only the atmosphere
above about 14 km at low latitudes has a net heating—less than 0.5°C per day. The
remaining, much more extensive part of the atmosphere shows, with one or two minor
exceptions, a net cooling.

From these charts one sees the overpowering effect of water vapor and clouds in
the lower atmosphere and carbon dioxide in the upper atmosphere in producing radiative cooling, but the ozone heating comes in strong in the heating of the upper layers

[1] T. G. Dopplick, Radiative Heating of the Global Atmosphere, *Journal of the Atmospheric Sciences*, vol. 29, pp. 1278–1294, 1972.

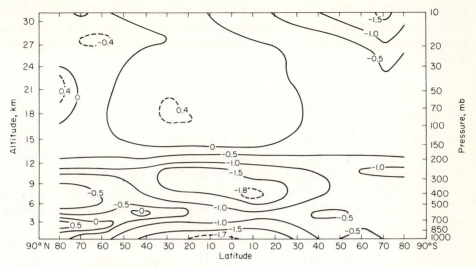

FIGURE 3-4
Rates of heating and cooling (−) of the atmosphere, June–August. (*After Dopplick.*)

over the tropics and into the brightly sunlit Southern Hemisphere in December–February and the equally sunny Northern Hemisphere in June–August.   It is seen that the negative effect of CO_2 at the top is most pronounced in the winter darkness of high latitudes.   The fact that there is a break in the distribution of cooling rates to small or near zero values at 2 to 6 km is due to the strong absorption of solar radiation by the water vapor there, as stated above.

These cross sections were prepared from computations based on the best data available in 1972.   Observed temperature distributions from balloon, rocket, and other soundings[1] provided a very important starting point.   Therefore, the results should not be interpreted as being aimed at determining the temperature distributions.   Also, average distributions and heights (therefore temperatures) of clouds were important factors in the computations.   Measurements of the absorbing gases made in various parts of the world were collected and certain approximations made to fill in gaps. Regardless of the incompleteness of the data and questions of validity of some of the approximations, the fact remains that if radiation were the only process acting, the atmosphere would give off more heat than it can absorb directly from the sun.

What balances these cooling rates?   In the first place, it must be remembered that the surface of the earth absorbs most of the solar radiation and transmits it back

---

[1] See Appendix B for a description of sounding methods and instruments.

to the atmosphere. Heat carried upward by convection, especially if there is a release of heat in condensation and precipitation, compensates for much of the radiation losses. The general circulation of the atmosphere helps, by advection, to preserve a steady state.

RADIATION CHARTS

Computer programs are used for routine or extensive determination of the radiation fluxes in the atmosphere. For instructional purposes and occasional use, charts are available for solving graphically the infrared radiation fluxes from a given vertical distribution of temperature and water vapor. Carbon dioxide and ozone absorption can be combined as constant contributions to the absorption, or separate charts for them can be used, provided their vertical distributions are known. A chart widely used in North America was developed by Elsasser[1] in 1942, with new data now provided.[2] It is shown in Fig. 3-5. The ordinate is water-vapor content (optical path, centimeters of precipitable water) increasing downward, and the abscissa is temperature, increasing to the left. The absorption is represented by areas on the diagram, and in the version shown here, a fixed area is enclosed at the top of the diagram to account for carbon dioxide absorption. The entire area to the right of a vertical line representing a temperature gives the blackbody emission for that temperature. For example, at the temperature T_s of the earth, the total emission in the absence of an atmosphere would be all the area of the diagram to the right of the line.

Starting from the surface with the CO_2 already assumed at B, one plots the accumulated vapor content as a function of temperature at the various heights through the atmosphere to produce the curve $BEDO$. The area above that curve bounded on the left by T_s is proportional to the infrared back-radiation from the atmosphere. The net loss from the surface is given by the difference between the whole area to the right of T_s and the area above the curve or, in other words, the area under the curve.

Taking another temperature level T_1 above the surface, one plots the accumulated water-vapor content downward to T_s to form $B'G$, and upward as before. The area under the curve to the right of T_1 represents the downward flux, and the area under $GB'C$ represents the upward flux at the level of T_1. In the case illustrated, the area under $GB'C$ is larger than that under $BEDO$; so more heat is going out at the level of T_1 than at T_s.

The optical path for water vapor (cumulative precipitable water) is plotted from the transmitted data from balloon or other soundings. If ρ_w is the vapor density, the

[1] W. H. Elsasser, Heat Transfer by Infrared Radiation in the Atmosphere, *Harvard Meteorological Studies*, no. 6, 1942.
[2] Elsasser and Culbertson, *op. cit.*

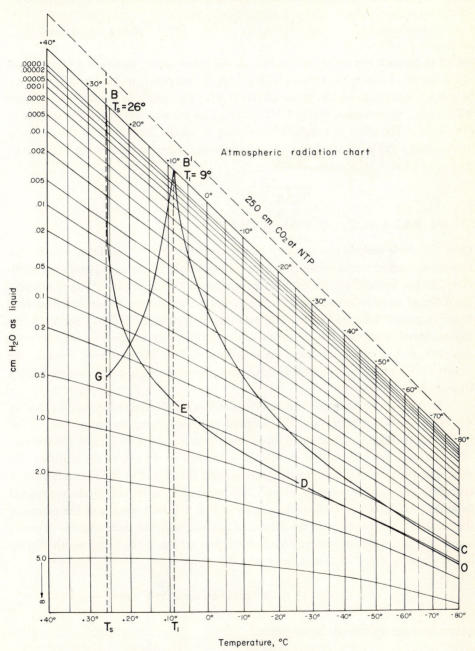

FIGURE 3-5
Elsasser diagram with plots as explained in text.

vapor content in a square centimeter column between heights z_1 and z_2 is given by

$$W = \int_{z_1}^{z_2} \rho_w \, dz$$

which in practice has to be computed from the mean vapor density in a layer (which the observer should select thin enough to suit the purposes) in the form $W = \bar{\rho}_w \, \Delta z$, where the bar refers to the mean of the layer. The result in grams is transferred directly to centimeters of precipitable water u_w on the assumption of unit density of the liquid. The effect of pressure broadening is usually introduced at this point by multiplying u_w by p/p_0, where p is the total pressure, taken as the average in the layer, and $p_0 = 1$ bar $= 1000$ mb.[1]

HEAT BALANCE OF THE EARTH

Over the years various investigators have pieced together all available information concerning temperatures, mean cloudiness, distributions of absorbing gases and particles, albedos, and emissivities of different surfaces and related data to figure out how the different areas or latitude belts of the earth contribute to the total balance between mean incoming solar radiation and outgoing terrestrial radiation. Naturally, as knowledge of absorption has increased, as worldwide observations have become more complete, and as computers have made heretofore hopeless problems solvable, the results of the studies have gradually improved. Earth-orbiting platforms are helping to provide better data.

In Fig. 3-6 are plotted the latitudinal distribution of solar radiation absorbed by the earth and atmosphere and the infrared radiation going out into space. The ordinate is in langleys per day and the abscissa is in a scale of the sine of the latitude. The area of a belt around the earth is proportional to the sine of the latitude. To determine the area of a belt, we integrate around the 2π rad of the circumference at latitude φ and over the width φ_1 to φ_2. The radius r of the latitude circle is measured from the polar axis, and a right triangle, as in Fig. 3-7, with apexes at the center of the earth and at latitude φ on the earth can be constructed to show that $r = R \cos \varphi$, where R is the radius of the earth considered as a sphere. The linear width of a $d\varphi$ belt is given by $R \, d\varphi$. The area integral is therefore

$$A = \int_{\varphi_1}^{\varphi_2} \int_0^{2\pi} R^2 \cos \varphi \, dL \, d\varphi = 2\pi R^2 \sin(\varphi_2 - \varphi_1)$$

where dL is the differential of longitude.

---

[1] For explanation of units, see Appendix C.

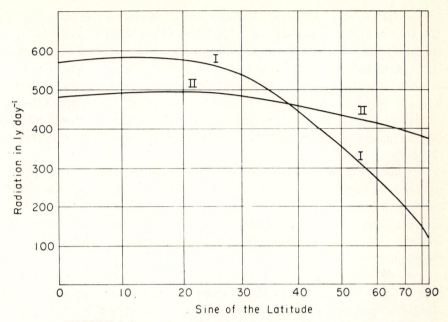

FIGURE 3-6
Distribution with latitude of absorbed and outgoing radiation as computed by
Houghton; curve I, solar radiation absorbed by earth and atmosphere; curve II,
long-wave radiation leaving the atmosphere. (*Ref. fn. 2, p. 50.*)

In Fig. 3-6 the total radiation energy is given by the areas under the respective
curves, and the areas between the curves show the excess or deficit of one form of
radiation compared with the other. If the curves are computed and plotted correctly,
the area between the curves poleward of the crossover should equal the area between
them equatorward of that point, thus showing a balance between income and outgo.

Now we can see why the atmosphere is never still. With the high-latitude deficit
and the low-latitude excess of energy accumulation, the middle to high latitudes could

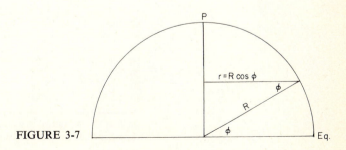

FIGURE 3-7

become colder and colder while the tropics and subtropics became hotter and hotter. This state of affairs is prevented by an exchange of heat through a circulation of the atmosphere and, to a lesser extent, of the oceans.   The fact that the circulation is highly perturbed rather than steady causes us to have weather instead of just climate as a day-to-day experience.   The general circulation, however unsteady or perturbed it may be, must produce climatically a net transport of heat energy across the latitude circles proportional to the accumulation of excess or deficit, and it should be expected that the greatest transport would be in the vicinity of the crossover of the two radiation curves.

Estimates of the poleward flux of heat required to overcome the radiation inequities have been made by various workers in the past 50 years.   An estimate by Sellers,[1] using data by Houghton[2] and by Budyko,[3] is given for the Northern Hemisphere in Table 3-1.   As expected, the maximum flux is in the vicinity of the radiation crossover.   It is noted that there is a southward (negative) flux across the equator into the Southern Hemisphere.

From average conditions for the Northern Hemisphere, computations of the heat budget of the earth-atmosphere system may be made.   Figure 3-8 presents Houghton's graphical summary—slightly modified—of the heat balance of the earth and atmosphere.   On the left side of the diagram the reflection, scattering, and absorption of the solar radiation are presented, and on the right the terrestrial effects are shown.

[1] W. D. Sellers, "Physical Climatology," pp. 66–68, The University of Chicago Press, Chicago, 1965.

[2] H. G. Houghton, On the Annual Heat Balance of the Northern Hemisphere, *Journal of Meteorology*, vol. 11, pp. 1–9, 1954.

[3] M. I. Budyko, Teplovoi Balans zemnoi Poverkhnosti, *Gidrometeorol. Izdatel'stvo.*, Leningrad, 1956. (English translation by N. A. Stepanova, Heat Balance of the Earth's Surface, Office of Technical Services, U.S. Department of Commerce, Washington, 1958.)

Table 3-1   REQUIRED POLEWARD FLUX OF HEAT ACROSS LATITUDE CIRCLES IN THE NORTHERN HEMISPHERE, IN UNITS OF 10^{19} KILOCALORIES PER YEAR

Latitude, °N	Flux	Latitude, °N	Flux
0	−0.26	50	3.40
10	1.21	60	2.40
20	2.54	70	1.25
30	3.56	80	0.35
40	3.91	90	0.00

SOURCE: W. D. Sellers, "Physical Climatology," The University of Chicago Press, Chicago, 1965.

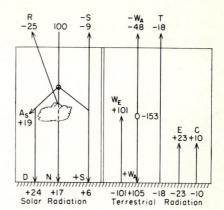

FIGURE 3-8
The heat balance of the earth and its
atmosphere as summarized by Houghton.

On the left, 100 units of solar energy are seen entering the top of the atmosphere and
34 going out again from reflection and scattering (albedo of 0.34).[1]   On the right
the remaining 66 units go out through the transparent region of the spectrum (18
units) and in the opaque region (48 units) to make up the external balance.   The
various symbols and quantities are as follows:

A_s = solar radiation absorbed in the atmosphere, 19 units

R = reflected solar radiation (depicted as coming from cloud, although some
comes from surface), 25 units

$-S$ = solar radiation scattered outward, 9 units

$+S$ = solar radiation scattered downward (sky radiation), 6 units

D = direct solar radiation reaching the earth, 24 units

N = diffuse solar radiation reaching the earth through clouds, 17 units

W_E = emission from the earth in the absorbing portion of the water-vapor
spectrum, etc., 101 units

$+W_A$ = downward flux of infrared radiation from the atmosphere in the
absorbing portion, 105 units reach the ground

$-W_A$ = upward flux of infrared radiation from the atmosphere in the absorbing
portion, 48 units leave at top

T = emission from the earth in the transparent region, 18 units

E = heat carried to the atmosphere in the hydrologic cycle (evaporation,
condensation, precipitation), 23 units

C = heat transported upward by convection or turbulence, 10 units to the
atmosphere

---

[1] This value of the albedo differs from the 0.36 used before, which was for the whole
earth, not just the Northern Hemisphere as considered here.

The balance is maintained as shown in Table 3-2. Thus, averaged over the entire year for the earth as a whole, there is a balanced heat budget.

Table 3-2

Balance to space			Surface balance				Atmospheric balance			
In	Out		In		Out		In		Out	
100	$-S$	9	D	24	W_E	101	W_E	101	$-W_A$	48
	R	25	N	17	E	23	E	23	$+W_A$	105
	$-W_A$	48	$+S$	6	C	10	C	10		
	T	18	$+W_A$	105	T	18	A_s	19		
100		100		152		152		153		153

EXERCISES

1 Treating the earth as a sphere with a radius of 6370 km, find the following earth dimensions:
   (a) The distance corresponding to 1° of latitude.
   (b) The length (circumference) of the 15th, 30th, 45th, 60th, and 75th parallels of latitude, and the distance corresponding to 1° of longitude on each.
   (c) The fraction of the surface of the earth contained within the tropics; within the two polar circles; between 30°N and 30°S.
2 A layer of water vapor in the atmosphere at 273 K radiates as a nonblackbody with an emissivity of 0.6. What is the radiation flux per unit area from the layer?
3 Convert the values in Table 3-1 to joules *per meter* per year along each of the given latitude circles.
4 On a replica of the atmospheric radiation chart, plot a current sounding from a nearby station, (a) from the surface upward, and (b) upward and downward from the base of a temperature inversion. Discuss the resultant radiation fluxes and net cooling. Note: The availability and interpretation of a sounding will depend on the instructor.

4

THE DISTRIBUTION OF TEMPERATURE

We now examine the averages or means of temperature in the atmosphere resulting from the different forms of heating and heat transport over and above the earth as described in the preceding chapter. The distribution of temperature over the earth's surface has been well documented from station records, some of which go back two or three hundred years. In the free atmosphere, data have been gathered from balloons and rockets mainly in the twentieth century, especially since the development of telemetering devices around 1940. Today upper-air soundings are made at least once daily throughout the civilized world. We shall examine first the data from upper levels.

Air density decreases exponentially with height, with the result that no upper boundary of the atmosphere can be defined. Half the weight of the atmosphere lies below about 6 km, three-fourths below about 10 km, and 99 percent below about 35 km. At about 90 km the gaseous composition of the atmosphere begins to change radically, audible sounds are not propagated, and finally the medium is describable only in terms of number concentrations of ions, energetic particles, or what is generally described as *plasma*.

That indefinable layer regarded as the outer limit of the atmosphere can best be described as the region where molecules cannot be held in the atmosphere but escape to space.   A height is reached where the collisions between molecules are extremely rare and a molecule from the denser atmosphere below has little likelihood of returning toward the earth by collision with molecules above.   Matter exists in interplanetary space, some of it originating outside the solar system.   The relationship of matter in near-earth space as influenced by the sun will not be discussed here.

Between, roughly speaking, 70 and 500 km are the layers of the ionosphere, well known as the important layers for long-distance radio transmission.   Radiation from the sun in the extreme ultraviolet ionizes the gases in these layers.   Cosmic rays also produce ions.   There are several ionized regions, but the zones of maximum ionization are the *E* layer and the *F* layer, centered at 150 and 250 km, respectively.   From 80 to 100 km the atmosphere of nitrogen and oxygen (O_2) is gradually transformed into one of nitrogen and monatomic oxygen (O).   The ultraviolet rays are responsible for dissociating the oxygen.   These heights are all approximate.

The study of the upper layers of the atmosphere is the subject of a branch of atmospheric and space physics and chemistry called *aeronomy*.   When speaking of meteorology, one normally does not mean to include aeronomy, but in its broadest sense meteorology may include all the atmospheres of all the planets.

In terms of temperature, the simplest divisions of the atmosphere in the vertical are the *troposphere*, 0 to about 10 km; the *stratosphere*, about 10 to 30 km; the *mesophere*, centered around 50 km; the *thermosphere*, above about 90 to perhaps 500 km, or to the region of the aurorae and airglow; then the *exosphere*.   The upper limits, or transition zones, of each of these layers bear the names *tropopause*, *stratopause*, *mesopause*, etc.   From Fig. 4-1 the logic in the naming of the different temperature layers is suggested.

THE VERTICAL DISTRIBUTION OF TEMPERATURE

Figure 4-1 shows the vertical temperature distribution taken as the standard for performance criteria of atmospheric and space vehicles.   It approximates the mean of balloon and rocket soundings in middle latitudes in the Northern Hemisphere.   The molecular-scale temperature above 91 km is the ordinary temperature corrected for the lower molecular weight of the air at those heights, given by $T \times m_0/m$, where m_0 is the molecular weight of 28.966, considered constant up to 91 km, and m is the molecular weight at the greater heights.   This is a region where the oxygen molecules are transformed into atomic oxygen, and a rapid increase in temperature with height is noted.

The *mesosphere* is characterized by high temperatures, reaching values as high as those at the ground.   The increase of temperature with height above the isothermal stratosphere to 47 km is explained by the intense absorption of ultraviolet radiation

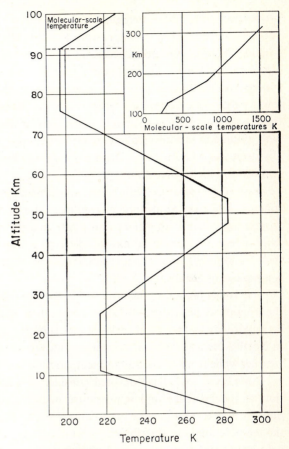

FIGURE 4-1
Standard atmosphere extended to about 300 km.

by the ozone which is concentrated there. The highest temperature is near the upper part of the ozone distribution where the solar ultraviolet is first intercepted and absorbed.

Leaving discussions of the upper atmosphere to more specialized texts, we shall proceed to describe the temperature conditions in the stratosphere and troposphere.

THE STRATOSPHERE

In a rough picturization of the stratosphere we usually think of a region in which the vertical distribution of temperature is isothermal. However, as more and more data are accumulated, it is brought out that it may in many cases consist of fairly pronounced temperature irregularities. In many situations, for reasons that are not yet fully understood, the temperature may increase with altitude in the stratosphere to great heights. On other occasions a number of small inversions and temperature lapses are indicated. Owing to the great difference in altitude of the base of the stratosphere over the earth, there is a wide variation in the temperatures observed in the lower part. The tropopause temperatures in the polar regions are very much higher than those over the tropics, but in the vicinity of the equator the temperature increases with altitude in the stratosphere, while in the polar stratosphere the temperature usually decreases with height. These two opposite temperature trends serve to diminish stratosphere meridional contrasts at levels near 20 km. In the polar stratosphere, the decrease of temperature with height sometimes approaches that of the upper troposphere. Since the stratosphere is defined and distinguished only by having a temperature decrease with height (lapse rate) markedly less than that of the underlying troposphere, one would be unable to locate a stratosphere in soundings made under these conditions. Such disappearance of the stratosphere over the antarctic region is a regular occurrence.

Between high altitudes over the tropics and low altitudes over the polar regions the tropopause has its greatest slope in middle latitudes. It is here also that the greatest day-to-day and seasonal variations in tropopause heights are found. For example, in North America, more or less polar stratosphere conditions accompany cold winter weather while tropical heights and temperatures accompany warm weather of summer.

Meteorologists have devoted considerable attention to a quantitative explanation of the existence and temperature distribution of the stratosphere. It is still difficult to present a concise explanation of why there must be a stratosphere or why the temperature is so distributed. We may approach the problem by starting with a cold isothermal atmosphere and allowing an accumulation of heat at the surface of the earth by absorption of solar radiation. This would cause a rapid increase of temperature at the surface in such a way that the temperature would be highest there and would decrease with altitude. Computations based on reasonable assumptions concerning the radiation fluxes produce a model in which the temperature decrease diminishes markedly at known stratosphere heights, but the frequently observed sharpness of the tropopause boundary does not appear. Vertical convection currents may arise that could carry heat upward or downward in the atmosphere, as the case might be; and as will be pointed out in more detail later, these vertical convection currents would accomplish a fairly complete stirring of the atmosphere and would produce within the stirred portion a uniform rate of temperature fall with height. There would be a certain maximum altitude of penetration of these convection currents. This height, which would also be the maximum altitude to which heating of the atmosphere would extend, would be the height of the tropopause. In the polar regions, where the amount of heating is small, this height would naturally be low, whereas in the tropics, where the heating is intense and convection can penetrate to high levels, the stratosphere would be found at much greater heights.

THE TROPOSPHERE

The temperature changes with which everyone is most familiar are those depending on latitude and altitude.   In general, temperatures decrease as one goes toward the poles and as one goes upward in the atmosphere.   The change of temperature with altitude is by far the most important, amounting, on the average, to about 6°C per km in the troposphere.   This is approximately 1000 times greater than the average rate of temperature change with latitude.   The presence of continents and oceans with their greatly contrasting temperatures modifies the latitudinal effect in a very pronounced way in certain localities.

The vertical distribution of temperature is likewise affected by the type of under-lying surface.   The temperature decreases most rapidly with height over continental areas in summer and oceanic areas in winter.   In cases of cooling by radiation losses at the surface or by contact of originally warm air with a cold surface, the vertical temperature lapse rate (rate of decrease) is often modified to the extent that the temperature actually increases with height through these lower layers, producing temperature inversions.   Inversions are most common over continents in winter and over oceans in summer; in other words they are characteristic of regions that are cool for their latitude because of surface conditions.

In meteorology the action in the troposphere is our main concern because storms and clouds are confined there.   However, studies of the tropopause and stratosphere have shown that the upper layers also have a direct bearing on the conditions that occur in the troposphere and at the surface of the earth.

Vertical cross sections of the mean temperature distribution in the troposphere and lower stratosphere along the 80th meridian from the equator to the North Pole in January and July are shown in Figs. 4-2 and 4-3.   The information is taken from charts by Landsberg and Ratner.[1]   The upper limit of the charts at 23 km represents the greatest height to which routine balloon soundings have provided enough data since World War II to arrive at good mean data.   The lines in the figures are *isotherms* connecting points of equal temperature in Celsius degrees.   The heavy line sloping upward toward the equator and showing a break in middle latitudes is the tropopause. It is noted that it forms a reasonably clear boundary between the region of persistent

[1] H. Landsberg and B. Ratner, Neue klimatologische Meridionalschnitte der Atmosphäre, in Linke–Baur, " Meteorologisches Taschenbuch," pp. 334–336, Akademische Verlagsgesellschaft Geest & Portig KG, Leipzig, 1964.

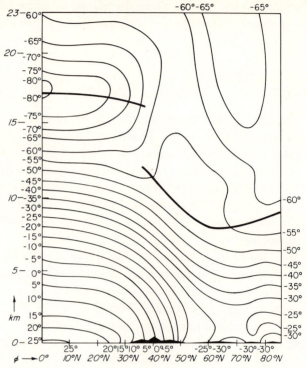

FIGURE 4-2
Climatic mean vertical cross section along the meridian of 80°W in the Northern Hemisphere—January.

temperature decrease with height in the troposphere and the region of slight or reverse temperature gradients in the stratosphere.

It is noted that the tropopause is lowest over the high-latitude stations and highest near the equator, with the maximum slope as well as breaks in middle latitudes. At all latitudes it is higher in summer than in winter and also has a more nearly constant and gradual slope in the warm season.

The troposphere exhibits stronger south-to-north temperature gradients in winter than in summer at all levels. Especially noteworthy on the winter cross section is the reversal of the isotherms north of lat 60° and below 3 km. This shows an increasing temperature with height, or temperature inversion, resulting from the strong winter cooling of the arctic and subarctic nights. Note also that the strongest horizontal temperature gradients are in middle latitudes, corresponding to the region of greatest slope of the tropopause. This is the region of greatest storm activity, and

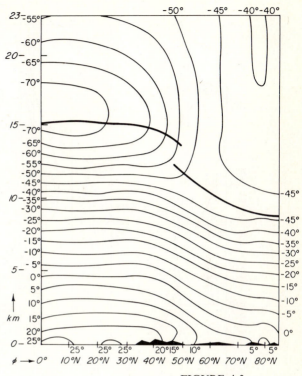

FIGURE 4-3
Same as Fig. 4-2 for July.

it should also be noted that this is near the crossover of the meridional solar and terrestrial radiation curves of Fig. 3-6.

These charts show that the coldest portion of the atmosphere lies over the equatorial region in the vicinity of the tropopause. For example, in winter the temperatures in the tropics at 17 km are below −75°C, whereas near the pole at the same altitude they are higher than −65°C with a maximum near −55°C at the tropopause. In summer the polar stratosphere has temperatures in excess of −45°C and the equatorial temperatures are lower than −70°C at 17 km, indicating a difference of nearly 30°C.

When the stratosphere was discovered toward the end of the nineteenth century in Europe, it was identified as an isothermal region. While this is seen to be true on the average in the latitudes of Europe, soundings in lower latitudes and near the poles show that this is not the case. The tropical stratosphere is characterized by a temperature inversion, as is also the case in the polar regions in summer. Until

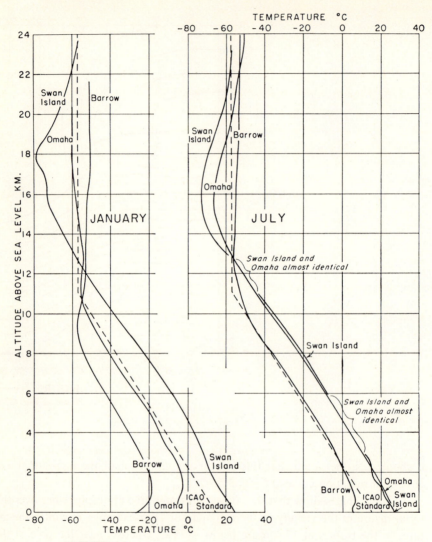

FIGURE 4-4
Graphs of mean winter and summer soundings in the North American region.

the 1920s meteorologists assumed that the isothermal stratosphere extended to the limits of the atmosphere, which as we have already seen, is far from true.

Another way to represent the temperature structure of the atmosphere is by means of soundings at individual stations plotted as a graph of temperature against height. Graphs of mean winter and summer soundings in North America as represented by three stations are shown in Fig. 4-4. The stations are Swan Island, 16°12'N, 83°55'W; Point Barrow, Alaska, 71°25'N, 156°30'W; and Omaha, Nebraska, a continental station at 41°25'N, 95°56'W. Plotted with these curves is the so-called "standard atmosphere," adopted for engineering and navigation purposes, which approximates a year-round average of all latitudes. The stations on these graphs are a few of the hundreds used for the construction of the cross sections in Figs. 4-2 and 4-3, and they show the features representative of their latitude. As a continental station, Omaha shows the effects of continental cooling of the lower atmosphere in winter. The average sounding of winter shows approximately isothermal conditions through the first 2 km. This average is made up of day-to-day changes from extreme inversions to conditions of appreciable lapse rates. In the summer mean soundings, Omaha has approximately the same troposphere temperatures as tropical Swan Island.

TROPOSPHERIC SUBDIVISIONS

Because of its exposure to the strong radiation, convection, and evaporation-condensation effects of the earth-air interface, the troposphere exhibits several zones or layers which are identifiable and classifiable according to their relationship to the energy exchanges at or near the surface.

At the very bottom is the *biosphere*, which is within the immediate environment of plants, insects, and other forms of life. The conditions are largely determined by the plant cover, soil covering, whether desert, grassy, bushy, open woods, forest, or thicket. It is a zone of relatively little air motion. Gases such as ammonia, methane, radon, and thoron, which are normally found in negligible trace amounts in the atmosphere, are sometimes important here, and the oxygen and carbon dioxide content is markedly affected by photosynthesis of plants and various forms of oxidation and bacterial action. The biosphere is of principal interest in agricultural applications. The designation usually is not applied to the air-sea boundary.

A layer described as the *boundary layer* or *microlayer* receives special attention in a special branch of meteorology known as *micrometeorology* or *microclimatology*. Frictional effects and low-level turbulence, including vertical flux of properties and materials (heat, momentum, water vapor, dust, and smoke) by turbulent eddies, are studied. Radiation exchanges and heat fluxes through the surface are also considered. The practical aims include measures of evaporation; wind stress on the sea surface and resulting wind-driven currents; diffusion of air pollution, military smoke screens, and gas barrages; understanding of the wind variation with height, in both speed and direction; and a variety of other problems associated with the microlayer. The layer is usually considered through a height of less than 100 m but may include a thicker layer, depending on the type of phenomenon studied.

It is convenient also to note a layer of diurnal influence and appreciable internal friction effects which is about 0.5 to 2 km deep, containing in many cases temperature inversions. There is a wide variation in the structure of these layers, depending upon the different relationships of heat transfer between the surface of the earth and these layers by radiation, convection, and eddy flux.

Variations with latitude are of considerable importance because of the transport of conditions acquired in one latitude to another, thus affecting the weather at various places by these advected properties. In high latitudes, where the outgoing terrestrial radiation is large, a cooling of the lower layers is pronounced. Therefore, we find that, in the polar regions, temperature inversions are of frequent occurrence. Over tropical regions, where the incoming solar radiation exceeds the loss by terrestrial radiation, pronounced temperature inversions do not occur.

Geographical variations in the lower layer are found because of different types of surfaces, such as land as opposed to sea surfaces. There is a tendency for the creation of temperature inversions over the land in winter and over the oceans in summer because air transported from oceans or other warm regions to the land in winter is chilled from below, and in the same way land air or tropical air carried by advection over the ocean in summer is cooled from below.

Over the continents pronounced diurnal variations occur in the temperature structure of the lowest layers. During the night there is a net loss of heat from the surface of the earth by terrestrial radiation when the sky is not overcast with clouds. One of the most common types of temperature inversion is the so-called "nocturnal inversion." On early-morning temperature soundings it almost invariably appears. This inversion usually disappears completely during the day, at least in the warmer part of the year, as the sun heats the earth's surface.

The Middle and Upper Troposphere

In most places during the course of a single day there is not enough heating of the surface of the earth to affect the temperatures above about 2 km. In terms of days and weeks the heating that occurs above this height, mainly from terrestrial radiation and convection, accumulates sufficiently to cause a gradual warming in spring, and conversely, the radiation losses as the days shorten and the sun gets lower cool the upper air in the fall. This effect is shown in Fig. 5-4, which indicates that in summer the temperatures in the troposphere above 2 km in middle latitudes average about 20°C higher than in winter.

The Tropopause

The tropopause is a transition layer between the troposphere and the stratosphere. Soundings made at single stations usually show a sharp delineation between the normal rate of temperature decrease occurring in the troposphere and the approximately isothermal distribution of the stratosphere. In some cases it is difficult to find such a well-marked discontinuity, especially when an attempt is made to locate a tropopause over a large area from a network of sounding stations. A tropopause delineated at one station may have a pressure and temperature combination such that no atmospheric processes can be found that would associate this tropopause layer with that noted at some other station. Consequently, meteorologists have come to recognize the *principle of multiple tropopauses*, in which it is realized that there is normally not one tropopause extending over a large area, but rather a series of overlapping steps in the tropopause, ranging from high altitudes and low temperatures over equatorial regions to low altitudes and high temperatures near the poles. These discontinuities may vary in location from day to day so that the mean data present only a smoothed picture.

HORIZONTAL TEMPERATURE DISTRIBUTION

Since ancient times people have been aware of differences in the average temperature conditions from place to place over the surface of the known earth. Likewise people have known that these temperatures were not wholly dependent on latitude. Furthermore the ancients knew that weather and temperature changes from day to day at a given locality are due to the transport by the winds of the temperature climates of warmer or colder regions—advection. Thus a connection between climatology—the average atmospheric conditions of the earth—and weather forecasting was established. Climatology is a special branch of meteorology and geography that treats of mean temperatures of the earth along with a number of other atmospheric properties. In discussing mean temperatures in this chapter, only a broad view of world temperature distribution will be considered, a more detailed treatment being more in the province of a book on climatology.

Only the data for the horizontal distribution of temperature at the immediate surface of the earth are available in abundance. In the upper air, there are large gaps in the data, but available observations indicate that local differences are much less pronounced than at the surface.

In representing the temperature distribution at the surface, one encounters the difficulty that, because of mountains and plateaus, one is obtaining a combined horizontal and vertical view of temperature variations. It is desirable to have a means of comparing temperatures and representing them on the same horizontal surface.

For this reason climatologists favor having all temperatures reduced to sea level by adding an empirically determined factor. This eliminates a tangled maze of isotherms that would otherwise be found in mountainous regions where no two stations are likely to be at the same elevation, but it produces some ridiculous results. For example, the reduction of temperature to sea level makes the relatively cool plateaus of Mexico and Abyssinia appear as the hottest regions on earth in summer.

The two charts in Figs. 4-5 and 4-6 show the world distribution of mean surface temperature in January and July, with isotherms in °F. Several features of these charts should be noted. In January the continents of the Northern Hemisphere show extreme winter conditions while the oceans exhibit mild conditions. The trend of the isotherms in western Europe and along the Pacific Coast of North America indicates that the oceanic temperature climate is transported by westerly winds; the eastern coasts show relatively little marine influence. The greatest meridional temperature gradient is over the continents, the least over the oceans. The intense winter cold of northern Siberia is especially striking, and the effects of the warm Gulf Stream are noted in the Atlantic extending even to the north of Norway. January is midsummer in the Southern Hemisphere. The continents are relatively less conspicuous than in the Northern Hemisphere, but they nevertheless stand out as warm spots. The low temperatures along the coast of Peru and Chile and the western coast of South Africa are caused by cold ocean currents in those regions.

The July map shows the Northern Hemisphere continents as considerably warmer than the oceans. The low temperatures along the Pacific Coast of the United States are due to the transported marine climate plus the effects of the cold California Current which is the counterpart of the cold current off the western coast of South America. A similar cool belt is noted on the northwestern coast of Africa. Eastern coastal locations of the Northern Hemisphere, particularly in the United States, show practically no marine influence, again suggesting the mechanism of a temperature climate transported from the continent.

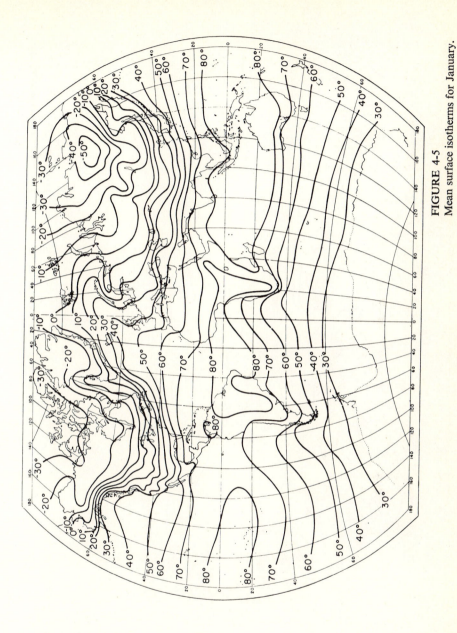

FIGURE 4-5
Mean surface isotherms for January.

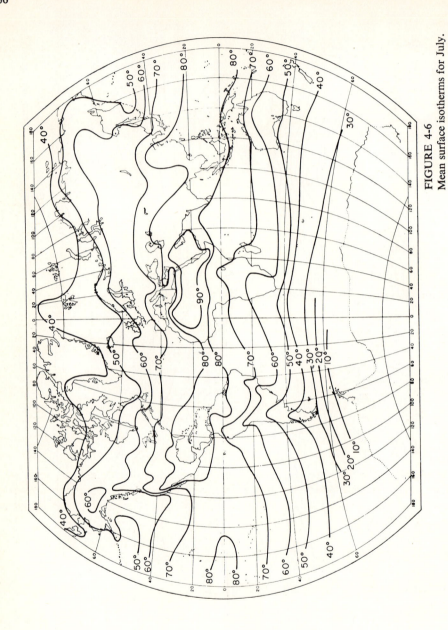

FIGURE 4-6
Mean surface isotherms for July.

EXERCISES

1 From Fig. 4-1 determine the temperature lapse rate of the standard atmosphere.

2 If the temperature at sea level at Miami, Florida, is 27°C and the tropopause is at 17 km, with a temperature of −68°C, what is the average tropospheric lapse rate in degrees Celsius per 100 m?

3 At Chicago, the temperatures as in the above problem are 15 and −57°C, respectively. With an average lapse rate of 0.6°C per 100 m, what is the height of the tropopause?

4 From San Francisco to Honolulu the California coast temperature inversion rises from its typical height of 500 m to become the trade-wind inversion at 2000 m. The ocean surface temperature at the San Francisco beach is 12°C, and at the Hawaiian Islands it is 26°C. In each locality the lapse rate up to the inversion is 0.6°C per 100 m.

(*a*) Find the temperature at the base of the inversion at each place.

(*b*) At the top of the inversion at 1700 m the temperature is 20° at San Francisco and 15° at 2800 m at Hawaii; and above, in both places, the lapse rate is 0.6°C per 100 m. What would the surface temperature have to be to destroy the inversion with the same lapse rate at San Francisco and at Hawaii? (Solve this problem by plotting on a sheet of graph paper.)

5 Given the following information for Miami and Milwaukee:

	Miami	Milwaukee
Station elevation	4 m	210 m
Surface temperature	27°C	15°C
Height of tropopause	17 km	?
Temperature at tropopause	−68°C	−57°C
Average lapse rate, °C/100 m	?	0.6

Supply the missing information indicated by the question marks.

6 The following places are in almost exactly the same latitude: Eureka, California; Omaha, Nebraska; New York, New York; Madrid, Spain; Naples, Italy; Istanbul, Turkey; Tashkent, USSR; Peking, Peoples Republic of China; and Aomori, Japan. From the January and July charts of surface isotherms (Figs. 4-5 and 4-6) determine the approximate temperature range between January and July at each of these places. Explain the differences in the temperature ranges.

5

THERMODYNAMICS AND STATICS

The atmosphere is a huge thermodynamic engine, driven by the energy received from the sun. The general circulation over the earth, all winds, storms, and clouds result from the differences in the amount and utilization of this energy. Since the radiant energy appears principally as heat, the student of meteorology must understand how the air reacts to heat changes; in other words, he must understand the *thermodynamics* of the atmosphere. Furthermore, since the thermal state of any portion of the atmosphere also determines its weight, a consideration of *statics* is required.

THE GAS LAWS

The atmospheric circulations mentioned above come about through changes in pressure, temperature, and density patterns of the air. Considering first the density, we note that it is defined as the mass per unit volume $\rho = M/V$, and that in a rigid body of simple configuration the density can be obtained easily with a pair of calipers or measuring tape and a weighing balance or scale, making use of the defining relationship. In gases, the volume changes with changes in pressure or temperature, so that this simple measurement based on the relationship between density, mass, and volume cannot readily be made although the relationship still holds.

The Laws of Boyle and Gay-Lussac

The relationship between pressure, temperature, and volume (therefore, also density) in gases has been determined by early experiments of classical physics. One of the fundamental concepts derived from these early experiments is Boyle's law, which states that, in a gas that is kept at constant temperature,

$$pV = p'V' = \text{const}$$

where p and V are the pressure and volume at one stage and p' and V' at another stage of the constant-temperature, or isothermal, process. This law is good for all common gases not too near their liquefaction temperature and is therefore applicable to the atmosphere.

The early experimenters recognized the property of nearly all matter that it expands on being heated. The experiment complementary to Boyle's was to determine the volume expansion of a gas with increasing temperature under constant pressure. Gay-Lussac related the change to the Celsius, or centigrade, temperature. If V_0 represents the volume at 0°C, then the change in volume as the temperature is raised or lowered to t°C is given by

$$V - V_0 = V_0 \alpha_p t$$

or, the new volume is given by

$$V = V_0(1 + \alpha_p t) \qquad (5\text{-}1)$$

where α_p is the volume coefficient of expansion with constant pressure. The coefficient was found to be constant for all gases at all temperatures not near the critical temperatures of liquefaction of the gases. Regnault determined the value of $\frac{1}{273}$ for α_p which, through Lord Kelvin's gas-thermometer experiment, became the accepted value for establishing the absolute, or Kelvin, scale of temperature. (See the section on units in Appendix C.)

We can combine the laws of Boyle and Gay-Lussac by considering two constant-pressure processes, one at pressure p and the other at a standard pressure p_s, and both with the same temperature t. At the pressure p the volume may be expressed by (5-1) and at the pressure p_s by $V_s = V_{0s}(1 + \alpha_p t)$, where the subscript s refers to standard pressure conditions. From Boyle's law, since p, V, p_s, and V_s all depend on the temperature t,

$$pV = p_s V_s = p_s V_{0s}(1 + \alpha_p t) = p_s V_{0s} \alpha_p \left(\frac{1}{\alpha_p} + t\right) \qquad (5\text{-}2)$$

Let $p_s V_{0s} \alpha_p$ be called C, since they are all constant, and $(1/\alpha_p) + t = 273 + t$ be called T, the Kelvin temperature; then

$$pV = CT \qquad (5\text{-}3)$$

Avogadro's Law and the Meaning of C

According to Avogadro's law the volume of a mole or gram-molecular weight (the number of grams of a gas equal to its molecular weight) is constant for all gases at the same temperature and pressure; therefore, the volume of a mole is a universal function of the temperature and pressure. At NTP it is 22,414 cm^3. If v is the volume occupied by 1 mole, then $v = V/n$, where n is the number of moles in the volume V (any volume). If we designate C/n as a constant R, then

$$pv = RT \qquad (5\text{-}4)$$

and R will be a universal constant, applicable to any gas. Also, since $C = nR$, we can write the relationship

$$pV = nRT \qquad (5\text{-}5)$$

The last two equations are forms of the so-called "equation of state for an ideal gas," expressed in terms most often used in physics.

Another way to consider the equation of state is to write it for a single molecule or by making use of Avogadro's number. Avogadro deduced that a mole of every gas contains the same number of molecules, called Avogadro's number N. If we divide (5-4) by N, we have the expression for a single molecule, or we can write

$$pv = NkT \qquad (5\text{-}6)$$

where $k = R/N$ is now a universal gas constant in terms of a molecule. It is known as Boltzmann's constant, and it is used in other contexts in physics, such as in Planck's radiation law in Chap. 2. In cgs units R is in ergs per degree per mole and k in ergs per degree per molecule. Pressure is expressed as force per unit area—ergs per square centimeter or newtons per square meter. Normal temperature and pressure (NTP) is 273 K and 1.0132×10^6 ergs cm^{-2}. (See Appendix C for a fuller discussion of units.) The various values can be obtained by substituting from a known relationship, such as v for NTP, thus: $pv/T = R$, or

$$\frac{1.0132 \times 10^6 \times 2.24 \times 10^4}{273} = 8.3143 \times 10^7 \text{ ergs K}^{-1} \text{ mol}^{-1}$$

Alternatively, the pressure can be in newtons per square meter (N m^{-2}) and v in cubic meters to produce R in joules per kelvin per kilogram-mole, thus eliminating the factor 10^7. Avogadro's number has been determined as 6.0248×10^{23} molecules per mole; so $k = R/N = 1.3806 \times 10^{-16}$ erg K^{-1} or 1.3806×10^{-23} J K^{-1}. Note that "per molecule" is not added here; it is understood when Boltzmann's constant is used.

Equation of State in the Atmosphere

If M is the mass of the volume V and m is the gram-molecular weight, then $n = M/m$ and therefore

$$pV = \frac{M}{m} RT \qquad (5\text{-}7)$$

but

$$\frac{V}{M} = \frac{1}{\rho} = \alpha \qquad (5\text{-}8)$$

where ρ is the density and α is the specific volume, or the number of cubic centimeters occupied by unit mass of the gas, and in these terms we can write the equation as

$$p\alpha = \frac{R}{m} T \qquad (5\text{-}9)$$

or

$$p = \rho \frac{R}{m} T \qquad (5\text{-}10)$$

These two forms are the ones in which the equation of state is expressed for meteorological purposes in terms of the universal gas constant. Sometimes the equation is written without the m, becoming $p\alpha = R'T$, where R' is the individual gas constant, different for every gas and equal to R/m.

Strictly speaking, air does not have a molecular weight, since it is a mixture of gases and there is no such thing as an air molecule; however, it is possible to assign a so-called "molecular weight" to dry air, which will make the equation of state operative. With sufficient accuracy we may take the value 28.9. For water vapor it has the value 18, which is the molecular weight of water.

Exact measurements show that the equation of state is only approximately valid. The higher the pressure and the lower the temperature, i.e., the nearer the gas approaches liquefaction, the greater the departure from the conditions of the equation. An ideal substance that follows the equation exactly is called an *ideal gas*. Real gases conform to the equation of state approximately, but the gases as they exist in the atmosphere may be considered as ideal gases. There is hardly a relationship in meteorology that is more basic and important than the equation of state. The student and professional meteorologist must make regular and frequent use of it either as a descriptive concept or for computations. It will often be referred to in this book.

Effects of Water Vapor—Virtual Temperature

All the gases in the atmosphere, including the water vapor, exert their own partial pressures. Concerning this, we have Dalton's law, which states that the total pressure of a gas mixture is equal to the sum of the partial pressures. Also, we have the corollary that, in a mixture of gases, every gas occupies the whole volume of the mixture as if the other gases were not present.   In meteorology we find it convenient to consider the mixture of "dry gases" forming one "gas" ("molecular weight" 28.9) and the water vapor (molecular weight 18) another.

The equation of state can be applied as in Eq. (5-10) to a mixture of gases provided that the relative quantities of the different gases in the mixture do not vary. Otherwise, the "molecular weight" m would not be a constant.   Water vapor is the only important gas in the atmosphere that varies in significant amounts, and it is found necessary to treat it separately or arrive at some means of taking into account the variable effects of its presence.   Since water vapor has a lesser molecular weight than that ascribed to dry air, its presence at a given temperature and pressure would tend to lower the density of the mixture.   We write the equation of state for water vapor in the form

$$e = \rho_w \frac{R}{m_w} T \qquad (5\text{-}11)$$

or

$$\rho_w = \frac{e m_w}{RT} \qquad (5\text{-}12)$$

where e is the partial pressure of the water vapor and the subscript w refers to water vapor.

We consider the other gases (dry air) separately and write

$$p_d = \rho_d \frac{R}{m_d} T \qquad (5\text{-}13)$$

or

$$\rho_d = \frac{p_d m_d}{RT} \qquad (5\text{-}14)$$

The two equations of state can be combined in such a way as to take into account the effects of water vapor.   To do this, meteorologists find it is not convenient to vary the "molecular weight," but instead they ascribe a fictitious temperature to the mixture, which would depend partly on the amount of water vapor present.   This can be interpreted from a recognition of the fact that an increase in water vapor decreases the air density, and an increase in temperature also decreases it.   We assign a fictitious increase to the temperature in just the right amount to correspond to the lowered density caused by the water vapor.   The new temperature is called the *virtual temperature*.   It may be defined as the temperature of dry air having the same total pressure and density as the moist air.

The combination of the equations of state for dry air and water vapor in terms of the virtual temperature and the total density ρ is accomplished as follows:

$$\rho = \rho_w + \rho_d = \frac{em_w + p_d m_d}{RT}$$

$$= \frac{em_w + (P - e)m_d}{RT} \tag{5-15}$$

where P is the total pressure ($p_d + e$). If we take Pm_d as a multiplier, we obtain

$$\rho = Pm_d \frac{\dfrac{e}{P}\dfrac{m_w}{m_d} + \left(1 - \dfrac{e}{P}\right)}{RT}$$

$$= \frac{Pm_d}{RT}\left(1 - \frac{3}{8}\frac{e}{P}\right) \tag{5-16}$$

since $\dfrac{m_w}{m_d} = \dfrac{18}{28.9} = \dfrac{5}{8}$ (approximately). If we call $T^* = \dfrac{T}{1 - \frac{3}{8}e/P}$, then

$$\rho = \frac{Pm_d}{RT^*} \tag{5-17}$$

where T^* is the virtual temperature. It will be pointed out later that e/P is a function of the *specific humidity q*, which is the number of grams of water vapor in a gram of air, so that we also have

$$T^* = \frac{T}{1 - \frac{3}{5}q} \tag{5-18}$$

By means of the virtual temperature, it is possible to use the equation of state in terms of the total pressure and the "molecular weight" of dry air. Since q or e/P seldom exceed 0.02, T^* seldom exceeds T by more than 2 or 3°. The correct equation of state for moist air is, then,

$$P\alpha = \frac{RT^*}{m_d} \tag{5-19}$$

or

$$P = \frac{\rho RT^*}{m_d} \tag{5-20}$$

THE FIRST LAW OF THERMODYNAMICS

The Concept of Internal Energy

The kinetic theory of gases relates the temperature of a gas to the rate at which the molecules are moving about, showing that the temperature is proportional to the mean kinetic energy of the moving molecules. The kinetic theory shows that the kinetic energy of the translatory motion of the molecules is proportional to the temperature.[1] If no heat is added but the gas is mechanically compressed, the kinetic energy will also be increased by the compression, and the temperature of the gas will rise.

Thus we define the *internal energy* of a given mass of a gas as the total energies of all the molecules in that mass, and note that it is proportional to the temperature.

Statement of the First Law

The first law of thermodynamics states that an increase in the internal energy can be brought about by the addition of heat or by performing work on the gas, or by a combination of both. If we represent the change in internal energy by dE, positive for an increase, and the heat added as $+dQ$, and if dW represents the work done *on* the gas, negative for work done *by* the gas, then we have the following relationship:

$$dE = dQ + dW \qquad (5\text{-}21)$$

Thus, in the case mentioned above, where the gas is compressed without adding heat, we would have $dE = dW$. In other words, the increase in internal energy would be caused by work (compression) performed on the gas. In the case of direct heating where no work is done, $dE = dQ$, and it is not difficult to conceive of a process in which both heating and mechanical work would be involved.

Work Done by External Forces

In considering the work done on a gas, we are mainly concerned with pressure. Pressure is defined as force per unit area.

$$p = \frac{F}{A}$$

or
$$F = pA \qquad (5\text{-}22)$$

Consider Fig. 5-1, which represents a parcel of air in the atmosphere. Any pressure p which may be applied in the atmosphere would act equally in all directions. Therefore, this pressure would be exerted on the parcel from all sides, causing it to be compressed to a smaller volume by an amount

$$-dV = A \, dn \qquad (5\text{-}23)$$

Here dn is the distance between the outer surface of the parcel in the two positions and A is the area of this surface, assumed to be the same in both cases because dn is

---

[1] In gases whose molecules contain two or more atoms, a part of the internal energy is stored as energy not only of translatory motion but also of rotation and vibration of individual molecules.

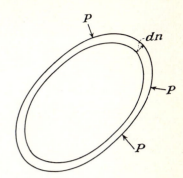

FIGURE 5-1

taken as infinitesimal.   The work done on the parcel in this compression is given, as always, by the product of the force acting on the parcel times the distance through which the force acts, measured in the direction of the force.

$$dW = F\,dn \qquad (5\text{-}24)$$

and by substituting for F from Eq. (5-22), we obtain the work done on the parcel in terms of pressure and volume,

$$dW = pA\,dn = -p\,dV \qquad (5\text{-}25)$$

Changes In Internal Energy

The temperature may be changed by holding the volume constant and varying the pressure, or by keeping the pressure constant and varying the volume.   This may be seen in Fig. 5-2, where the equation of state is represented in a pV diagram, with temperatures shown in the curved lines (rectangular hyperbolas).   It is apparent that the temperature may be changed by following along a horizontal line $p = $ const, or along a vertical line $V = $ const.   We shall arrive at the change in internal energy by considering these two processes.

Having obtained the expression for the work done dW, we may now write the equation for the first law as

$$dE = dQ - p\,dV \qquad (5\text{-}26)$$

If we consider a constant-volume process, $p\,dV = 0$.   The heat added dQ, corresponding to a change in temperature dT, will be given by the mass times the heat required to raise unit mass $1°C$ at constant volume, i.e., the specific heat, multiplied by dT, or

$$dE = MC_v\,dT \qquad (5\text{-}27)$$

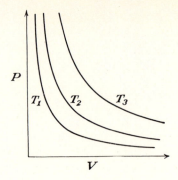

FIGURE 5-2
pV diagram.

The quantity C_v is the specific heat.   It is determined by experiment and is found to be constant.   The expression (5-27) is taken as the definition of the change in internal energy and can be substituted for dE whether one is dealing with a constant-volume process or not.

Equation (5-27) is a statement of Eq. (5-26) for the case of heat added at constant volume so that $dV = 0$ and no external work is involved.   The heat added all goes toward changing the internal energy.   If heat is added at constant pressure, $dV \neq 0$ and the gas performs work against the environment through expansion as it is heated.   We get two different results depending upon which of these two conditions is imposed during the heating.   The amount of heat required to raise the temperature of a given mass of the substance by one degree will be different in the two cases. That is another way of saying that there are two specific heats of a gas, the specific heat at constant volume C_v and the specific heat at constant pressure C_p, both constant.

To see how the two specific heats differ, we may consider a constant-pressure process involving the same change in internal energy dE as in the constant-volume process of Eq. (5-27).   We substitute for this dE in the fundamental Eq. (5-26) and apply the heat dQ at constant pressure:

$$MC_v\, dT = dQ - p\, dV \qquad (5\text{-}28)$$

At constant pressure, the heat added would be $MC_p\, dT$, where C_p is the specific heat at constant pressure.   The change in internal energy in the constant-pressure process would be

$$dE = MC_v\, dT = MC_p\, dT - p\, dV \qquad (5\text{-}29)$$

This may be written in the form

$$M(C_p - C_v)\, dT = p\, dV \qquad (5\text{-}30)$$

The equation of state (5-5) in differential form is

$$p \, dV + V \, dp = nR \, dT \qquad (5\text{-}31)$$

In this constant-pressure process, $dp = 0$; so

$$p \, dV = nR \, dT \qquad (5\text{-}32)$$

or

$$C_p - C_v = \frac{nR}{M} = \frac{R}{m} \qquad (5\text{-}33)$$

since, from Eq. (5-7), $n/M = 1/m$. The difference $C_p - C_v = R/m$ is sometimes called the *molecular heat difference* because it is inversely proportional to the molecular weight m. The ratio of the specific heats, which can be obtained from Eq. (5-33) by dividing by C_v, is

$$\frac{C_p}{C_v} = 1 + \frac{R}{mC_v} \qquad (5\text{-}34)$$

It is nearly the same in all gases in the same class, as defined by the number of atoms contained in their molecules. For example, in monatomic gases like helium it has the value 1.67; in diatomic gases (N_2, O_2, etc.) about 1.40; in gases with from three to five atoms per molecule (H_2O, CO_2, NH_3, etc.) about 1.30.

Application to the Atmosphere

In the atmosphere, volume measurements are difficult, and it is preferable to use Eq. (5-28), after dividing by M, as

$$dq = C_v \, dT + p \, d\alpha \qquad (5\text{-}35)$$

where dq is the heat added to unit mass and $d\alpha$ is the change in specific volume. This equation can be reduced to terms of pressure and temperature by writing the equation of state in differential form

$$p \, d\alpha + \alpha \, dp = \frac{R}{m} \, dT$$

$$p \, d\alpha = \frac{R}{m} \, dT - \alpha \, dp \qquad (5\text{-}36)$$

and substituting for $p \, d\alpha$ with the result

$$dq = C_v \, dT + \frac{R}{m} \, dT - \alpha \, dp \qquad (5\text{-}37)$$

But since
$$C_p - C_v = \frac{R}{m}$$

and therefore
$$C_v + \frac{R}{m} = C_p$$

we note that
$$dq = \left(C_v + \frac{R}{m} \right) dT - \alpha \, dp$$

and therefore
$$dq = C_p \, dT - \frac{R}{m} T \frac{dp}{p} \qquad (5\text{-}38)$$

the last term on the right being obtained by solving the equation of state for α.

Equation (5-38) is another of the half dozen or so highly important basic equations used to describe the properties, motions, and transformations of the atmospheric medium. It is a statement of the first law of thermodynamics in terms of the easily measured atmospheric variables, temperature and pressure, and requires no consideration of measured or controlled volumes or masses. It will now be shown how in the atmosphere we often find Eq. (5-38) simplified by the prevalence of processes in which $dq = 0$.

Adiabatic Process

The adiabatic process is defined as one in which there is no heat added or taken away, in which $dq = 0$. In an adiabatic process the change in internal energy of the gas would, therefore, be due entirely to the work performed on it, particularly by forces compressing the gas, or by the work done by the gas in the form of an expansion. In the atmosphere both adiabatic and nonadiabatic processes are occurring. The addition of heat to the atmosphere near the ground when the surface is warmer than the overlying air and the removal of heat in the same manner when the surface is colder are continuous nonadiabatic processes. In the free atmosphere where there is no solid or liquid surface to give off or remove heat and where the amount of energy absorbed from the sun's rays or lost by direct radiation is insignificant by comparison, we are justified in assuming that all short-period processes are essentially adiabatic. Since in the atmosphere the pressure decreases rapidly with elevation, adiabatic temperature changes, which as is seen above are determined solely by pressure changes, occur most readily when portions of the air undergo motions having a vertical component. Thus, a parcel of air carried upward cools adiabatically with the decreasing pressure that it experiences, and conversely, a descent with increasing pressure causes an increase in the temperature.

From the first law of thermodynamics it is possible to obtain an equation by means of which, if given the temperature and pressure of a parcel or sample of air at the beginning of an adiabatic process, its temperature at any known pressure in which it is found later on during that process can be calculated.   The derivation of such an equation affords the best opportunity to introduce the concept of *entropy*.   The complete explanation of entropy is derived from the second law of thermodynamics. However, since the second law involves mainly concepts that are not generally applied in meteorology at the present time, no effort will be made to enter into a general discussion of it here.   For our purposes we shall merely define entropy.   We shall make use of the specific entropy, which is the entropy applicable to unit mass of the substance.   Expressing it in differential form $d\varphi$, meaning the change in entropy, we have the relationship applicable to any reversible process

$$d\varphi = \frac{dq}{T} \qquad (5\text{-}39)$$

In other words, the increase in entropy is given by the ratio of the heat added to the temperature at which it is added.   We see from this relationship that a reversible adiabatic process is one in which the entropy does not change; therefore, it is what we call an *isentropic* process.

By making use of the entropy differential, we have for an adiabatic process

$$d\varphi = \frac{dq}{T} = C_p \frac{dT}{T} - \frac{R}{m} \frac{dp}{p} = 0 \qquad (5\text{-}40)$$

and

$$C_p \frac{dT}{T} = \frac{R}{m} \frac{dp}{p} \qquad (5\text{-}41)$$

We start out in an adiabatic process with the temperature T_0 and the pressure p_0 and wish to find the temperature T at some other pressure p.   To do this, we integrate the last equation from T_0 to T and from p_0 to p as follows:

$$\int_{T_0}^{T} C_p \frac{dT}{T} = \int_{p_0}^{p} \frac{R}{m} \frac{dp}{p} \qquad (5\text{-}42)$$

which becomes

$$C_p \ln T - C_p \ln T_0 = \frac{R}{m} \ln p - \frac{R}{m} \ln p_0$$

or

$$C_p \ln \frac{T}{T_0} = \frac{R}{m} \ln \frac{p}{p_0} \qquad (5\text{-}43)$$

which is the same as

$$\left(\frac{T}{T_0}\right)^{C_p} = \left(\frac{p}{p_0}\right)^{R/m} \qquad (5\text{-}44)$$

$$\frac{T}{T_0} = \left(\frac{p}{p_0}\right)^{R/mC_p} \qquad (5\text{-}45)$$

This is known as Poisson's equation.   The exponent R/mC_p is usually given the value $\frac{2}{7}$, or 0.286.

A similar expression in terms of temperature and specific volume can be obtained from Eq. (5-35), where the adiabatic process would be represented by

$$C_v \, dT = -p \, d\alpha \quad \text{or} \quad C_v \frac{dT}{T} = -\frac{p}{T} d\alpha = -\frac{R}{m} \frac{d\alpha}{\alpha} \quad (5\text{-}46)$$

$$C_v \ln \frac{T}{T_0} = \frac{R}{m} \ln \frac{\alpha_0}{\alpha}$$

$$\frac{T}{T_0} = \left(\frac{\alpha_0}{\alpha}\right)^{R/mC_v} \quad (5\text{-}47)$$

In physics and engineering, the thermodynamic processes of gases, such as isothermal processes and adiabatic processes, are represented on the so-called "pV diagram," having pressure and volume as coordinates. This diagram has the property that the work involved in any closed cycle, made up of any series or combination of processes bringing the gas back to its starting pressure and volume, is directly proportional to the area enclosed by the plot of the cycle on the diagram. From Eq. (5-25), it is seen that the work done on the gas in compression to change its volume by an amount $-dV$ would be $dW = -p \, dV$. The total work done in a cyclic process would then be

$$W = -\int_c p \, dV \quad (5\text{-}48)$$

where the integral sign with the subscript c indicates integration around a closed path (line integral). This is then the area enclosed by the closed path on the pV diagram. The negative sign means work done *on* the gas and the positive sign would denote work done *by* the gas, according to the convention adopted in Eq. (5-21).

In the atmosphere, where we are not dealing with controlled volumes, it is more convenient to represent the processes in terms of pressure and temperature. It is desirable to retain, however, the work-measuring property of the pV integral. This can be done in terms of p and T with a suitable transformation. Since in the atmosphere it is more convenient to deal with unit mass and specific volume, we measure the work on a unit mass,

$$dw = -p \, d\alpha \quad (5\text{-}49)$$

Again substituting from the equation of state in differential form as in the development of Eq. (5-36), we note that

$$dw = \alpha \, dp - \frac{R}{m} dT \quad (5\text{-}50)$$

and the integral becomes

$$w = \int_c \alpha \, dp - \int_c \frac{R}{m} dT \quad (5\text{-}51)$$

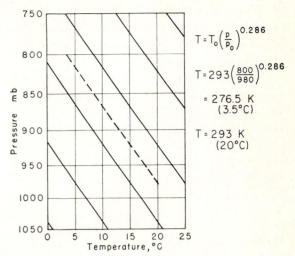

FIGURE 5-3
Adiabatic chart illustrating the graphical solution of Poisson's equation.

But the second integral on the right is equal to zero because it takes two variables to make an area.

The α in the first term on the right can be changed to an expression of p and T by means of the equation of state; so we have

$$w = \frac{R}{m} \int_c T \frac{dp}{p} = \frac{R}{m} \int_c T \, d(\ln p) \qquad (5\text{-}52)$$

Thus a diagram with T and $\ln p$ as coordinates will have the desired property of representing work by an area.

This is particularly fortunate for atmospheric studies, since height is very nearly a function of the logarithm of pressure, and a diagram with temperature as abscissa and logarithm of pressure, decreasing upward, as ordinate has the appearance of a temperature-height graph. These points will be demonstrated in the next section of this chapter. Another fortunate circumstance is that the exponent in Eq. (5-45) has such a value (0.286) that the differential of p to that power is proportional to $d(\ln p)$ at atmospheric values and the ordinate in the diagram can therefore be p to the power 0.286 with close approximation. This transformation produces adiabats—lines representing the adiabatic process—as straight lines. A portion of an adiabatic chart of this nature is shown in Fig. 5-3. The sloping lines are the adiabats.

Poisson's equation (5-45) can be solved graphically with this diagram. Suppose we have a sample of air at a temperature of 20°C ($T_0 = 293$) and a pressure p_0

of 980 mb, as in the figure.   If we want to find the temperature after the sample has been carried through an adiabatic process in which it has been brought to a pressure p, let us say of 800 mb, the required temperature T can be read off the scale below, which in this example turns out to be 3.5°C.

Equation (5-41) can give the rate at which the temperature would change with pressure during an adiabatic process, and therefore the slopes of the adiabatic lines. The expression is

$$\frac{dT}{dp} = \frac{RT}{mC_p p} \tag{5-53}$$

or

$$\frac{dT}{d(\ln p)} = \frac{RT}{mC_p} \tag{5-54}$$

Hydrostatic Equation

Ordinarily the only force operating vertically on the whole atmosphere is that force produced by the acceleration of gravity.   Transient upward or downward accelerations of the air are noted in certain situations which may on occasion have noticeable effects on readings of sensitive barographs, but except under special conditions in mountainous areas, the effects are trivial.   The force produced by the acceleration of gravity g on a mass M is $F = Mg$.   This, of course, is the definition of *weight*, and for the atmosphere it would be the weight of the overlying air.   By definition, the pressure is the force per unit area, and in cgs units the unit of area is 1 cm². The pressure at any point in the atmosphere would be equivalent to the weight of an air column of 1 cm² cross section extending to the top of the atmosphere.   If the atmosphere were of constant density with height, we would have $M = \rho V$ and $p = \rho g H$, where H would be the height of the air column extending to the outside of the atmosphere.

But because the density varies with height, we express the relation in terms of a derivative $\partial p / \partial z$, where z is vertical distance, positive upward.   The partial derivative is used because (1) pressure may change also in the horizontal direction and with time, and (2) in later, more advanced considerations the partial derivative will connote changes between geometric points in the atmosphere while the total derivative will be used for changes within a moving sample of air.   Through a height δz the pressure changes from p_1 to $p_2 = p_1 + (\partial p / \partial z)\,\delta z$; so the difference is

$$-\frac{\partial p}{\partial z}\,\delta z = g\rho\,\delta z$$

and

$$\frac{\partial p}{\partial z} = -g\rho \tag{5-55}$$

The minus sign indicates that z is measured upward in the direction of decreasing pressure.

The total differential of the pressure takes the form

$$dp = \frac{\partial p}{\partial t} dt + \frac{\partial p}{\partial x} dx + \frac{\partial p}{\partial y} dy + \frac{\partial p}{\partial z} dz$$

where t is time, x, y are the two horizontal distances, and z is vertical.   In an atmospheric column at rest, with pressure changing only with height, $\partial p/\partial t$, $\partial p/\partial x$, and $\partial p/\partial y$ are zero; so

$$dp = \frac{\partial p}{\partial z} dz = -g\rho \, dz \qquad (5\text{-}56)$$

Substituting for ρ from the equation of state, we have

$$dp = -g\frac{pm}{RT} dz \qquad (5\text{-}57)$$

It should be emphasized again that the pressure is exerted equally in all directions.   It is apparent from the hydrostatic relationships that the pressure depends only on the density and thickness of the overlying atmosphere.   Thus, if there are two small volumes or parcels of air next to each other but each has a different density, the pressure in each will nevertheless be the same.   It will be shown in this chapter that two such parcels will not remain together at the same level.

Equation (5-56) is called the *hydrostatic equation* and is applicable to bodies of fluid and gaseous atmospheres.   It is the third equation of great basic importance which we have derived in this chapter.   We now are in a position to extend the three basic relationships (equation of state, first law of thermodynamics, and hydrostatic equation) to a variety of problems of thermodynamics and statics in the atmosphere.

Adiabatic Temperature Change with Height

The adiabatic rate of change with pressure expressed in Eq. (5-53) can be substituted in the hydrostatic equation (5-57) to obtain the adiabatic rate of temperature change with height.   It is assumed that the atmosphere is in complete hydrostatic equilibrium so that the total differential can be used.   The substitution is

$$C_p \, dT = -\frac{RT}{mp} g\rho \, dz \qquad (5\text{-}58)$$

but from the equation of state $\rho = mp/RT$; so the result is

$$dT = -\frac{g}{C_p} dz \qquad (5\text{-}59)$$

where $-g/C_p$, having the dimensions of kelvin per centimeter in *cgs* units, is the dry adiabatic[1] rate of temperature change. The negative sign indicates that the temperature of a sample of air would decrease with ascent in the atmosphere. The term g/C_p has values ranging with the range of sea-level values of the acceleration of gravity from $978.036/(1.005 \times 10^7)$ at the equator to $983.208/(1.005 \times 10^7)$ at the pole, or 9.632×10^{-5} to 9.783×10^{-5} °C cm^{-1}. The rounded value 10^{-4} °C cm^{-1} or 1°C per 100 m is generally used. This is sometimes referred to as the *dry-adiabatic lapse rate*, although the term lapse rate is usually reserved for observed vertical distributions of temperature. We have defined here a *process* rate for a sample of air moving up or down through an atmosphere at rest and in hydrostatic equilibrium.

Determination of Altitude

On the adiabatic chart, altitudes can be represented conveniently because all three factors, pressure, temperature, and humidity, determine the altitude. The relationship between pressure and height at a given temperature may be determined from the hydrostatic equation

$$dp = -\rho g \, dz$$

or, by substituting from the equation of state,

$$dp = -\frac{pmg}{RT^*} \, dz$$

where T^* is the virtual temperature. This equation is then integrated from sea level, where the pressure is p_0, to the height z, having the pressure p, as follows:

$$\int_{p_0}^{p} \frac{dp}{p} = -\frac{mg}{R} \int_0^z \frac{1}{T^*} \, dz = \ln \frac{p}{p_0} \qquad (5\text{-}60)$$

$$p = p_0 \, e^{-\frac{mg}{R} \int_0^z \frac{1}{T^*} \, dz} \qquad (5\text{-}61)$$

It is convenient and feasible to divide the atmosphere into layers of about 100 mb in thickness, using the mean virtual temperature of each layer to determine its depth. The various depths are then added together to obtain the total height. Hence we have for a layer between z_1 and z_2, having the mean virtual temperature $\overline{T}^*$,

$$\int_{p_1}^{p_2} \frac{dp}{p} = -\frac{mg}{R\overline{T}^*} (z_2 - z_1) \qquad (5\text{-}62)$$

$$\ln p_2 - \ln p_1 = -\frac{mg}{R\overline{T}^*} (z_2 - z_1) \qquad (5\text{-}63)$$

---

[1] The significance of the term "dry" will be discussed on a subsequent page.

$$z_2 - z_1 = \frac{R\bar{T}^*}{mg}(\ln p_1 - \ln p_2) \qquad (5\text{-}64)$$

This is known as the *hypsometric formula*, and the total height is the sum of the thicknesses of the layers.

These relationships show that, if temperature in a sounding is plotted against pressures on a logarithmic scale with pressure decreasing upward, we have, at least for a certain mean virtual temperature $\bar{T}^*$, an exact plot of temperature against height on a linear scale.

Geopotential Altitude

A horizontal surface on the earth or in the atmosphere is one that everywhere parallels the surface of the sea, or in other words, a surface that is everywhere at the same distance above (or below) mean sea level. One learns in elementary physics that a particle in straight steady frictionless horizontal motion neither performs nor requires work in its motion. In the gravitational field of the earth, the work performed in lifting a mass M from a height z_1 to a greater height z_2 is said to be $Mg(z_2 - z_1)$. To be more exact, we should say that the work involved in a displacement of a mass M from any point 1 to any other point 2 is given by

$$W = \int_{z_1}^{z_2} Mg \, dz \qquad (5\text{-}65)$$

or, for unit mass,

$$= \int_{z_1}^{z_2} g \, dz \qquad (5\text{-}66)$$

If the acceleration of gravity were constant in a horizontal plane all over the earth, then a surface of $z = \text{const}$, in other words, a horizontal surface, would be one along which air particles could move in straight frictionless flow without work. Such a surface would then be one of constant potential energy in the earth's gravitational field ($Mgz = \text{const}$). We should call such surfaces *equipotential surfaces* or *geopotential surfaces* or *level surfaces*. However, we know that g is not constant at any given height over the earth, that it varies especially with latitude, being at a maximum near the pole and at a minimum near the equator. Therefore, *horizontal surfaces are not level surfaces*. In order for an air particle to move in a horizontal surface over the earth, work must be performed, for, although in so doing it goes from point 1 to point 2 with $z_1 = z_2$, there is a difference in g_1 and g_2.

If we start with zero at sea level, we may define a geopotential given by

$$\Phi = \int_0^z g \, dz \qquad (5\text{-}67)$$

$$d\Phi = g \, dz \qquad (5\text{-}68)$$

In meteorology and oceanography, heights or depths are given in terms of this geopotential under the name *geopotential height* or *depth*. In order to make this appear in units similar to meters, the value with z in centimeters is divided by 980×10^2 and the geopotential height is given in *geopotential meters* (gpm). These would correspond exactly with geometric meters at a location where $g = 980$, which is true at sea level at latitude 38°. If z is expressed in meters, the geopotential height in meters is given by

$$\Phi = \frac{1}{9.8} \int_0^z g \, dz \qquad (5\text{-}69)$$

The geopotential is conveniently substituted in the hydrostatic equation, for

$$dp = -g\rho \, dz = -\rho \, d\Phi = -\frac{pm}{RT} d\Phi \qquad (5\text{-}70)$$

and

$$\int_{p_1}^{p_2} \frac{dp}{p} = -\frac{m}{R} \int_{z_1}^{z_2} \frac{1}{T^*} g \, dz = -\frac{m}{R} \int_{\Phi_1}^{\Phi_2} \frac{1}{T^*} d\Phi \qquad (5\text{-}71)$$

where the virtual temperature T^* is introduced to provide a more accurate measurement. However, in practice the *mean* virtual temperature defined by

$$\overline{T}^* = \frac{\int_{p_2}^{p_1} T^* \, d(\ln p)}{\int_{p_2}^{p_1} d(\ln p)}$$

is used. Then

$$\ln p_2 - \ln p_1 = -\frac{m}{\overline{T}^* R} (\Phi_2 - \Phi_1) \qquad (5\text{-}72)$$

and, with the previously stated constant to convert from centimeters to geopotential meters, we have the height between the two pressures, or geopotential thickness between the two pressures, given by

$$\Phi_2 - \Phi_1 = \frac{R\overline{T}^*}{m} \frac{\ln p_1 - \ln p_2}{9.8 \times 10^4} \qquad (5\text{-}73)$$

Tables for obtaining the geopotential thicknesses between standard-pressure surfaces with various mean virtual temperatures have been compiled by the World Meteorological Organization.

To illustrate the nature of the tables, a small sample portion is reproduced in Table 5-1. In part a the geopotential differences between the 1000-mb and the 950-mb surfaces are given as a function of the mean virtual temperatures in the layer as read on the left in whole degrees and in tenths of a degree across the top. For example, with a mean virtual temperature of 26°C the 950-mb surface is 449 gpm above the 1000-mb surface. The tables are printed in this form for 50-mb layers and lesser pressure differences at high levels. For residual layers at the bottom and top of a sounding, another table, seen in part in Table 5-1b, is used. For example, if the pressure at the ground is 1016.4 mb the height of the 1000-mb surface above the ground has to be added to the totals from the first table. In part b the value for 0°C is found to be 130 gpm. Table 5-1c shows the correction to be added for the actual mean virtual temperature. If, in our example, the observed mean is 30°C, a correction of 14 gpm is added to the 130 gpm to give a separation of 144 gpm between 1016.4 and 1000 mb. This value is added to the 1000- to 950-mb separation to obtain the height of the 950-mb surface above the ground (144 + 449 = 593 gpm). In this way the heights are built upon, layer by layer.

Altimeters

The hypsometric formula shows that in order to determine the height, the pressure at a known height (the ground or sea level) and at the level in question, and also the mean virtual temperature of the intervening air column must be known. The quantities

Table 5-1 PORTION OF WMO TABLES FOR DETERMINING GEOPOTENTIAL DIFFERENCES BETWEEN ISOBARIC SURFACES

a. Differences in geopotential meters (gpm) between 1000 and 950 mb as function of mean virtual temperature of the layer

mb °C	0	1	2	3	4	5	6	7	8	9
	gpm	gpm	gpm	gpm	gpm	gpm	gpm	gpm	gpm	gpm
950.....										
−70	305	304	302	301	299	298	296	295	293	292
−60	320	319	317	316	314	313	311	310	308	307
−50	335	334	332	331	329	328	326	325	323	322
−40	350	349	347	346	344	343	341	340	338	337
−30	365	364	362	361	359	358	356	355	353	352
−20	380	379	377	376	374	373	371	370	368	367
−10	395	394	392	391	389	388	386	385	383	382
−0	410	409	407	406	404	403	401	400	398	397
0	410	412	413	415	416	418	419	421	422	424
10	425	427	428	430	431	433	434	436	437	439
20	440	442	443	445	446	448	449	451	452	454
30	455	457	458	460	461	463	464	466	467	469
40	470	472	473	475	476	478	480	481	483	484
50	486	487	489	490	492	493	495	496	498	499
1000.....										

Table 5–1 (*continued*)

b. Geopotential differences between 1000-mb surface and surfaces of given pressure below it, with mean virtual temperature of 0°C

p_1	p_2	0.0	0.1	0.2	0.3	0.4	0.5	0.6	0.7	0.8	0.9
mb	mb	gpm	gpm	gpm	gpm	gpm	gpm	gpm	gpm	gpm	gpm
	1000	0	1	2	2	3	4	5	6	6	7
	1001	8	9	10	10	11	12	13	14	14	15
	1002	16	17	18	18	19	20	21	22	22	23
	1003	24	25	26	26	27	28	29	30	30	31
	1004	32	33	34	34	35	36	37	38	38	39
	1005	40	41	41	42	43	44	45	45	46	47
	1006	48	49	49	50	51	52	53	53	54	55
	1007	56	57	57	58	59	60	61	61	62	63
	1008	64	65	65	66	67	68	69	69	70	71
	1009	72	72	73	74	75	76	76	77	78	79
	1010	80	80	81	82	83	84	84	85	86	87
	1011	88	88	89	90	91	91	92	93	94	95
	1012	95	96	97	98	99	99	100	101	102	103
	1013	103	104	105	106	106	107	108	109	110	110
	1014	111	112	113	114	114	115	116	117	118	118
1000	1015	119	120	121	121	122	123	124	125	125	126
	1016	127	128	129	129	130	131	132	133	133	134
	1017	135	136	136	137	138	139	140	140	141	142
	1018	143	144	144	145	146	147	147	148	149	150
	1019	151	151	152	153	154	155	155	156	157	158
	1020	158	159	160	161	162	162	163	164	165	165
	1021	166	167	168	169	169	170	171	172	173	173
	1022	174	175	176	176	177	178	179	180	180	181
	1023	182	183	183	184	185	186	187	187	188	189
	1024	190	191	191	192	193	194	194	195	196	197
	1025	198	198	199	200	201	201	202	203	204	205
	1026	205	206	207	208	208	209	210	211	212	212
	1027	213	214	215	215	216	217	218	219	219	220
	1028	221	222	223	223	224	225	226	226	227	228
	1029	229	230	230	231	232	233	233	234	235	236
	1030	236	237	238	239	240	240	241	242	243	243

c. Correction to geopotential differences as a function of departure of mean virtual temperature (+ or —) from 0°C (to be added for positive temperatures, subtracted for negative temperatures)

gpm

°C	20	40	60	80	100	120	140	160	180	200 etc.
2	0	0	0	1	1	1	1	1	1	1
4	0	1	1	1	1	2	2	2	3	3
6	0	1	1	2	2	3	3	4	4	4
8	0	1	2	2	3	4	4	5	5	6
10	1	1	2	3	4	4	5	6	7	7
12	1	2	3	4	4	5	6	7	8	9
14	1	2	3	4	5	6	7	8	9	10
16	1	2	4	5	6	7	8	9	11	12
18	1	3	4	5	7	8	9	11	12	13
20	1	3	4	6	7	9	10	12	13	15
22	2	3	5	6	8	10	11	13	14	16
24	2	4	5	7	9	11	12	14	16	18
26	2	4	6	8	10	11	13	15	17	19
28	2	4	6	8	10	12	14	16	18	21
30	2	4	7	9	11	13	15	18	20	22

are p_0, p, and $\overline{T}^*$. The airplane altimeter is an aneroid barometer[1] which measures p only. By setting the zero of the altitude scale at p_0 as given by that instrument or a similar one on the ground, the altitude at p will be obtained, provided that the scale can be adjusted for $\overline{T}^*$.

Altimeters have no provision for adjusting the altitude scale for variations in the mean virtual temperature of the air column underneath. The conditions of the so-called "standard atmosphere" are always assumed.[2] This standardized temperature distribution, which is internationally accepted, is based on what are thought to be the average conditions over the earth as a whole. A temperature of 15°C is assumed at sea level with a lapse rate of 0.65°C per 100 m. As a result of this simplified assumption, altimeters indicate altitudes that are too high in cold weather and too low in warm weather. The error is negligible when flying near the ground, provided that a correct setting for the ground elevation is obtained. Manufacturers sometimes supply with their altimeters corrections based on the temperature at the flying level. This is not a complete correction, since it must assume a certain lapse rate. It should be kept in mind that the mean virtual temperature of the air column between the airplane and the ground determines the temperature factor. If there is a reasonably uniform lapse rate, this mean can be determined accurately enough if the temperatures at the ground and at the flying level are known. If one or more important temperature inversions exist, the mean temperature cannot thus be obtained from the two end points of the curve.

POTENTIAL TEMPERATURE AND STABILITY OF DRY AIR

Potential Temperature Defined

A quantity θ, designated the *potential temperature*, is defined from Poisson's equation (5-45) as

$$\theta = T\left(\frac{1000}{p}\right)^{R/mC_P} \qquad (5\text{-}74)$$

where θ takes the place of T_0 in Eq. (5-45), with p_0 at 1000 mb. This means that a sample of air would achieve an actual temperature equal to its potential temperature when brought dry-adiabatically to a pressure of 1000 mb.

---

[1] See Appendix A for definition and description.

[2] The complete specifications of the standard atmosphere call for a sea-level pressure of 1013.25 mb, a surface temperature of 59°F (15°C), and a lapse rate of 3.566°F per 1000 ft or 0.65°C per 100 m until a temperature of -67°F (-55°C) is reached, thence isothermal. The altimeter is based on the observed ground pressure, but the other factors are the same.

Constant θ, Adiabatic, and Isentropic

Since on the $T \ln p$ or similar diagram a dry-adiabatic line can intersect the 1000-mb line at only one point, every point on an adiabatic line must be characterized by one and the same potential temperature. Therefore, dry-adiabatic lines are constant potential-temperature lines, and each can be designated according to its potential temperature. In Fig. 5-3 we may label the sloping lines, starting from the left, as 280, 290, 300, etc., in terms of potential temperature. A dry-adiabatic process in the atmosphere is a constant potential-temperature process. Furthermore, since under atmospheric conditions a dry-adiabatic process does not involve a change in the specific entropy, the terms *constant potential temperature*, *dry-adiabatic*, and *isentropic* are used to describe one and the same process.

At this point it might be explained that the word "dry" applied to an adiabatic process does not require that the air should contain no water vapor. The process operates up to the saturation point of the water vapor or, in other words, at all relative humidities below 100 percent. The effect of water vapor on the air density is essentially negligible in the atmosphere, but it can be considered by using the virtual temperature and the virtual potential temperature derived from it.

Distribution of Potential Temperature

In an atmosphere with a dry-adiabatic lapse rate, the potential temperature would not change with altitude. Except under certain circumstances near the ground, lapse rates are nearly always less than this value, which means that the potential temperature must normally increase with altitude. If the potential-temperature values are entered on the dry-adiabatic lines of Fig. 5-3 this point can be demonstrated, for any normal lapse rate must produce a line with a smaller decrease of temperature with decreasing pressure than the adiabats and thus show increasing potential temperature with decreasing pressure (upward in the atmosphere).

In considering the three-dimensional distribution of potential temperature, it is to be noted that, since the potential temperature depends on pressure and temperature, it would, at any given pressure (and therefore, roughly, at any given level), be greater the higher the temperature. Thus, we find that the potential temperature increases toward the equator. These normal latitudinal and vertical distributions require, then, that, in a north-to-south vertical cross section through the atmosphere, the potential-temperature lines or isentropic lines should slope upward toward the north, as shown in Fig. 5-4.

The potential-temperature or isentropic lines such as in the illustration become potential-temperature or isentropic surfaces when extended east and west to complete the three-dimensional picture. Since all processes in the free atmosphere tend to be largely adiabatic as long as no condensation occurs, air particles should remain on

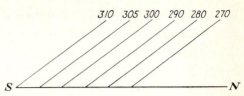

FIGURE 5-4
Slope of potential-temperature surfaces from south to north in the Northern Hemisphere.

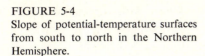

the same isentropic surface until saturation is reached. The true paths of air particles can best be traced on the surfaces along which they move—the isentropic surfaces. This tracing of paths or "flow patterns" is accomplished by mapping an isentropic surface to produce from the upper-air data a so-called "isentropic chart." Contour lines of the surface in terms of either altitudes or pressures are drawn, and the flow patterns are traced by means of identification of air particles through some quantity that measures the absolute water-vapor content. Isentropic charts will be discussed in detail in a later chapter.

The Importance of Vertical Motions

The modern application of the principles of physical hydrodynamics to the weather has given conclusive evidence that the vertical components of air motions are the principal factors in producing the more important meteorological phenomena. Their chief significance comes from the rapid changes in temperature and moisture of the air caused by these upward and downward movements. For, when compared with the rate of increase or decrease of temperature and moisture in horizontally moving air, the rapidity of change of these properties in vertical currents is tremendous.

The weather phenomena of perhaps greatest practical interest are those having to do with *condensation*—clouds, fog, rain, snow, ice, hail, and sleet. Condensation occurs when the amount of water vapor contained in a given volume in the atmosphere reaches the capacity of that volume. Since the vapor capacity of the volume depends only on the temperature, it is possible to produce condensation by lowering the capacity (i.e., lowering the temperature) or by adding more and more water vapor. Observations show that condensation can be produced more easily and quickly in the atmosphere by the first method (cooling) than by the addition of moisture, especially at altitudes above the surface where no moisture supply is readily available.

The principal processes that bring about this cooling of an element of air are *radiation, contact with cold bodies, mixing with colder air masses,* and *adiabatic cooling with expansion* of the air.[1] It is not difficult to show that the last (adiabatic expansional cooling) is the most effective process, a fact that will become more apparent

[1] For an early discussion, see J. Bjerknes and H. Solberg, Meteorological Conditions for the Formation of Rain, *Geofysiske Publikasjoner*, vol. 2, no. 3, 1921.

in the following discussions.   Adiabatic cooling of a volume of air results from an expansion under decreasing pressure, that is, from movement from high to lower pressure, unaccompanied by any exchange of heat with the environment.   Here again, comparing the pressure decrease in horizontally moving air, which is seldom more than 5 mb in 100 miles, with the pressure decrease with height, which is normally about 500 mb in the first 3 or 4 miles above the surface, it is easily seen that *the principal cause of adiabatic cooling is upward motion.*   Precipitation seldom, if ever, occurs without upward-moving air as its direct cause.   For clouds that do not produce precipitation, and especially for fogs, the other cooling processes are of importance, but these phenomena will be discussed in a later chapter.   In view of the fact that condensation, as well as many other important atmospheric processes, depends on rapid changes in temperature of a mass of air and that adiabatic temperature changes are of the greatest consequence, a thorough familiarity with adiabatic processes in the atmosphere is necessary for the student of meteorology.

In the atmosphere, we can regard vertical motions and the accompanying expansion or compression as occurring so rapidly that the absorption or loss of heat by radiation, conduction, or convection is negligible; thus the process is adiabatic in accordance with the definition.   It is apparent, then, that as long as air is unsaturated it will always cool at the rate of about 1°C for every 100 m that it is lifted; and on the other hand, if it is carried downward in the atmosphere, the moving air will be heated about 1°C for every 100 m of its descent.

The Adiabatic Rate of Cooling and the Prevailing Lapse Rate

The student is likely to confuse the adiabatic rate of cooling for upward-moving air with the rate of temperature decrease with altitude ordinarily observed in the atmosphere.   Temperature lapse rates are observed from data recorded on balloons or airplanes passing through a great many different parcels or samples of air.   If it were possible to select a condition when upward motion is prevalent, and if we could take an individual element of air and follow it along as it went upward, recording the temperature of this and no other air element, we would find that it would cool approximately 1°C for every 100 m of its ascent.   During this time, the surrounding air through which it was rising might have had quite a different temperature distribution.

A simple case to illustrate the distinction between lapse rate in the atmosphere and cooling rate of vertically moving air will be helpful.   Consider a small island on a windy day.   Suppose that the air is moving perfectly horizontally over the water, and that within this air the temperature decrease with height is 0.5°C per 100 m.   The surface air has a temperature of 10°C.   Some of the air will strike the island and

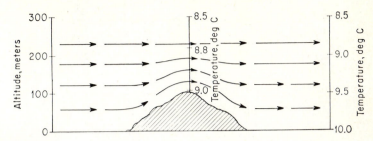

FIGURE 5-5
Adiabatic cooling by forced ascent compared with observed lapse rate in surrounding air.

in doing so will be forced upward over it, to produce lines of flow as shown in Fig. 5-5. Elsewhere, the air will pass undisturbed. The air forced up over the island will cool at the rate of 1°C per 100 m so that the air just above the crest of the island, having been lifted exactly 100 m, will have a temperature 1° lower than it had over the surface of the water. It has been cooled from 10 to 9°. The air surrounding the island, however, will still have a temperature of 9.5°C at this height, because it has been moving horizontally.

Thus, in meteorology, it is the practice to speak of small, isolated elements of air lifted or lowered through the atmosphere. These elements during this motion are heated or cooled adiabatically, depending on whether they are displaced downward or upward, while the temperature decrease with altitude in the surrounding air through which they are moving remains unchanged.

The data on the temperature distribution in the atmosphere presented in Chap. 4 show that normally the temperature decrease with height is less than the rate at which air would be heated or cooled in passing downward or upward through the same height. The example of Fig. 5-5 represents the normal state of the atmosphere. This means that an upward motion would usually make the air in the rising current colder than the air at the same height in the surrounding, horizontally moving air. If the temperature lapse rate is the same as the rate for adiabatic changes in dry air, we say that the air has a *dry-adiabatic* lapse rate. If it is greater, the lapse rate is called *superadiabatic*.

Temperature Lapse Rate and Stability

Stability is sometimes defined as that condition in the atmosphere in which vertical motions are absent or definitely restricted; and, conversely, instability is defined as the state wherein vertical movement is prevalent. The effects are demonstrated from

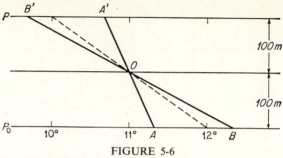

FIGURE 5-6
Graphical determination of stability.

a consideration of the temperature distribution. By noting at any given level the difference in temperature between an upward-moving element and the surrounding atmosphere, definite conclusions can be drawn as to the stability or instability. The surrounding atmosphere is described as stable or unstable depending on whether its temperature lapse rate brings about a decrease or an increase of the buoyancy forces on an upward-moving element. In the normal case, such as in Fig. 5-5, the rising air has at a corresponding level a lower temperature than the surroundings. That this represents stability will be seen in the following generalization of the principles of vertical motions through the atmosphere.

Suppose that an unsaturated layer in the atmosphere is 200 m thick and has the prevailing lapse rate represented by the curve AA' of Fig. 5-6, which, it is seen, is less than the dry-adiabatic lapse rate. Suppose that an element of air at O, midway between the top and bottom of the layer, is displaced upward, encountering steadily lowering pressures and therefore cooling at the adiabatic rate. Its course through the layer in terms of temperature and pressure is shown by the upper half of the *broken line* passing through O. It will be noted that along its path, as soon as it leaves the original level, its temperature is less than that of its surroundings (represented by AA') and that this temperature difference between moving element and surrounding atmosphere becomes larger the greater the displacement. Knowing from the equation of state that at the same pressure and, therefore, at the same height in the atmosphere the density of the air depends inversely on the temperature, we recognize that the moving element becomes heavier than its surroundings; and as the temperature is lowered more and more by expansion, this difference in density will become greater. Therefore, unless a strong mechanical force pushes it strongly upward, it will sink back to the original level. Similarly, if the element moves downward from O, represented by the lower part of the broken line, it will become increasingly warmer than its surroundings and therefore lighter, so that the buoyancy forces will tend to return it to the original position. Thus we say that the curve AA' represents a stable lapse rate or that the atmosphere in this case is stable, meaning that its density distribution suppresses any vertical motions that may be started.

The lapse rate BB' is characteristic of an unstable atmosphere. This is seen by again moving an element from O through this unstable air, for now it is continually becoming lighter than its surroundings as it moves upward, and heavier (colder) as it sinks. Given an impetus upward from O, this air will continue to rise of its own accord; and if pressed downward, it will continue to sink at an accelerated rate.

We have yet to consider the case where the lapse rate of the surrounding atmosphere is dry-adiabatic. In such a condition, the rising or descending element would always have the temperature of its surroundings; and while the atmosphere would offer no resistance to vertical motions, neither would it favor them.

The necessary condition, then, for the support of vertical motions, up or down, in dry air is a superadiabatic temperature lapse rate.

If we represent the actual lapse rate by γ and the adiabatic rate by γ_0, then we may write the following conditions:

$$\gamma < \gamma_0 \quad \text{stable}$$
$$\gamma = \gamma_0 \quad \text{neutral}$$
$$\gamma > \gamma_0 \quad \text{unstable}$$

Also, since a dry-adiabatic line is a line of constant potential temperature θ, we may refer to Fig. 5-3 and note the following conditions:

$$\frac{\partial \theta}{\partial z} > 0 \quad \text{stable}$$

$$\frac{\partial \theta}{\partial z} = 0 \quad \text{neutral}$$

$$\frac{\partial \theta}{\partial z} < 0 \quad \text{unstable}$$

The requirements for stability and instability[1] under conditions of water-vapor saturation, that is, with 100 percent relative humidity, are quite different from these. We then speak of *saturation-adiabatic* equilibrium, a condition which will be considered in the next chapter after we have become familiar with the properties and thermodynamics of water vapor.

Potential Temperature as an Expression of State

At this point it is important to note the usefulness of potential temperature as an expression of state in the atmosphere.

---

[1] Note that in English the adjective is *un*stable and the noun is *in*stability.

The most practical way to represent an equation of state in the atmosphere is in terms of temperature, pressure, and potential temperature. The state is then derived from two easily measured quantities—temperature and pressure. The potential temperature represents that state. With p expressed in bars, making the reference pressure unity, we have

$$\theta = Tp^{-R/mC_p} = T\left(\frac{\alpha}{\alpha_0}\right)^{R/mC_v} \qquad (5\text{-}75)$$

In these terms we can tabulate the different thermodynamic processes as shown in Table 5-2.

From the logarithmic form of the potential-temperature equation, we write

$$\ln \theta = \ln T - \frac{R}{mC_p}\ln p + \frac{R}{mC_p}\ln p_0 \qquad (5\text{-}76)$$

where it is to be noted that the last term is constant with $p_0 = 1000$ mb. In differential form the expression becomes

$$\frac{d\theta}{\theta} = \frac{dT}{T} - \frac{R}{mC_p}\frac{dp}{p} \qquad (5\text{-}77)$$

The differentials can be related to the derivatives in space taken in the z-direction only, showing that

$$d\theta = \frac{\partial \theta}{\partial z}dz \qquad dT = \frac{\partial T}{\partial z}dz \qquad dp = \frac{\partial p}{\partial z}dz \qquad (5\text{-}78)$$

Then, recognizing that $\partial p/\partial z = -\rho g = -pmg/RT$, we substitute for the differentials in (5-77) to obtain

$$\frac{\partial \theta}{\partial z} = \frac{\theta}{T}\left(\frac{\partial T}{\partial z} + \frac{g}{C_p}\right) \qquad (5\text{-}79)$$

Table 5-2

Process	θ	Constant	Thermodynamic equilibrium
Constant p (isobaric)	$\theta = k_1 T$	$k_1 = p^{-R/mC_p}$	$dq = C_p\, dT$
Constant T (isothermal)	$\theta = k_2 p^{-R/mC_p}$	$k_2 = T$	$dq = -\dfrac{RT}{m}\dfrac{dp}{p} = p\, d\alpha$
Constant α (isosteric)	$\theta = T$	$\alpha = \alpha_0$	$dq = C_v\, dT$
Constant θ (isentropic or adiabatic)	$\theta = k_3$	$k_3 = Tp^{-R/mC_p}$	$dq = 0;\ C_p\dfrac{dT}{T} = \dfrac{R}{m}\dfrac{dp}{p}$

and note that the second term in the parentheses is the adiabatic rate while the first is the observed rate of temperature change with height. Unless a negative $\partial T/\partial z$ exceeds g/C_p in absolute value (superadiabatic lapse rate), θ increases with height. Thus the conditions for stability and instability previously demonstrated graphically are derived from an application of mathematics to the problem.

Nature of Vertical Accelerations

The force causing warm parcels to rise in the air and cold ones to sink is the *buoyancy force*. This is sometimes called the *Archimedean force*. According to the principle of Archimedes, a body floating or immersed in a fluid is subjected to an upward-directed buoyancy force equal to the weight of the amount of fluid that the body displaces. The body will rise, sink, or remain at the same level depending on whether this force is greater than, less than, or equal to, respectively, the downward force on the body due to the acceleration of gravity (the weight of the body).

Instead of a fixed body, we are here dealing with a parcel of air, which would have the weight $\rho'Vg$, where ρ' is its density, V its volume, and g the acceleration of gravity. The weight of the displaced fluid (air) is ρVg, where ρ is the density of the surrounding, displaced air. The volumes are the same. The resultant force F, positive upward, would be

$$F = M'a = \rho Vg - \rho'Vg = (\rho - \rho')Vg \qquad (5\text{-}80)$$

where a is the acceleration upward and M' is the mass of the accelerated parcel. However, $M' = \rho'V$, so the upward acceleration is

$$a = g\frac{\rho - \rho'}{\rho'} \qquad (5\text{-}81)$$

It is more convenient to express this acceleration in terms of the temperature of the parcel and of its surroundings. This may be done by substitution from the equation of state,

$$\rho = \frac{pm}{RT}$$

and

$$\rho' = \frac{p'm}{RT'}$$

Realizing that the parcel is under the same pressure as its immediate surroundings $p = p'$, we may write

$$a = g\frac{T' - T}{T} \qquad (5\text{-}82)$$

This shows that, as long as the temperature of the displaced element is higher than that of the air it is displacing, the element will be accelerated upward. When the surroundings are warmer, the element is accelerated downward. It should be noted that the acceleration, not the speed, is proportional to this temperature difference. Ideally the speed will increase until temperature equilibrium is again reached, but friction and turbulence work against it.

Autoconvective Lapse Rate: Homogeneous Atmosphere

Under ordinary conditions the density decreases rapidly with height, even when the prevailing lapse rate is considerably greater than the dry-adiabatic rate. When the density is the same throughout, we have what is called a *homogeneous atmosphere*, which is of only theoretical interest, for such a condition never exists. However, there are evidences of its occurrence in a restricted layer next to the ground on days of intense solar heating, and it is believed that conditions of this kind sometimes exist in tornadoes.

The lapse rate of a homogeneous atmosphere may be calculated from the hydrostatic equation and the equation of state in differential form. The hydrostatic equation is

$$dp = -\frac{g}{\alpha} dz$$

The equation of state in differential form is

$$dT = \frac{m}{R}(p \, d\alpha + \alpha \, dp)$$

In terms of change with height

$$dT = \frac{\partial T}{\partial z} dz \qquad \text{and} \qquad d\alpha = \frac{\partial \alpha}{\partial z} dz$$

In the homogeneous atmosphere $\partial \alpha / \partial z = 0$, so

$$\frac{\partial T}{\partial z} dz = \frac{m}{R} \alpha \, dp \qquad (5\text{-}83)$$

or, substituting for dp from the hydrostatic equation, we have

$$\frac{\partial T}{\partial z} = -\frac{mg}{R} \qquad (5\text{-}84)$$

With numerical values in the equation, we have, assuming g to be 980,

$$\frac{\delta T}{\delta z} = \frac{-28.9 \times 980}{8.314 \times 10^7} = -3.41 \times 10^{-4} \, °C \text{ per cm}$$

or $-3.41°C$ per 100 m.

This is called the *critical autoconvective lapse rate* or often simply the *autoconvective lapse rate*, because if it is exceeded no displacement of a parcel from its initial level is necessary for upward acceleration to occur. The parcel would already be less dense than the air immediately above it, so that vertical overturning must take place spontaneously. Over desert regions, large dust whirls or "dust devils" almost the size of small tornadoes are apparently caused by autoconvective lapse rates near the ground. Smaller dust whirls seen in almost any locality may arise from similar conditions on a very local scale.

A homogeneous atmosphere, having no decrease in density with height, would have a finite depth. While this is not a reasonable assumption, the concept of a homogeneous atmosphere finds usefulness in many theoretical problems. The height of the homogeneous atmosphere can be calculated by using the hydrostatic equation and substituting for the density from the equation of state, giving

$$dz = - \frac{RT}{mg} \frac{dp}{p}$$

But the density at all heights is the same, so that

$$\rho = \frac{pm}{RT} = \frac{p_0 m}{RT_0} = \rho_0$$

and

$$p = p_0 \frac{T}{T_0}$$

where the subscript 0 refers to the bottom of the atmosphere. We therefore have

$$dz = - \frac{RT_0}{mgp_0} dp$$

which we may integrate through the depth of the atmosphere to the top H, where the pressure would be zero.

$$\int_0^H dz = - \frac{RT_0}{mgp_0} \int_{p_0}^0 dp = H = \frac{RT_0}{mg} \qquad (5\text{-}85)$$

For a surface temperature of $T_0 = 273°$, we would have

$$H = \frac{8.315 \times 10^7 \times 273}{28.9 \times 980} = 8 \times 10^5 \text{ cm (approx)}$$

$$= 8000 \text{ m (approx)}$$

It is evident that the temperature at the top of this atmosphere would be absolute zero. We could have obtained, therefore, the top of the atmosphere by dividing the surface temperature by the lapse rate.

Layer Stability

In addition to the ascent or descent of isolated parcels of air through their atmospheric environment, there are numerous occasions when whole layers of air many hundreds of meters thick undergo lifting or sinking (subsidence). The lifting may occur through the upslope motion of a large-scale current in a mountainous region or when a warm current ascends over a colder one. Subsidence may occur to compensate for the lateral spreading out of the air of the lower part of an area of high pressure. These processes will be described in detail in later chapters, but at present we shall examine only the thermodynamic effects of lifting and subsidence.

The weight or mass of a layer per vertical column of unit area is given by the pressure difference between the top and bottom of that layer. The weight is then given by the hydrostatic formula, which may be expressed in the following manner:

$$W = Mg = p_1 - p_2 = \bar{\rho}g(z_2 - z_1)$$

where the subscript 1 refers to the bottom, 2 to the top of the layer, and $\bar{\rho}$ is the mean density of the layer. If the changes in g with height are neglected, this pressure difference will also be a measure of the mass. The mass of air contained, let us say, between two isentropic surfaces as boundaries can change only by the lateral addition or removal of mass. The addition of mass in this way is called *horizontal convergence*, and the removal is called *horizontal divergence*. If no convergence or divergence occurs, the pressure difference between top and bottom will remain the same during lifting or subsidence. If there is convergence, the pressure difference will become greater, and if there is divergence it will become less.

The changes in lapse rate involved in lifting and subsidence are best illustrated on the adiabatic chart. Suppose that a layer of very dry air lying between the 1000- and 900-mb surfaces (Fig. 5-7) is lifted in such a way that there is no lateral convergence or divergence. The pressure difference between the top and the bottom would remain the same, but the altitude difference corresponding to a given pressure difference is greater at high altitudes; so the condition as shown in Fig. 5-7 results. During the lifting, which is an adiabatic process, the lower part of the layer at A will cool to A' and the upper part with the temperature at B will cool to B'. From the diagram, it is obvious that the stability is decreasing with the lifting.

By taking the layer under the condition $A'B'$ and noting the effects of subsidence, the reverse process is followed, and it is apparent that the lapse rate becomes more stable.

If the layer were superadiabatic in the beginning, the lifting would make the lapse rate less superadiabatic, and the subsidence would make it more superadiabatic. If the lapse rate were in adiabatic equilibrium in the beginning, no change in lapse rate could occur.

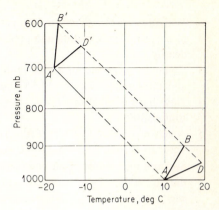

FIGURE 5-7
Effects on lapse rate of lifting, sinking, divergence, and convergence.

We have the rule: *Lifting causes the lapse rate to approach the dry-adiabatic and subsidence causes it to depart more and more from the adiabatic.*

In lateral divergence, the horizontal dimension of the volume occupied by the air is enlarged at the expense of the vertical thickness. The result of the divergence is seen in Fig. 5-7 by the line AD corresponding to the lapse rate of AB or $A'B'$ under the combined effects of subsidence and divergence. $A'D'$ shows the type of temperature distribution resulting from lifting of AB, accompanied by divergence.

In horizontal convergence, the effect would be the reverse; i.e., convergence would tend to make the lapse rate AB become less stable. For superadiabatic lapse rates convergence would tend to make the lapse rate less superadiabatic and divergence would decrease the instability.

We have the rule: *Horizontal convergence causes the lapse rate to approach the adiabatic, and lateral divergence causes it to depart more and more from the adiabatic.*

Thus it is seen that convergence and lifting work in the same direction in affecting the lapse rate, and divergence and subsidence work together. It is possible in the atmosphere to have these processes working in opposition, as for example, lifting and divergence or subsidence and convergence. It is possible to compute the relative magnitudes of these opposing effects that would make them just cancel each other.[1]

EXERCISES

1 With the universal gas constant R in ergs per kelvin per mole, give the units or dimensions of all terms in the equation of state for an ideal gas written as $pv = RT$, $pV = nRT$, and $p\alpha = RT/m$.

[1] H. R. Byers, Combined Effects of Ascent and Divergence on the Lapse Rate, *Bulletin of the American Meteorological Society*, vol. 23, pp. 319–320, 1942.

2   With pressure of 1000 mb and temperature of 30°C, compute the dry-air density.   If 1 kg of air at this temperature and pressure contains 30 g of water vapor, what is the total density?

3   Compute the density of dry air at the tropopause in the standard atmosphere (-56.5°C and 226.3 mb), at 5 km (-17.5°C and 540.2 mb) and at sea level (15°C and 1013.25 mb).

4   How big (in cubic meters) is 1 kg of dry air at 1000 mb, 15°C?   At the standard tropopause?

5   A classroom is 30 by 24 by 9 ft.   Assume the average density of the air in the room is 1.2×10^{-3} g cm^{-3}.   What is the weight of this volume in pounds?

6   Suppose the volume of air considered in the last problem is at the earth's surface and during the daytime is warmed isobarically by 20°C.   Let us assume that the amount of energy represented by this warming could be extracted and used at 100 percent efficiency to accelerate a 2-ton automobile for a distance of 100 m.   Compute the average acceleration of the car in meters per second per second.   Assume that the car starts from rest and compute the time needed to cover the 100 m.   What is the speed at this distance?

7   A parcel of dry air at a temperature of 30°C and 1000 mb is cooled isobarically during the night to a temperature of 20°C in the early morning.   Compute the following: (*a*) the initial specific volume; (*b*) the specific volume after the cooling; (*c*) the change in internal energy; (*d*) the work performed; (*e*) the heat exchanged.

8   Expressing the first law of thermodynamics in terms of dT and $d\alpha$ (where α is specific volume), obtain Poisson's equation in terms of T, T_0, α, α_0.

9   What would be the depth of a layer in the atmosphere having a uniform density of 10^{-3} g cm^{-3} that would exert a pressure equal to that at a depth of 1 m in a lake where the water has a density of 1?

10   Construct isentropic lines on a p, α diagram for potential temperatures of 273, 283, 293, 303, and 313.   Construct your diagram in the range 100 to 1000 mb and α in a range compatible with p and θ.

11   Compute the work done on or by 1 g of air in going isobarically at 1000 mb from a temperature of 273 to 283 K, then isothermally at 283 K to 900 mb, then isobarically back to 273 K and isothermally to the starting point.

12   Construct on a p, α diagram isotherms of 273 and 293 and two adiabatic lines connecting them at 900 mb, 293 and at 1000 mb, 273.   Compute the work around the closed path inscribed by these lines.

13   Find the thickness in geopotential meters of the layer between 975 and 880 mb having a mean virtual temperature of (*a*) 20°C; (*b*) -20°C.

14   Compute the geopotential heights for a pressure-temperature-humidity sounding in your locality through use of either the WMO or similar tables or by your own computation.

15   Plot temperature versus pressure on an inverted semilog graph sheet for an available sounding.

16   A layer 100 mb thick has a potential temperature of 293 at 600 mb and 297 at 500 mb. If this layer, preserving its mass and therefore pressure difference of 100 mb, sinks adiabatically to 900 to 800 mb, how will the temperature difference between bottom and top of the layer change as a result of the sinking?   Find the lapse rate (°C per 100 gpm)

before and after sinking. If the pressure thickness of the layer were reduced to 50 mb by horizontal spreading (divergence) during the sinking, what would the new lapse rate be? Note: This problem is most easily solved with the help of a graph.

17 Express in terms of temperature and pressure the heat that would have to be removed to preserve isothermalcy in compressing an ideal gas from 2 to 1 m^3 kg^{-1} at 273 K (i.e., along the 273 isotherm in the p, α diagram).

6

WATER VAPOR AND ITS THERMODYNAMIC EFFECTS

The presence of water either in the invisible vapor form or as visible cloud has been shown in previous chapters to have a pronounced effect on the transmission of radiation. Equally important are the phase-change effects. Water is present on our planet at a temperature far below the boiling point, but it is sufficiently volatile to change from the liquid or ice state to the gaseous state by the process of evaporation (liquid to gas) or sublimation (ice to gas) at atmospheric temperatures and pressures. Conversely, it changes from gas to liquid by condensation and from gas to solid by crystallization (sometimes also called sublimation) in the free air to form clouds and on surfaces to form dew or frost. Melting and freezing also occur at temperatures commonly found in the atmosphere. The precipitation of water from the clouds in the form of rain, snow, and hail affects the heat distribution in the atmosphere. Electric as well as thermal energy is transformed by these phenomena, as shown by the occurrence of lightning.

Water is supplied to the atmosphere by evaporation from the surface of the earth, from the land as well as from the oceans. Precipitation returns water to the surface. Some of the water precipitated over the land flows to the oceans in rivers to complete the so-called "hydrologic cycle." For balance, this means that precipitation must exceed evaporation when averaged over an entire continent, while over the oceans as a whole evaporation must be greater than precipitation.   The continental excess and the oceanic deficit are equalized by the discharge of rivers and glaciers into the oceans.

Although the percentage of water vapor in the atmosphere is quite small compared with the other gases, about six times more water is transported in the atmosphere over North America, for example, than is transported by all the rivers combined.[1] This water in its various actions of cloud, fog, and precipitation formation is responsible for what we usually call *weather*.   For these reasons we are much concerned with measuring the water-vapor content of the atmosphere and understanding its thermodynamic effects.

VAPOR PRESSURE AND SATURATION

Fundamentally, all measures of water vapor or atmospheric humidity are based upon quantities related to evaporation and condensation over a flat surface of pure water. From a water surface that is evaporating, the excess of molecules of water leaving the surface over those coming back in is expressed and measured as a pressure.   This pressure, called the *vapor tension*, depends only on the temperature of the water surface. This concept is based on measurements made in a closed space.   Let us consider a closed container about half-filled with water.   We shall assume that the water and the air space above it are kept at the same temperature.   If the air is relatively dry to begin with, the water will start to evaporate and introduce additional water vapor into the space above.   If the pressure in this space is measured during the process, it will be noted that the pressure is increasing slightly as the new vapor is added.   This increasing pressure is due to an increase in the partial pressure of the water vapor, known as the *vapor pressure*.   The vapor pressure in the space will increase until it is exactly equal to the vapor tension of the water surface.   When this balance of pressure is reached, no further evaporation will occur, and if the overlying gases and the water are at the same temperature, the closed space above is then *saturated* with water vapor.   If the water had been exposed to the open air, equilibrium would not have been reached and all the water would eventually evaporate.   If the temperature of the overlying gases is different from that of the water, quite another equilibrium will be required.

The *saturation vapor pressure* of pure water vapor is defined as the pressure of the vapor when in a state of neutral equilibrium with a plane surface of pure water at the same temperature.   It is then equal to the vapor tension of the water surface at

[1] G. S. Benton and M. A. Estoque, Water-Vapor Transfer over the North American Continent, *Journal of Meteorology*, vol. 11, pp. 462–477, 1954.

this temperature.　It varies with the temperature exactly as does the vapor tension of the water, so that the higher the temperature, the greater the vapor pressure required for saturation.　We can look at this from another point of view and say that any volume has a certain water-vapor capacity depending upon the temperature of the water vapor and other gases in that volume.　When this capacity is reached, we have saturation.

The dependence of the saturation vapor pressure on temperature is a specific characteristic; it is independent of the pressure of the other gases at all atmospheric pressures.　It has been determined empirically for both the liquid and the solid phases of water over a wide range of temperature.　The values in Table 6-1 include the pressures over both ice and liquid water at temperatures below freezing.　This is important in meteorology, because there is a marked tendency for water in the atmosphere, especially in clouds, to remain in liquid form at temperatures many degrees below the freezing point of pure bulk water.　Water in the liquid form at these temperatures is said to be *supercooled* or *undercooled*.　Tables for use with dry- and wet-bulb thermometer readings embody the vapor pressure with respect to ice at subfreezing temperatures because the wet-bulb thermometer gives a more constant reading if the water on it is allowed to freeze (see Appendix C).

Table 6-1　SATURATION VAPOR PRESSURE OVER WATER AND OVER ICE, IN MILLI-BARS (Values over water at subfreezing temperatures in italics)

Temperature, °C										
Units										
Tens	0	1	2	3	4	5	6	7	8	9
40	73.777	77.802	82.015	86.423	91.034	95.855	100.89	106.16	111.66	117.40
30	42.430	44.927	47.551	50.307	53.200	56.236	59.422	62.762	66.264	69.934
20	23.373	24.861	26.430	28.086	29.831	31.671	33.608	35.649	37.796	40.055
10	12.272	13.119	14.017	14.969	15.977	17.044	18.173	19.367	20.630	21.964
+0	6.1078	6.5662	7.0547	7.5753	8.1294	8.7192	9.3465	10.013	10.722	11.474
−0	6.1078	5.623	5.173	4.757	4.372	4.015	3.685	3.379	3.097	2.837
	6.1078	*5.6780*	*5.2753*	*4.8981*	*4.5451*	*4.2148*	*3.9061*	*3.6177*	*3.3484*	*3.0971*
−10	2.597	2.376	2.172	1.984	1.811	1.652	1.506	1.371	1.248	1.135
	2.8627	*2.6443*	*2.4409*	*2.2515*	*2.0755*	*1.9118*	*1.7597*	*1.6186*	*1.4877*	*1.3664*
−20	1.032	0.9370	0.8502	0.7709	0.6985	0.6323	0.5720	0.5170	0.4669	0.4213
	1.2540	*1.1500*	*1.0538*	*0.9649*	*0.8827*	*0.8070*	*0.7371*	*0.6727*	*0.6134*	*0.5589*
−30	0.3798	0.3421	0.3079	0.2769	0.2488	0.2233	0.2002	0.1794	0.1606	0.1436
	0.5088	*0.4628*	*0.4205*	*0.3818*	*0.3463*	*0.3139*	*0.2842*	*0.2571*	*0.2323*	*0.2097*
−40	0.1283	0.1145	0.1021	0.09098	0.08097	0.07198	0.06393	0.05671	0.05026	0.04449
	0.1891	*0.1704*	*0.1534*	*0.1379*	*0.1239*	*0.1111*	*0.09961*	*0.08918*	*0.07975*	*0.07124*
−50	0.03935	0.03476	0.03067	0.02703	0.02380	0.02092	0.01838	0.01612	0.01413	0.01236

ABSOLUTE HUMIDITY

The equation of state for water vapor is written

$$e = \rho_w \frac{R}{m_w} T \qquad (6\text{-}1)$$

The ρ_w is the vapor density or *absolute humidity*. It is usually multiplied by 10^6 to express it in grams per cubic meter. At saturation, ρ_{ws} is given by

$$\rho_{ws} = \frac{e_s m_w}{RT_s} \qquad (6\text{-}2)$$

where the subscript s refers to saturation conditions.

Values of the saturation absolute humidity are given in Table 6-2. These values, which represent the water-vapor capacity of a cubic-meter volume, are dependent only on the temperature. This is because e_s is also a function of the temperature only.

SPECIFIC HUMIDITY

The specific humidity is defined as the mass of water vapor contained in a unit mass of air (dry air plus water vapor), expressed in grams per gram or grams per kilogram. Designating it as q, we have

$$q = \frac{M_w}{M_a}$$

Table 6-2   WATER-VAPOR   CAPACITY   OF   A CUBIC-METER VOLUME $(\rho_{ws} \times 10^6)$ AT VARIOUS TEMPERATURES

Temp., °C	Grams	Temp., °C	Grams
−40	0.120	0	4.847
−35	0.205	5	6.797
−30	0.342	10	9.401
−25	0.559	15	12.832
−20	0.894	20	17.300
−15	1.403	25	23.049
−10	2.158	30	30.371
−5	3.261	35	39.599
		40	51.117

where M_w is the mass of the water vapor and M_a that of the air in which it is contained. Since all the gases occupy the same volume V, and since $M_w = \rho_w V$ and $M_a = (\rho_d + \rho_w)V$, the V's cancel and

$$q = \frac{\rho_w}{\rho_w + \rho_d} \qquad (6\text{-}3)$$

the denominator being the total density of the dry gases ρ_d plus the water vapor ρ_w. To express this in terms of easily measurable quantities, we substitute from the equations of state for dry air and for water vapor

$$\rho_d = \frac{p_d m_d}{RT}$$

$$\rho_w = \frac{e m_w}{RT}$$

to form the equation

$$q = \frac{m_w e}{m_d p_d + m_w e} \qquad (6\text{-}4)$$

Table 6-3   QUANTITY OF WATER VAPOR REQUIRED FOR SATURATION AT VARIOUS TEMPERATURES AND PRESSURES
(In grams per kilogram of moist air, saturation specific humidity)

Temp., °C	Pressure, mb						
	1000	900	800	700	600	500	400
−40	0.118	0.131	0.147	0.168	0.196	0.235	0.294
−35	0.195	0.217	0.244	0.279	0.326	0.391	0.488
−30	0.317	0.353	0.397	0.453	0.529	0.635	0.793
−25	0.503	0.559	0.629	0.719	0.839	1.007	1.259
−20	0.784	0.871	0.980	1.120	1.307	1.569	1.962
−15	1.20	1.33	1.49	1.71	1.99	2.39	2.99
−10	1.79	1.99	2.23	2.55	2.98	3.58	4.48
−5	2.63	2.92	3.29	3.76	4.39	5.27	6.59
0	3.80	4.23	4.76	5.44	6.35	7.62	9.54
5	5.44	6.05	6.81	7.79	9.09	10.92	13.67
10	7.67	8.53	9.60	11.0	12.8	15.4	
15	10.7	11.9	13.4	15.3	17.9		
20	14.7	16.3	18.4	21.1			
25	20.0	22.2	25.0				
30	26.9	29.9	33.7				
35	35.8	39.8					
40	47.3						

the RT canceling out.   We then divide both the numerator and the denominator by m_d to obtain

$$q = \frac{m_w}{m_d} \frac{e}{p_d + (m_w/m_d)e} \qquad (6\text{-}5)$$

The total pressure is $P = p_d + e$; therefore $p_d = P - e$, which we may substitute in the equation to find

$$q = \frac{m_w}{m_d} \frac{e}{P - [1 - (m_w/m_d)]e} \qquad \text{g per g} \qquad (6\text{-}6)$$

which gives us the specific humidity in terms of the easily obtainable quantities e and P.
    Using numerical values, we have

$$\frac{m_w}{m_d} = \frac{18}{28.9} = 0.622$$

Then $$q = 0.622 \frac{e}{P - 0.378e} \qquad \text{g per g} \qquad (6\text{-}7)$$

For ordinary values of the vapor pressure, we can say without appreciable error that

$$q = 0.622 \frac{e}{P} \qquad \text{g per g (approx)} \qquad (6\text{-}8)$$

The specific humidity at saturation may be written as

$$q_s = 0.622 \frac{e_s}{P - 0.378e_s} \qquad (6\text{-}9)$$

A table of saturation specific humidities, similar to Table 6-2, can be written to illustrate the saturation phenomena.   These data are given in Table 6-3.   In the previous table, no mention was made of pressure because the *saturation absolute humidity* does not change with pressure.   For a constant mass (1 kg) of air, a change in volume takes place with changing pressure, thus requiring a different quantity of water vapor to produce saturation.   Values for every even hundred millibars pressure are given in Table 6-3.   Using a graph of pressure on a logarithmic scale and temperature on a linear scale as rectangular coordinates, one finds that the saturation specific humidities become very nearly straight lines.

THE MIXING RATIO

The mixing ratio is defined as the mass of water vapor contained in mixture with a unit mass of dry air, expressed in grams per gram or grams per kilogram.   It differs from specific humidity only in that it is related to dry air instead of to the total of dry air plus vapor.   Designating it as w, we have

$$w = \frac{\rho_w}{\rho_d} \qquad (6\text{-}10)$$

and from the equations of state, we get

$$w = \frac{m_w e}{m_d p_d}$$

or

$$w = 0.622 \frac{e}{P - e} \qquad (6\text{-}11)$$

The relationship between q and w can be seen by taking the expression for q,

$$q = \frac{\rho_w}{\rho_d + \rho_w}$$

and dividing both the numerator and the denominator by ρ_d, to obtain

$$q = \frac{\rho_w/\rho_d}{1 + (\rho_w/\rho_d)}$$

However, ρ_w/ρ_d is the mixing ratio w; so

$$q = \frac{w}{1 + w} \qquad (6\text{-}12)$$

Since w seldom exceeds 0.02 g per g, it can be seen that, in an extreme case,

$$\frac{w}{1 + w} = \frac{0.02}{1 + 0.02} \cong 0.02$$

or approximately, $q = w$, and no appreciable error results if q and w are used interchangeably.

RELATIVE HUMIDITY

By international agreement, the relative humidity is defined as the ratio of the observed mixing ratio to that which would prevail at saturation at the same temperature. Expressed as a percentage, it is given as

$$f = \frac{w}{w_s} \times 100 \qquad (6\text{-}13)$$

The relative humidity is also given by

$$f = \frac{e}{e_s} \times 100 \qquad (6\text{-}14)$$

$$= \frac{\rho_w}{\rho_{ws}} \times 100 \qquad (6\text{-}15)$$

and, with more than ample accuracy,

$$f = \frac{q}{q_s} \times 100 \qquad (6\text{-}16)$$

TEMPERATURE OF THE DEW POINT

The dew-point temperature is defined as the temperature at which saturation barely would be reached if the air were cooled *at constant pressure* without the removal or addition of moisture.   In other words, it is the temperature at which the quantity of water vapor actually present in the atmosphere would be the capacity amount.

It should be pointed out that there are two ways to produce saturation: (1) by decreasing the temperature and thereby reducing the capacity for water vapor and (2) by increasing the amount of water vapor.   The dew-point temperature is defined in terms of the first process.

ISENTROPIC-CONDENSATION TEMPERATURE

The isentropic-condensation temperature is the temperature at which saturation barely would be reached if the air were cooled *adiabatically* without the removal or addition of moisture.   It differs from the dew-point temperature in that it is defined in terms of an adiabatic instead of a constant-pressure process.   It is always less than the dew-point temperature, because in an adiabatic process, owing to expansion, the water vapor has an increasingly large volume to saturate.

In the adiabatic or isentropic process, condensation will occur at a level or pressure specified by the potential temperature and the specific humidity or mixing ratio.   Thus we speak of the *isentropic-condensation pressure* of an air sample, which for a given isentropic surface, will depend only on the specific humidity or mixing ratio.

VARIABILITY OF HUMIDITY QUANTITIES

Vapor pressure   If the quantity of water vapor present does not vary, that is, if there is no addition by evaporation or removal by condensation, the vapor pressure will remain constant *provided* that the total pressure of all the gases does not change. In an adiabatic expansion, for example, the partial pressures of all the gases, including water vapor, decrease. Thus the vapor pressure is an absolute measure of the quantity of water vapor present so long as we do not have to consider vertical displacements.

Relative humidity The relative humidity is a highly variable quantity. It involves the ratio of two vapor pressures, the actual and the saturation. The actual vapor pressure changes with the pressure, as noted above, and the saturation vapor pressure varies with temperature. Therefore, relative humidity changes markedly in an adiabatic process and in all other processes involving a change in temperature or pressure. At most stations, relative humidity has a wide diurnal variation, changing inversely with the temperature.

Absolute humidity The absolute humidity varies with all volume changes, since it is expressed in terms of the amount of moisture contained in a unit volume. It is used mainly in physics and engineering where fixed volumes are considered.

Specific humidity and mixing ratio Since these quantities are based on measurements of the masses of water vapor and air, the law of conservation of matter requires that they cannot be changed unless water vapor is actually added or removed. During an adiabatic process the specific humidity and mixing ratio are unchanged except after saturation is reached. In an adiabatic cooling at saturation these quantities show a decrease because water vapor is then being changed into liquid that cannot be included in the measurement.

Dew-point temperature The temperature of the dew point changes in the same way as does the vapor pressure. It is conservative for nonadiabatic processes involving no pressure change provided that there is no evaporation or condensation. It is useful for comparing the absolute quantities of moisture present at stations having roughly the same elevation.

Isentropic-condensation pressure (or temperature) This quantity remains unchanged in the same sense as specific humidity or mixing ratio as long as the potential temperature does not change.

WATER-VAPOR CONTENT OF AN AIR COLUMN

It is frequently useful to know the total water-vapor content of the atmosphere over a given place. Such a determination gives an idea of the possible amount of rain that may be expected from the overlying air if conditions become favorable for precipitation of the moisture.   Also, in radiation studies it is necessary to know this quantity, called *the precipitable water*, expressed in centimeters.

Consider an air column having a cross-sectional area of 1 cm². The total mass of water vapor contained between the height 0 and the height z would be

$$W = \int_0^z \rho_w \, dz \qquad (6\text{-}17)$$

where ρ_w is the vapor density or absolute humidity.   Substituting from the hydrostatic equation $dp = -\rho g \, dz$, where ρ is the total air density, we have

$$W = \int_p^{po} \frac{\rho_w}{\rho} \frac{1}{g} \, dp$$

$$= \frac{1}{g} \int_p^{po} q \, dp \qquad (6\text{-}18)$$

where q is the specific humidity, defined as ρ_w/ρ.   This can be integrated in steps or by making a plot of the specific humidity against pressure and finding the area between the plotted curve and $q = 0$ and the lines p and p_0.   Actually, it is simpler to express the quantity in terms of vapor pressure and pressure.   Since, approximately, $q = 0.622e/p$, we can write

$$W = \frac{0.622}{g} \int_p^{po} e \, \frac{dp}{p} \qquad (6\text{-}19)$$

$$W = \frac{0.622}{g} \int_p^{po} e \, d(\ln p) \qquad (6\text{-}20)$$

where W is in grams per square centimeter of area.   Taking the density of the precipitable water as unity, we see that, since ρ and A are both unity, $W = \rho V = AD = D$, where D is the depth of precipitable water.

Figure 6-1 shows maps of the distribution of the mean precipitable water, in centimeters, for the months of January and July over the United States.   The values are given only for that part of the atmosphere lying below 8 km; it is found that in the mean the contribution above that altitude is of the order of hundredths of a centimeter in summer and thousandths of a centimeter in winter.   At locations near sea level the principal concentration of water vapor is in the lowest 2 km.

For comparison purposes, it might be pointed out that for a *saturated* atmosphere with a sea-level temperature of 28°C and in saturation-adiabatic equilibrium, the precipitable water would be 10.54 cm; under the same conditions for a sea-level temperature of 20°C it would be 5.28 cm, while at 0°C it would be 0.84 cm.

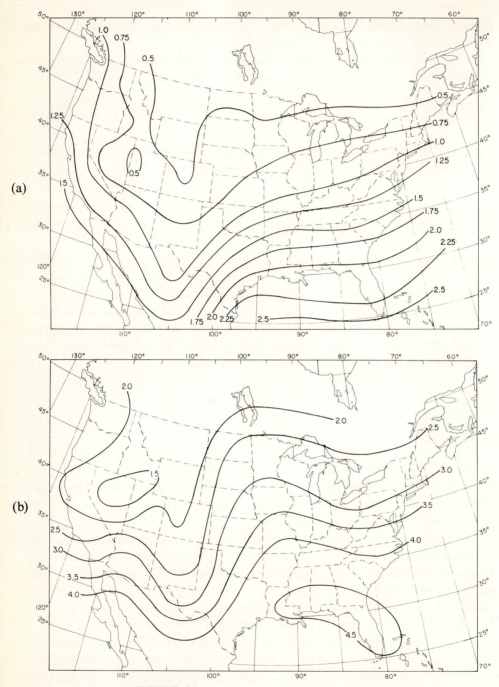

FIGURE 6-1
Precipitable water, in centimeters, for the months of (a) January and (b) July.

CHANGE OF PHASE

Water is the one substance in the atmosphere that occurs commonly in all three of its phases—vapor, liquid, and solid. This has many consequences of great importance, not the least of which is the condensation and precipitation in the form of rain or snow. At this point, however, we are concerned with the thermodynamic effects.

It can be shown experimentally that heat energy is added or removed in these changes of phase. In the case of change from liquid to vapor (evaporation), the energy involved is called the *heat of vaporization*, and in the change from liquid to solid (freezing) it is called the *heat of fusion*. It is also possible to have a change from solid to vapor directly, as we often see in the case of the disappearance of snow without melting on a clear dry day. This process, called *sublimation*, involves energy in the form of the *heat of sublimation*, which is the sum of the heat of vaporization and the heat of fusion.

In the case of evaporation, there is a considerable increase in the volume occupied by the water. Since this occurs against a definite vapor pressure in the air, external work must be done by the vapor as it leaves the liquid surface. Also, the molecules and atoms themselves have their potential energy increased. Both these processes involve a removal of energy from the evaporating surface, and the remaining liquid must be cooled unless heat is added from the outside.[1] In nature, evaporation has many noticeable cooling effects, such as the cooling caused by the evaporation of the rain that falls in a summer shower, or the cooling a person notices when he steps dripping out of a warm bath. Certain other liquids, which are more volatile than water, such as ether, alcohol, and carbon tetrachloride, exhibit a much more pronounced cooling effect.

The energy removed from the liquid surface in the evaporation is carried latent in the water vapor. When the vapor condenses again, each gram of vapor gives off the energy that it acquired from the evaporating surface. Thus we have the reverse of the heat of vaporization in the *heat of condensation*. We may say, therefore, that evaporation on a surface or in a sample of air has a cooling effect, while condensation gives off heat to the surface or air sample.

The heat of vaporization is defined as the heat removed by the evaporation of 1 g of the substance. For water at normal atmospheric temperatures, it has a value close to 600 cal per g but decreases with increasing temperature as shown in Table 6-4.

The heat of fusion is defined as the heat removed in the freezing of 1 g of the substance. For water, it has the value of approximately 80 cal per g. Within the limited ranges of pressure in the atmosphere, the freezing point of water may be

---

[1] In the case of water being boiled, heat is added from the stove or flame as rapidly as it is removed by evaporation. The more rapidly the heat is supplied, the greater the rate of evaporation.

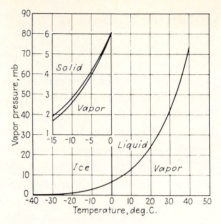

FIGURE 6-2
Saturation vapor pressure. *Inset:* saturation vapor pressure over ice (lower
curve) and over liquid (upper curve) at
subfreezing temperatures.

considered to be always at 0°C, and the removal of heat energy to cause freezing or its addition to produce melting must occur isothermally at this temperature.

In Fig. 6-2 is a *pT* diagram of the saturation vapor-pressure curve for water at temperatures above and below freezing. It is seen that below the curved line the conditions support the vapor only, whereas above the line one finds conditions for liquid at temperatures above freezing and solid at temperatures below freezing. For saturation at the freezing point, the three phases, vapor, liquid, and solid, may theoretically occur in equilibrium together. This point is called the *triple point*. It occurs in water at a temperature of 0°C and with a vapor pressure of 6.11 mb.

In the inset, two curves of the saturation vapor pressure are shown for subfreezing temperatures. The lower one is for saturation with respect to a plane surface of ice, and the upper one is for saturation with respect to a plane liquid surface supercooled to these temperatures. All these values are the same as those given in Table 6-1.

Table 6-4   LATENT HEAT OF VAPORIZATION OF WATER

Temp., °C	Latent heat, cal/g	Temp., °C	Latent heat, cal/g
−40	621.7	0	597.3
−30	615.0	10	591.7
−20	608.9	20	586.0
−10	603.0	30	580.4
		40	574.7

SOURCE: Smithsonian Meteorological Tables, 6th ed.
In work units 600 cal g^{-1} would be $600 \times 4.186 \times 10^7 = 2.512 \times 10^{10}$ ergs g^{-1} or 2.512×10^6 J kg^{-1}.

Without the presence of supercooled water, there would be direct sublimation at temperatures below freezing.

At all points in the vapor phase, below the line in Fig. 6-2, the relative humidity is less than 100 percent. At all points above the line (or lines, in the case of subfreezing temperatures) there is greater than 100 percent relative humidity, or *supersaturation*. On the line, the water is exactly at saturation. In the atmosphere, the natural processes of condensation and evaporation occur in such a way that appreciable supersaturation does not exist. In the atmosphere, then, one probably never finds conditions represented by points appreciably above the line or lines, and for all practical purposes one may consider that saturation is not exceeded and that the conditions along the lines exactly represent those in water or ice clouds.

As pointed out previously, meteorologists have discovered that supercooled water is prevalent in clouds at subfreezing temperatures down to about $-15°C$. Therefore, the upper of the two curves in the inset diagram of Fig. 6-2 would represent the conditions often found in clouds at those temperatures.

THE ADIABATIC PROCESS AT SATURATION

After air has been cooled sufficiently, either by adiabatic expansion or otherwise, a temperature is reached where there is no longer room for all the water contained in it to remain in vapor form, and so some of it starts to condense into fog or cloud. This temperature is called the *saturation* or *dew point* for cooling at constant pressure, and the *temperature of the condensation level* when we speak of adiabatic cooling. As the air is further cooled, more and more of the vapor turns to liquid droplets or solid particles, until finally, at some low temperature that is probably never reached, all the vapor has changed to solid or liquid.

The humidity relationships derived thus far do not include the liquid-water droplets in the clouds. In a parcel of air in which there is no exchange with the surroundings, all the water vapor that condenses goes to increasing the liquid-water content such that $-dw_s = dx$, where w_s and x represent the saturated cloud water vapor and the liquid-water content, respectively, per gram of air. The mixing ratio or specific humidity decreases steadily with ascent of an air parcel after saturation because it measures only the vapor and not the liquid water. Note that $w_s = 0.622\, e_s/p_d$, and it is found that e_s is very sensitive to temperature, while the air pressure has a much smaller effect. While w_s decreases in ascent, it increases in descent; however, the required cloud saturation value of w_s may not occur in descent unless there is a supply of evaporating water droplets or ice particles with which saturation equilibrium can be maintained.

The heat produced in condensation of the amount $-dw_s$ of saturated vapor is $dq = -d(Lw_s)$, where L is the latent heat released by condensation of 1 g of the vapor. The factor L decreases slowly with increasing temperature (Table 6-4); so it is often considered constant through a small range of temperature, and the heat differential is written as $-L\,dw_s$.

Consider 1 g of dry air mixed with w_s g of water vapor making a $(1 + w_s)$-g ascending parcel. The parcel will have a certain mass of liquid water x carried along from the warmer levels below. The droplets will give off the sensible heat[1] $-C\,dT$, where C is their specific heat. Meanwhile the $(1 + w_s)$ g of air are expanding in a manner that is adiabatic except for the two forms of heat—latent and sensible—so the balance is

$$-d(Lw_s) - Cx\,dT = (1 + w_s)\left(C_p'\,dT - \frac{RT}{m}\frac{dp}{p}\right) \qquad (6\text{-}21)$$

C_p' is the value for moist air, equal to $C_p(1 + 0.90q)$, where q is the specific humidity in grams per gram, for C_p expressed in calories per gram. With q normally 0.01 or less, the added value is less than 1 percent of C_p. A further refinement would be the use of virtual temperature in the equation of state in the last term on the right, which might add as much as 1 percent to it. It is assumed that the heat exchange is fast enough to keep all components at the same temperature as they cool.

An examination of the various terms shows that the w_s on the right adds less than 1 percent to the difference which follows. We can take C_p and m as representing only the dry gases and use the partial pressure p_d; thus with w_s neglected, the terms on the right would represent dry air. The liquid-water term, second on the left, might have a maximum value of 5 percent of $C_p\,dT$ in an extreme case. In exact calculations the effect of the liquid water should not be overlooked if there are large amounts of liquid water, such as $10^3 x = 5$ g per kg or approximately 5 g per m^3. We may then find suitable accuracy in the following equation:

$$(Cx + C_p)\,dT - \frac{RT}{m_d}\frac{dp_d}{p_d} + d(Lw_s) = 0 \qquad (6\text{-}22)$$

The maximum value of x might be expected in a parcel in which all the condensed water is conserved and carried along. We would then have $x = w_{s_0} - w_s$, where w_{s_0} is the saturation mixing ratio at the condensation level, or cloud base. In such a parcel x would increase steadily with height.

For low liquid-water content we may write

$$C_p\,dT - \frac{RT}{m_d}\frac{dp_d}{p_d} + d(Lw_s) = 0 \qquad (6\text{-}23)$$

---

[1] Sensible heat is heat which can be sensed or felt as contrasted with latent heat, which cannot be sensed directly.

Thus it is seen that the process in most cases is adequately described by adding a term $d(Lw_s)$ to the dry-adiabatic process.

REVERSIBLE ADIABATIC AND PSEUDOADIABATIC CONDENSATION PROCESSES

Although Eq. (6-23) is represented as a simplification of (6-22), the physical processes represented by the two expressions are very different.   In the first case we are dealing with a process in which all the liquid water that is accumulated by condensation during the ascent of the parcel is retained.   If the process is reversed, the liquid water will evaporate with increasing total pressure at the same rate as it condensed with ascent, and the heat exchange will be the same but with the opposite sign, dw_s positive.   In the second case the water will precipitate out as rapidly as it is formed.   If a descent is started, there is no water to evaporate and the temperature will increase at the dry-adiabatic rate.   The first process is reversible, the second, called the *pseudoadiabatic* process, is not.   In the first case, we have all cloud with no precipitation; and in the second, all precipitation with no cloud.   Each is an extreme case; the real situation lies somewhere between the two.   Yet in terms of temperature change in ascent, the two cases produce essentially the same result.

As a parcel ascends, it reaches a low temperature at which the vapor forms and deposits upon ice crystals.   The heat of deposition (sublimation) is given off to the air and replaces the heat of condensation in the equations.   It is greater than the heat of condensation, being the sum of that quantity and the heat of fusion, adding approximately 80 cal per g.   However, at subfreezing temperatures the vapor mixing ratio is small, and in applied simplifications only the heat of condensation is used.   This practice is partly justified by the observed fact that water exists in clouds in the undercooled-liquid form at temperatures down to $-15°C$ and lower.   In the reversible case the liquid water that is carried along in the parcel reaches a temperature level where it freezes, releasing to the air the heat of fusion.   Whether undercooled or not before the freezing, the temperature of freezing remains at $0°C$ until all the liquid water is changed to ice—the so-called " hail stage."

EQUIVALENT-POTENTIAL TEMPERATURE

The dry-adiabatic process is defined by a constant potential temperature.   A quantity can also be derived to define the pseudoadiabatic process.

In pseudoadiabatic ascent the water vapor keeps condensing until there is practically none left and the process becomes dry-adiabatic.   This means that each pseudoadiabatic line approaches a dry-adiabatic or potential-temperature line. The pseudoadiabat can be defined by the potential temperature achieved after all the vapor is condensed and precipitated out. This final potential temperature is called the *equivalent-potential temperature* for the entire process, having one constant value for each pseudoadiabatic line. A formula for the equivalent-potential temperature θ_E can be obtained in the following way.

From the logarithmic form of the potential-temperature equation given in (5-76), one may write, for p and p_0 in bars,

$$\ln T = \frac{R}{mC_p} \ln p_d + \ln \theta_d \qquad (6\text{-}24)$$

where θ_d is determined from the partial pressure of the dry air. If the simplified pseudoadiabatic equation (6-23) is formed into an indefinite integral, it becomes, after rearranging,

$$\ln T = \frac{R}{mC_p} \ln p_d - \frac{Lw_s}{C_p T} + \text{const}$$

Substituting this expression for $\ln T$ in Eq. (6-24), we may write

$$\ln \theta_d + \frac{Lw_s}{C_p T} = \text{const} \qquad (6\text{-}25)$$

We may drop the subscript s on the mixing ratio, since it will always be at saturation in this process. As $w \to 0$, $\theta_d \to \theta_E$, the equivalent-potential temperature, so that

$$\ln \theta_E = \text{const}$$

At nonzero values of w the constant is still the same, and

$$\ln \theta_E = \ln \theta_d + \frac{Lw}{C_p T} \qquad (6\text{-}26)$$

or

$$\theta_E = \theta_d \exp \frac{Lw}{C_p T} \qquad (6\text{-}27)$$

The graphical demonstration of the relationship may be seen in the thermodynamic diagrams sketched on subsequent pages in this chapter or, better still, from the more complete charts used in practice which should be available in meteorological offices and laboratories.

An equivalent temperature T_E at any atmospheric pressure level is the actual temperature achieved by the same complete irreversible process brought back to the original pressure instead of to 1000 mb.

It is useful to note that at the condensation level

$$\theta_E = \theta_d \exp \frac{Lw}{C_p T_c} \qquad (6\text{-}28)$$

where w is the observed mixing ratio in the parcel ascending dry-adiabatically up to this point and T_c is the temperature of the condensation point. Since in a dry-adiabatic parcel ascent, w and θ are both constant and together uniquely determine T_c, it is apparent that θ_E is also constant. The condensation point is referred to as a *characteristic point* because it defines the pseudoadiabatic path to be followed at saturation and the dry-adiabat below as well as being characterized by the constant water-vapor content contained in the unsaturated parcel. It is useful in describing and classifying the characteristics of air masses.

WET-BULB TEMPERATURE

At constant pressure, there are two separate ways of producing saturation: (1) by lowering the temperature to the dew point by means of cooling without any change in the water-vapor content, (2) by evaporating water into the air without adding or removing sensible heat. The process of evaporation in a closed container described at the beginning of this chapter involves a combination of both these methods, since in order to preserve a constant temperature in the enclosure, heat must be added while the evaporation is occurring. The first of the two processes is easy to understand, but in the second it is necessary to determine the cooling effect involving the heat of vaporization. In particular, it is necessary to know how much water has to be evaporated to produce saturation and what the temperature at saturation will be. It is not a simple matter of raising the moisture content to that required for saturation at the initial temperature of the air, because the temperature will decrease owing to the evaporative cooling. In other words, we cannot raise the dew point to the initial temperature because the latter will be falling and they will meet somewhere at an intermediate temperature point. We will now find out at what temperature this saturation will occur.

At constant pressure $dp = 0$, the equation for the condensation process [Eq. (6-23)] reduces to

$$C_p \, dT = -L \, dw \qquad (6\text{-}29)$$

Let us start with temperature T and mixing ratio w and find the temperature T_w and mixing ratio w' at which the added water vapor will produce saturation. The equation becomes

$$C_p(T - T_w) = -L(w - w') = L(w' - w) \qquad (6\text{-}30)$$

Solving for the new temperature T_w, we have

$$T_w = T - \frac{L}{C_p}(w' - w) \qquad (6\text{-}31)$$

Note that since water vapor has been added, $w' > w$; so $T_w > T$.

At the beginning of this process we know only T and w. It is apparent that at the end T_w is the actual temperature and w' is the actual mixing ratio, which at this point is at saturation. Since w' is a function of T_w (saturation mixing ratio at a given pressure depends on temperature only), we can solve for T_w or use tables to find a w' that fits the relation, or better still, we can solve graphically.

Actually, T_w is essentially the wet-bulb temperature as described in Appendix A. The *psychrometer*, a combination of an ordinary thermometer and one with its bulb covered by a thin dampened cloth, is the common instrument used to obtain humidity measurements. The cooling of the wet bulb to its equilibrium temperature is represented by processes similar to those just described. At equilibrium, the transfer of sensible heat from the air to the cooler bulb is exactly equal to the heat transferred from the bulb by evaporation. At a given temperature, a low mixing ratio means a high rate of evaporational cooling of the bulb, and a large temperature difference requires a correspondingly large transfer of sensible heat from the air to the bulb. In a somewhat analogous sense we can say that for the cooling of air by evaporation, equilibrium is reached when the temperature has been reduced by an amount corresponding to the heat removed from the air by the evaporation. The drier the air initially, the greater must be the amount of water evaporated to produce saturation equilibrium and the greater must be the temperature reduction.

The wet-bulb temperature of a mass of air may be defined as the temperature at which the air reaches saturation when water is evaporated into it without any other heat exchange. It is the lowest temperature to which the air can be cooled by evaporating water into it. These statements apply to a constant-pressure process.

It can be shown also that a wet-bulb temperature can be achieved through a saturation-adiabatic process. This temperature differs so little from that obtained by a constant-pressure process that for practical purposes it may be considered the same. Consider a parcel that has ascended to its characteristic (condensation) point. As already shown, this point determines the pseudoadiabatic line to be followed in further ascent. To relate this to the wet-bulb temperature, one may imagine a process in which enough liquid water is added at the characteristic point in the form of cloud droplets to keep the air parcel saturated by evaporation of these droplets during descent to the original level. If there is more than the required amount of liquid water, the result will not be affected. With the water continually evaporating, the parcel will descend along a reversible saturation-adiabatic line, which is essentially the same numerically as a pseudoadiabatic line. Where this line intersects the original pressure level, the wet-bulb temperature is achieved.

In this process, instead of adding water vapor by evaporation at constant pressure as described previously, we have added the same amount of water during descent from the condensation level.   By definition, the mixing ratio w at the condensation point is the same as the initial mixing ratio of the parcel.   In obtaining the wet-bulb temperature adiabatically, one must use Eq. (6-23), which differs from (6-31) used in the isobaric case in that the term $(RT/m)(dp/p)$ must be taken into account.   When we say, as above, that the results of the two processes are essentially the same, we are saying that this term integrated up to the condensation point and back to the original pressure is essentially zero.   Actually it represents an amount of work expended in the process, but this amount is too small to bother about in practice.

The wet-bulb potential temperature is defined by this process carried to the 1000-mb pressure.   It is seen that each pseudoadiabatic line corresponds to one and only one wet-bulb potential temperature, as indicated at its intersection with the 1000-mb line.   Since there can be only one pseudoadiabat going through a given condensation point and since the condensation point is uniquely determined by the pressure, temperature, and mixing ratio or, more specifically, by the potential temperature and mixing ratio of the parcel, it follows that the wet-bulb potential temperature is a function of p, T, and w or of θ and w or θ, p_c (the condensation point).

CONDITIONAL INSTABILITY

In tracing the altitude-temperature curve of a moving element that contains moisture but is not at saturation, the lapse rate followed is practically the dry-adiabatic until the condensation level is reached, after which the air will cool at the slower saturation rate. If the prevailing lapse rate in the surrounding atmosphere is between the dry- and saturation-adiabatic rates, an element lifted upward would first encounter resistance to its vertical motion because it would cool with altitude at the faster dry-adiabatic rate.   After condensation, it would follow the saturation-adiabatic rate, which is slower than that of the surrounding atmosphere, so that the line traced on the altitude-temperature graph by the moving element would finally intersect the curve for the surrounding atmosphere and pass from stable to unstable, that is, from heavier to lighter than its surroundings.   The process is illustrated in Fig. 6-3, where the solid line represents the prevailing lapse rate, and the other parts of the figure are as indicated.

This type of temperature distribution (lapse rate between dry- and saturation-adiabatic rates) is called *conditional instability*.   The conditions that must be fulfilled before the element may become unstable with respect to its surroundings are

1 A sufficient amount of moisture in the air so that the moving element becomes saturated soon enough to follow a saturation-adiabatic line which intersects the prevailing lapse-rate curve

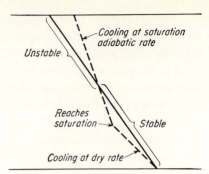

FIGURE 6-3
Conditional instability.

2 A mechanically produced lifting of sufficient strength to overcome the stabilizing forces at the lower levels and to carry the element to the equilibrium point

ENTRAINMENT

The most serious departure from reality in the process we have described comes from the assumption of an isolated parcel of air. In the type of convection which gives rise to cumulus and cumulonimbus clouds (defined in Appendix B), a certain amount of mixing between the cloud air and the environment occurs. It has been shown that in order not to be decelerated in its upward growth, the ascending parcel must be warmer than its environment. The mixing of colder environment air into the cloud has the effect of reducing the temperature of the rising air and thus of reducing or destroying its buoyancy. Because of the dry air being mixed into it, the cloud is unable to maintain itself unless some of the liquid water already present can be evaporated to make up the water-vapor deficit, thus preserving saturation. The result-ant cooling by evaporation of cloud droplets further reduces the temperature in the cloud. The greater the proportion of outside air in the mixture and the lower its initial water-vapor content, the greater must be the evaporation of cloud water. The limiting case would be that in which the liquid content in the mixture is equal to or less than the saturation deficit in the mixture; in this case the cloud would dissipate in that portion.

Without introducing appreciable errors, we may describe the ascent and mixing as taking place in the following three steps:

1 Take 1 kg of air at saturation and lift it through a pseudoadiabatic expansion some chosen small distance to a new T', p.

2   Mix this with M kg of environment air having the temperature T and the specific humidity q to make a mixture of $1 + M$ kg.   In this step the *thermodynamic* effects of changing the specific humidity will not be considered.   The temperature of the mixture would then be the weighted mean

$$T_m = \frac{T' + MT}{1 + M} \qquad (6\text{-}32)$$

and the specific humidity of the mixture would be the weighted mean

$$q_m = \frac{q' + Mq}{1 + M} \qquad (6\text{-}33)$$

Here q' is the saturation specific humidity which the isolated parcel would have at T', p.   Since q, the environment specific humidity, is less than the saturation value, q_m is not sufficient to saturate the mixture.   We must evaporate some of the cloud water to provide the necessary amount of water vapor for saturation. This we do in the next step.

3   Evaporate enough of the cloud water to keep the mixture at saturation. This involves the exchange of a certain amount of latent heat of vaporization. This heat can be calculated by using Eq. (6-29) for constant pressure ($dp = 0$), since no pressure change is involved in this step.   The change in temperature is then

$$dT = \frac{-L}{C_p} \, dw \qquad (6\text{-}34)$$

Here we recognize that by evaporating an amount of liquid water $-dx$ we produce an amount of vapor $+dw$.   Also, since we are dealing with a mixture, it is more convenient to use the specific humidity change dq in place of the change in mixing ratio dw.   The new temperature, then, is

$$T'' = T_m + dT = T_m - \frac{L}{C_p} \, dq = T_m - \frac{L}{C_p} (q'' - q_m) \qquad (6\text{-}35)$$

where T'' and q'' are the actual temperature and specific humidity, respectively, resulting in the cloud.

In making these computations it always is necessary to determine whether or not there is enough liquid water in the cloud at the end of step 1 to accomplish the saturation in step 3.   The amount of water in liquid form x_2 contained in an isolated parcel going from point 1 to point 2 is given by

$$x_2 = x_1 + (q_{s1} - q_{s2})$$

where the quantities in the parentheses are the saturation specific humidities at points 1 and 2, respectively.   As shown on page 108, they can be determined directly from the temperatures and pressures of the two points.

The name *entrainment* has been given to this process of incorporation and mixing of environment air into the ascending cloud mass.   Its precise thermodynamic significance was first pointed out by Stommel.[1]   He used the term because he recognized that the ascending current in the cumulus cloud was essentially like a jet of air whose physical analysis always requires a consideration of entrainment of some of the still air into which it is injected.

In considering the ascent of air in convection currents in the atmosphere, it is customary to refer to the entrainment in terms of the percentage increase of mass by mixing in a given pressure interval.   In cumulus clouds of restricted development Stommel found entrainment rates of 100 percent in 100 mb, while Byers and Braham[2] obtained a rate of 100 percent in 500 mb for bulging cumulus growing into thunderstorms.

After a further discussion of thermodynamic diagrams and thermodynamic properties of the atmosphere in the next few pages, a simple graphical method of following the ascent of air with and without entrainment will be shown.

THE ADIABATIC CHART AND THERMODYNAMIC DIAGRAM

Data from soundings of pressure, temperature, and humidity are entered and computed on some form of adiabatic chart.   In its simplest form this is a diagram with temperature in degrees Celsius as abscissa and pressure in millibars as ordinate. The temperature is on a linear scale, but the pressure is usually on a logarithmic scale, decreasing upward.   There is some advantage in having the ordinate in terms of $p^{.286}$, making the dry-adiabatic or potential-temperature lines straight.   The potential temperature is given by

$$\theta = T \left(\frac{1000}{p} \right)^{.286}$$

as shown in Eq. (5-74).

A true thermodynamic diagram is one which is a transformation of the pV diagram, which permits the computation of energy exchanged in a closed cycle.   As

[1] H. Stommel, Entrainment of Air into a Cumulus Cloud, *Journal of Meteorology*, vol. 4, pp. 91–94, 1947.
[2] H. R. Byers and R. R. Braham, Jr., "The Thunderstorm," Government Printing Office, Washington, 1949.

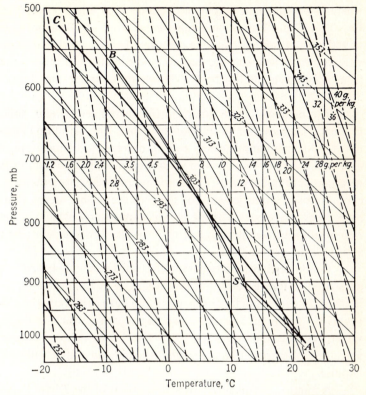

FIGURE 6-4
Stüve diagram.

shown on page 81, this transforms into $T \ln p$ where, as in the atmosphere, we desire
to use pressure and temperature in place of pressure and volume. Coordinates of T
and $p^{.286}$ are a close approximation to T and $\ln p$. A useful diagram based on the
latter approximation is the *Stüve diagram*, named after the German meteorologist who
had much to do with its development.

A Stüve diagram is shown in Fig. 6-4, with various lines for atmospheric computations. The potential-temperature or dry-adiabatic lines are the straight lines
sloping upward from right to left and labeled 273, 283, 293, etc. The lines of saturation mixing ratios are the dashed lines labeled 8, 10, 12, 14, ..., 28 g per kg. They
are based on data similar to that given in Table 6-3 and connect points in terms of
temperature and pressure where the amount of water vapor represented by each line
will just produce saturation. From these, the isentropic-condensation temperature or
pressure may be obtained. For example, if an air sample starting from A and
having a mixing ratio of 10 g per kg ascends, it will reach condensation at the point

where its dry-adiabatic or potential-temperature line intersects the line for saturation for 10 g per kg.   This point is represented by *S*—an isentropic-condensation pressure of 900 mb and condensation temperature of 12°C.   If the air sample starting out at *A* were cooled at constant pressure, it would reach saturation where a horizontal line from *A* intersects the 10-g line.   This would be the temperature of the dew point, or about 14°C.

The pseudoadiabatic lines are the curved set of light, solid lines that are not labeled.   The variation in slope of these lines with variation in temperature should be noted.   At high temperatures, a given decrease in temperature of a saturated parcel results in a greater reduction in saturation mixing ratio than in the case of low temperatures.   In other words, condensation proceeds more rapidly in warm air than in cool air for a given decrease of temperature.   This means that more heat will be released by condensation at the higher temperatures, which, in turn, will decrease the rate of adiabatic temperature drop.   At very low temperatures the pseudoadiabats approach the dry adiabats.

Conditional instability is also illustrated in Fig. 6-4.   After the rising air parcel goes from *A* to the condensation level at *S*, it follows along a pseudoadiabat *SB*. If *AC* is the prevailing lapse-rate curve, we have conditional instability indicated as in Fig. 6-3.   The point where the two curves cross, which in Fig. 6-4 is at 770 mb, is sometimes called the *level of free convection*, because above that level the rising parcel is unstable with respect to its surroundings and convection will proceed without added impulse.

A number of equal-area transformations of the $T \ln p$ diagram have been made and are used regularly by meteorological services.   They are aimed at easier plotting, reading, and computation of the various quantities.   Since adiabatic equilibrium or departures from it are important features to be recognized, some of the diagrams feature the dry adiabats as horizontal lines or vertical lines, so that departures from adiabatic equilibrium stand out more clearly.   It is argued that neither isothermal nor constant-pressure conditions are of special interest in the atmosphere and that the vertical lines and horizontal lines on the conventional diagram are of no great help in recognizing critical situations.   In one transformation the important pseudoadiabats for summer convection stand approximately vertical.   In addition to the $T \ln p$ diagram, two true energy diagrams, the *tephigram* and the *skew T* ln *p diagram*, have gained wide acceptance.

Tephigram   This derives its name from its coordinates of temperature and entropy (T, φ).   It was introduced by Sir Napier Shaw.[1]   A line of constant entropy

[1] N. Shaw, "Manual of Meteorology," vol. 2, p. 36; vol. 3, pp. 223–224, Cambridge University Press, London, 1926, 1930.

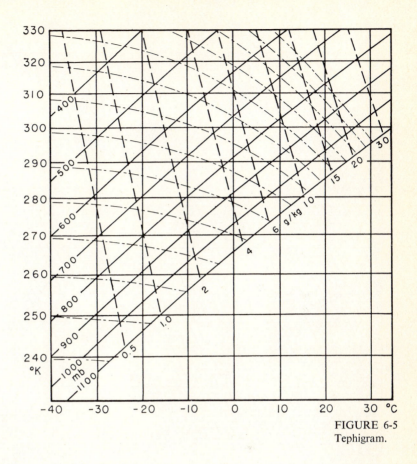

FIGURE 6-5
Tephigram.

is a line of constant potential temperature; so the ordinate in Fig. 6-5 is labeled in terms of potential temperature. The pressure lines are diagonals with the lowest values in the upper left. Again, the moisture lines are the straight, dashed lines and the pseudoadiabats are curved. This chart has been favored by British, Canadian, and some American meteorologists.

Skewed diagram Certain desired characteristics are obtained by making the coordinates nonrectangular while preserving the $T \ln p$ areal representation. A diagram of this type, suggested by Herlofson,[1] has the conventional $\ln p$ scale as the ordinate, but the temperature lines are rotated clockwise somewhat less than 45°. This produces the maximum angle of intersection (90° in the standard atmosphere at

[1] N. Herlofson, *Meteorologiske Annalen* (Oslo), vol. 2, no. 10, 1947.

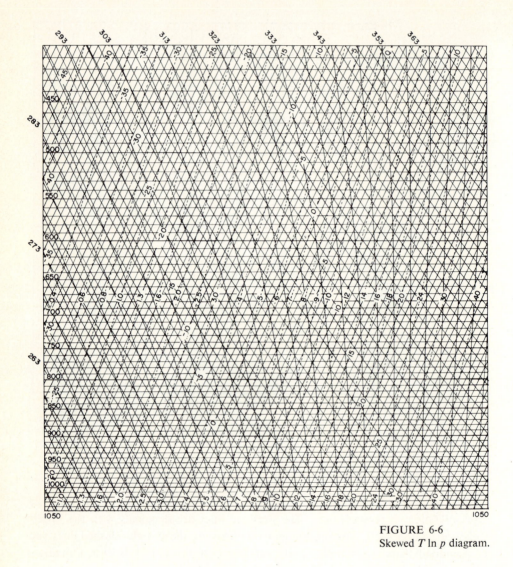

FIGURE 6-6
Skewed $T \ln p$ diagram.

500 mb) between the lines of temperature and of potential temperature. The angle between dry-adiabatic (potential temperature) and pseudoadiabatic lines is also large. At temperatures in the low atmosphere characteristic of summer convection the pseudoadiabats are nearly vertical. A skewed $T \ln p$ diagram is shown in Fig. 6-6.

GRAPHICAL COMPUTATIONS ON THE DIAGRAMS

In practical meteorological work the thermodynamic diagrams are used for computing a variety of quantities from the given data. Radiosondes provide measurements of pressure, temperature, and relative humidity. The soundings are plotted on the diagram, and from the plot other variables can be read off or plotted on related curves. Some of the data obtainable in this manner are:

Potential temperature is obtained by following the dry adiabat to 1000 mb and reading the temperature there, or, more easily, by noting the values of potential temperature printed on the dry-adiabatic lines.

Saturation mixing ratio is read immediately at any point on the temperature-pressure plot by interpolation from the mixing-ratio lines.

Mixing ratio is obtained by multiplying the saturation mixing ratio by the measured relative humidity.

Dew-point temperature is obtained as the temperature at the point where the actual mixing-ratio line or interpolated line intersects the observed pressure line. It is useful to plot this for every significant level of the sounding to produce a curve that is a plot of both the mixing ratio and the dew-point temperature. From such a curve, working in a reverse sense, one can get the relative humidity from the ratio of the mixing ratio to the saturation mixing ratio.

Condensation pressure, temperature (*isentropic*) is found at the intersection of the potential-temperature line and the mixing-ratio line which corresponds to the value at the starting point. For certain purposes it is convenient to plot this quantity for the various significant levels.

Wet-bulb temperature, wet-bulb potential temperature are obtained by following the pseudoadiabatic line from the condensation point to the original pressure and to 1000 mb, respectively.

Equivalent temperature, equivalent-potential temperature are found by determining the potential-temperature line which is approached asymptotically by the pseudoadiabat that passes through the condensation point. This potential-temperature line has the value of the equivalent-potential temperature. By following it back to the original pressure, the equivalent temperature is read on the temperature scale.

PROPERTIES USED IN TRACING AIR PARCELS

Some of the thermodynamic properties are useful in tracing, both horizontally and vertically, the paths of air parcels through the atmosphere. Those properties, which under ordinary circumstances do not vary appreciably, called "conservative" properties, can be used for this purpose. In Table 6-5 the relative variability of thermodynamic properties which can be measured or derived easily from the observations of the conventional radiosondes is given. Those properties which are constant, at least before saturation, are classified as conservative. It will be noted that wet-bulb-potential temperature and equivalent-potential temperature are the only two properties that are constant before and after saturation. As tracers, however, they are not as useful as specific humidity or mixing ratio, because over large sections of the atmosphere, and even over large horizontal distances, the values are observed to be unchanging with altitude. This means that a parcel displaced from its original location will not carry a distinctive value to differentiate it from its surroundings. The water-vapor mixing ratio has normally a highly nonuniform distribution, and any parcel carried away from its initial position is more easily recognized. Over the large regions studied daily on synoptic charts, the areas of saturation are relatively small and their effects on the mixing ratio as a tracer are not usually serious.

ENERGY DIAGRAMS

The determination of conditional instability and the calculation of the energy that might be released may be made graphically from the thermodynamic diagram. This is a step toward predicting the occurrence of strong vertical currents, especially thunderstorms, from upper-air soundings. On the $T \ln p$ diagram or equal-area transformations of it the buoyant energy that can be realized in cumulus-cloud convection may be determined by measuring the areas between the prevailing lapse-rate

Table 6-5 VARIABILITY OF AIR-MASS PROPERTIES IN ASCENT

Property	Before condensation (dry-adiabatic)	After condensation (saturation-adiabatic)
Temperature	Decreases	Decreases
Virtual temperature	Decreases	Decreases
Vapor pressure	Decreases	Decreases
Relative humidity	Increases	Constant, 100%
Dew-point temperature	Decreases	Decreases
Absolute humidity	Decreases	Decreases
Specific humidity or mixing ratio	Constant	Decreases
Potential temperature	Constant	Increases
Isentropic-condensation pressure or temperature	Constant	Not defined
Equivalent temperature	Decreases	Decreases
Wet-bulb temperature	Decreases	Decreases
Equivalent-potential temperature	Constant	Constant
Wet-bulb-potential temperature	Constant	Constant

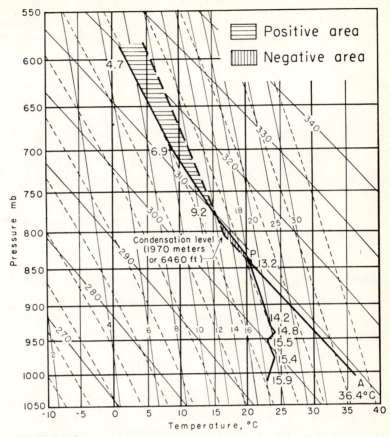

FIGURE 6-7
Energy diagram prepared from an early-morning sounding at Washington, D.C.

curve and the adiabats followed by an ascending parcel.   Areas on the diagram where the rising air is warmer than its surroundings are *positive areas*, and the parts where the moving parcel is colder are *negative areas*.   Figure 6-7 shows an energy diagram in terms of $T \ln p$ for a selected sounding, and Fig. 6-8 shows the same sounding on a tephigram.

From the buoyancy equation (5-82) the force per unit mass is

$$a = g \frac{T' - T}{T} \qquad (6\text{-}36)$$

where T' is the temperature of the parcel and T that of the ambient air.   The differential of work per unit mass, that is, the force per unit mass times the distance displaced, is

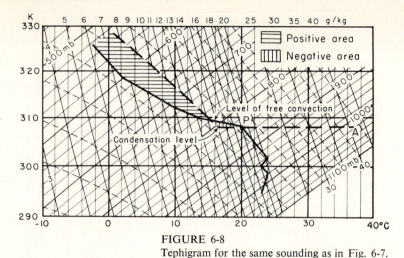

FIGURE 6-8

Tephigram for the same sounding as in Fig. 6-7.

written as $dw = a\,dz$ for the height differential dz; so the work done *by* the air parcel is

$$dw = \frac{T' - T}{T}\,g\,dz = -\frac{T' - T}{T}\frac{dp}{\rho} = -\frac{T' - T}{T}\frac{dp}{p}\frac{RT}{m} \qquad (6\text{-}37)$$

$$dw = \frac{R}{m}(T - T')\,d(\ln p) \qquad (6\text{-}38)$$

which on a $T \ln p$ diagram would represent the differential area dA such as might be envisaged as a very thin slice across a shaded area in Fig. 6-7. The work by unit mass through a finite interval from p_0 to p is obtained by integrating:

$$w = \frac{R}{m}\int_{p_0}^{p} T\,d(\ln p) - \frac{R}{m}\int_{p_0}^{p} T'\,d(\ln p) \qquad (6\text{-}39)$$

This expression, with $\ln p$ decreasing upward in the usual diagram, gives the area between p_0 and p above the T curve minus that above the T' curve, both multiplied by R/m, as may be visualized in Fig. 6-7 starting from the level of free convection. This is equivalent to R/m times the area between the two curves in the pressure interval.

The air through which the parcel is ascending offers a frictional resistance referred to as the *form drag*, which is neglected here. For small parcels not exceeding a few cubic meters in size it might be possible to approach the problem of form drag, at least for a simple form such as a sphere. In the buoyancy of convection in the atmosphere where elements the size of a growing cumulus cloud are involved, the surrounding air develops an accommodating circulation. Even more important is the loss of parcel identity through mixing with the air of the cloud-free environment.

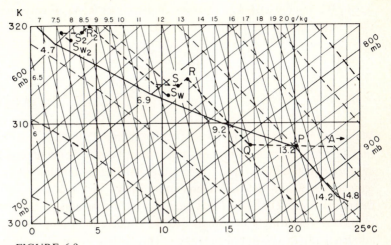

FIGURE 6-9
Graphical computation of entrainment in 100-mb steps, magnified from Fig. 6-8.

The thermohydrodynamic modeling of convection is the subject of extensive treatment in the literature of meteorology and fluid mechanics.

As a forecasting aid, the energy diagram is usually constructed from an early-morning or late-evening sounding, and the conditions for the time of maximum temperature the ensuing day are analyzed. In Figs. 6-7 and 6-8 the calculations are based on expected afternoon conditions, reasoning that at the time of maximum temperature a dry-adiabatic lapse rate will prevail below the point P, so that the temperature at A will result at the ground. The adiabatic mixing in this layer is expected to reduce the moisture content of the air at the ground to 15 g per kg, or to about the average of the values (entered to the right of the curve) between the ground and P. A sample of this ground air is then lifted. In this example, it is noted that a small negative area needs to be overcome before the ascending sample of air in the cloud reaches the *level of free convection* whence the positive area and instability begin.

The type of convection just described is based on the full-parcel method. Entrainment modifies this picture. To illustrate this, let us take a magnified portion of the plot in Fig. 6-8 between 950 and 600 mb and perform an entrainment computation, as shown in Fig. 6-9. For simplicity, the computation is made in 100-mb steps assuming an entrainment rate of 50 percent per 100 mb of ascent.

The 1-kg parcel at Q (pressure 810 mb) in Fig. 6-9 is carried along the pseudo-adiabat to 710 mb, point R, where its temperature T' becomes 11.8°; the environment temperature T is 9.9° while the corresponding specific humidities q' and q are 12.6 and 7.2 g per kg, respectively. The environment q is interpolated between the two points having 9.2 and 6.9 g per kg.   Adding 50 percent or 0.5 kg of environment air gives for the mixture, from Eqs. (6-32) and (6-33),

$$T_m = \frac{T' + 0.5T}{1.5} = \frac{11.8 + 5.0}{1.5} = 11.2°C$$

$$q_m = \frac{q' + 0.5q}{1.5} = \frac{12.6 + 3.6}{1.5} = 10.8 \text{ g kg}^{-1}$$

We obtain the point S for this mixing by finding $T_m \equiv 11.2$ at 710 mb.   It is noted that q_m is less than the saturation value at this point.   Some of the liquid water must evaporate to produce a new temperature and specific humidity.   Actually, this new temperature is the wet-bulb temperature of the air at S, as is apparent from the thermodynamic definition of that quantity.   It can be obtained by the usual graphical method of following a dry adiabat to the condensation point and tracing a pseudo-adiabat back to the initial pressure.   This procedure produces the point S_w, with temperature 10.3°C and specific humidity of 11.3 g per kg (read from the diagram).

Table 6-6   VALUES OBTAINED IN COMPUTATION OF ENTRAINMENT IN FIG. 6-9

a. Temperature and water vapor, assuming adequate supply of liquid water

Point	T'	q_m	q_{sat}	Environment T	Environment q
Q	16.6		15.0	17.4	11.2
R	11.8		12.6	9.9	7.2
S	11.2	10.8	12.0	9.9	7.2
S_w	10.3		11.3	9.9	7.2
R_2	4.5		8.7	2.6	4.9
S_2	3.9	7.5	8.3	2.6	4.9
S_{w2}	3.1		7.8	2.6	4.9

b. Check as to adequacy of liquid-water contents, grams per kilogram

From points	Added by condensation	Result of dilution by mixing	Removed by evaporation	Net carried forward
Q to R	$15.0 - 12.6 = 2.4$			2.4
R to S		$\frac{2.4}{1.5} = 1.6$		1.6
S to S_w			$10.8 - 11.3 = -0.5$	1.1
S_w to R_2	$11.3 - 8.7 = 2.6$			3.7
R_2 to S_2		$\frac{3.7}{1.5} = 2.5$		2.5
S_2 to S_{w2}			$7.5 - 7.8 = -0.3$	2.2

All net values positive; therefore, liquid-water contents are adequate for all required processes.

The necessary bookkeeping, tabulated in Table 6-6, shows that from Q to R, 2.4 g of water vapor, representing the difference between the saturation specific humidities (mixing ratios) at the two points, passed into liquid form in the 1-kg parcel of air.   Putting this amount into 1.5 kg of air by the mixing reduces its concentration to 2.4/1.5 or 1.6 g per kg as the liquid-water content.   After evaporating the 0.5 g (11.3 − 10.8) necessary to bring the mixture to saturation at the wet-bulb temperature, there is still 1.1 g per kg of liquid water left.   This is added to the liquid-water load of the parcel through the next step of the ascent.

Taking S_w as the new starting point, we go up another 100 mb, repeating the procedure to obtain R_2, S_2, and S_{w2} as shown in the figure and recorded in the table. The entrainment moist-adiabatic ascent curve would be obtained by constructing straight-line segments between Q and S_w and between S_w and S_{w2}.   For a clearer explanation of the process, this was not done in Fig. 6-9.

The results indicate that the buoyancy in the cloud is greatly reduced by the entrainment.   In the actual atmosphere this reduced buoyancy represents the true state of affairs in cumulus clouds, except in the cores of giant ones where little environment air is able to penetrate.

LAYER STABILITY, CONVECTIVE INSTABILITY

In Chap. 5 the consequences of lifting, sinking, stretching, and shrinking of layers in dry-adiabatic processes were described.

The results may apply to saturation as well as to dry conditions if the saturation or pseudoadiabatic is used in place of the dry adiabatic.   In nature, the saturation conditions would not appear to be applicable for descent.   Superadiabatic lapse rates with respect to saturation are common in the atmosphere, and it is under these conditions that the effects on a superadiabatic lapse rate are important.   For example, divergence in an ascending layer at saturation but superadiabatic with respect to saturation would preserve steep lapse rates during ascent.

Frequently the vertical distribution of moisture within the lower layers of the atmosphere is such that, when they are lifted, the bottom part reaches its condensation level before the upper.   This condition arises most often when the relative humidity decreases with height within the layer in question.   The result of lifting in such cases is a very rapid steepening of the lapse rate, for now, in addition to the gradual change brought about in any lifting process as previously described, the top is cooling more rapidly than the bottom of the layer.   If the air is very moist in the low levels, the difference between the saturation and dry adiabatic rates is large so that steep temperature drops with height within the air layers can develop with great rapidity. The process is shown in Fig. 6-10, where AB is the original lapse rate in the layer. A

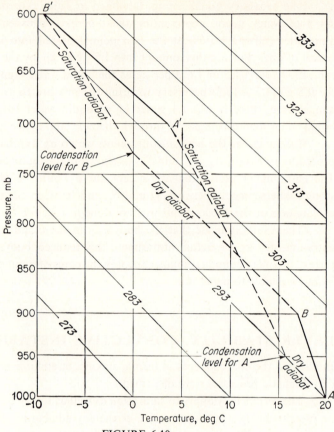

FIGURE 6-10
Graphical representation of convective instability.

reaches saturation much earlier than *B*, and the results of the different rates of cooling at the two points are shown by *A'B'*.

Air layers that exhibit this property of sharp lapse-rate changes due to different condensation levels are designated as convectively unstable or as possessing the property of convective instability. Layers that are originally quite stable or that may even consist of temperature inversions are often convectively unstable because of the pronounced stratification of moisture.

Rossby[1] studied the problem of convective instability in considerable detail and showed that a layer will be convectively unstable if the equivalent-potential tem-

---

[1] C. G. Rossby, Thermodynamics Applied to Air Mass Analysis, *Massachusetts Institute of Technology, Meteorology Papers*, vol. 1, no. 3, 1932.

perature decreases with height through the layer. This is the accepted definition of convective instability, although wet-bulb-potential temperature may be substituted in the definition.

With the introduction of convective instability, confusing terminology appears. We generally test convection by the parcel method, considering the conditional instability, whereas convective instability refers to layer conditions and has little reference to convection as we generally know it. It is a form of *potential* instability in which a layer that may at the present time be stable can become unstable with the appropriate amount of lifting. On the other hand, conditional instability refers to the character of the layer *now*. It is possible for a layer to be convectively unstable and not conditionally unstable.

An interesting case arises when a fairly deep layer that is convectively unstable becomes saturated not by lifting but by the evaporation of water into it from rain falling from a higher cloud layer. Once this layer becomes saturated, it is absolutely unstable. It has been noted on some occasions that thunderstorms develop from rain falling out of an altostratus layer. The saturation of the convectively unstable layer by evaporation of the falling rain is thought to be the cause. The student can verify the instability of this situation by plotting decreasing wet-bulb-potential temperatures with height and recognizing that this will be the actual lapse rate if saturation is reached.

EXERCISES

1 In a cloud that remains in liquid form at temperatures below 0°C the saturation vapor pressure is higher than the equilibrium over ice. What would be the percent of super-saturation with respect to ice in a liquid-water cloud at saturation at -12°C? (Table 6-1.)

2 From tables and equations given in this chapter, obtain values to replace the question marks. The symbols are those employed in this chapter.

T, °C	e, mb	e_s	q or w, g/kg	q_s or w_s	ρ_w, g cm^{-3} ρ_{ws}	P, mb	f, %	T_d, °C
20	?	?			?	?	50	?
			6	10			?	
0	?	?	2.8	?	?	?	800	?	?
	?	?					900		14
5	?	?	?	?	?	?	850	?	0
10						11	?	

3 A 1-kg parcel at saturation ascends from a condensation level at 850 mb with an initial temperature of 22°C and mixing ratio of 20 g kg^{-1}. It reaches 800 mb with a temperature of 20°C.

(*a*) Calculate the mixing ratio at 800 mb from the simplified pseudoadiabatic equation, assuming a mean temperature of 21°C in the pressure-change term. $L = 2.454$ J kg^{-1}; $R/m = 287.05$ J kg^{-1} K^{-1}; $C_p = 1.005 \times 10^3$ J kg^{-1} K^{-1}.

(*b*) How much heat is released in this 50-mb ascent by condensation?

(*c*) How high would this energy, converted into work, lift a 220-lb man, if it is operating at 100 percent efficiency?

(*d*) Continuing in this pseudoadiabatic process, one finds that at 600 mb $w_s = 13$ g kg^{-1}. Compute the temperature at that point, using a mean temperature of 15°C during the ascent.

4   From a thermodynamic diagram (skew T log p, tephigram, Stüve, or similar diagram) determine the amount of liquid water condensed into an isolated parcel of air ascending from a condensation level at 900 mb and 20°C to the 500-mb level without mixing or entrainment of environment air. Convert to grams per cubic meter, taking the air density at 500 mb to be 6.4×10^{-4} g cm^{-3}.

5   On a thermodynamic diagram as in the last question, enter and label the following point in the atmosphere:

$$p \quad 890 \text{ mb} \qquad T \quad 26°C \qquad w \quad 10 \text{ g kg}^{-1}$$

Find:

(*a*) The p and T at the condensation level

(*b*) Saturation mixing ratio at the 890-mb, 26°C point

(*c*) Potential temperature

(*d*) Pseudo-wet-bulb temperature

(*e*) Wet-bulb-potential temperature

(*f*) Temperature of the dew point

(*g*) Relative humidity

(*h*) The mixing ratio at the wet-bulb temperature

(*i*) The amount of water vapor condensed in going from the condensation level to 620 mb

(*j*) The vapor pressure at the 890 point

6   From a condensation point at 810 mb and 16.5°C, 1 kg of air ascends to 710 mb with an entrainment at 710 mb of 0.5 kg of environment air having a temperature of 10°C and a mixing ratio of 3 g kg^{-1}. From a 100-mb single-step graphical-numerical computation, determine whether or not there will be enough liquid water to preserve saturation after mixing with the entrained air. Give actual values.

7   (A major work or laboratory exercise.) From a blank piece of paper construct a thermodynamic diagram. For simplicity make it a Stüve diagram ($p^{.286}$). Enter dry and pseudoadiabatic lines and lines of w_s as well as temperature and pressure (-40 to $+40$ and 1050 to 400 mb). Assume liquid condensation at all temperatures.

7

HORIZONTAL MOTION IN THE ATMOSPHERE— THE WINDS

The wind is air in motion, usually measured only in its horizontal component. It is an essential part of the total thermodynamic mechanism of the atmosphere, serving as a means of transporting heat, moisture, and other properties from one region of the earth to another. While vertical motions, when they occur, are extremely important for the production of clouds, precipitation, thunderstorms, etc., their total sustained magnitudes are quite small in comparison with the total horizontal flows of the atmosphere.

The physical-mathematical treatment of the flow of air particles over the surface of the earth is complicated by the fact that both the earth and the atmosphere are rotating. Before discussing the action of forces in the wind motions, it is necessary to understand some of the principles of rotational motion derived from elementary physics. For a more complete statement of these fundamentals the reader is referred to physics textbooks.

ANGULAR VELOCITY

The rate of rotation of a body is called its *angular velocity*. It is expressed as an angle through which a body turns in unit time; for a particle moving in the arc of a circle, it is the angular displacement in a unit time. It is customary to state the angular velocity in radians per second.

Rotating motion may also be expressed in terms of linear velocity according to a simple relationship. Consider a particle moving in an arc the distance Δs with speed v in time Δt. In terms of angular speed ω, it moves through the angle $\Delta\varphi$ subtended by the arc at radius r. From elementary kinematics it is seen that

$$\Delta s = v\,\Delta t \qquad \text{and} \qquad \Delta\varphi = \omega\,\Delta t \qquad (7\text{-}1)$$

The total circumference of a circle is $2\pi r$, and the distance Δs is the $(\Delta\varphi/2\pi)$th part of the circumference; so $\Delta s = r\,\Delta\varphi$. Therefore, in terms of the calculus

$$\underset{\Delta t \to 0}{\text{Limit}} \frac{\Delta s}{\Delta t} = \frac{ds}{dt} = r\frac{d\varphi}{dt} \qquad \text{or} \qquad v = \omega r \qquad (7\text{-}2)$$

ANGULAR-VELOCITY VECTOR

The rotation of a volume, a system of points, or a frame of reference can be represented by a vector. The vector is along the axis of rotation, perpendicular to the plane of rotation. The convention has been adopted that the vector should point in the direction of advance of a screw being turned in the same direction of rotation. Thus a vector representing the rotation of the hands of a clock would point into the clock from the face because the screw turned with the hands would advance toward the back of the clock. The length of the vector is made proportional to the angular speed. The minute hand of the clock has an angular speed of 2π rad per hr or $\pi/1800 = 1.745 \times 10^{-3}$ rad per sec. Taking a centimeter of length as representing 10^{-3} rad per sec, one would draw a vector 1.745 cm long pointing inward from the front of the clock to represent the angular velocity. It should be noted that dimensionally angular velocity is expressed simply as $\sec^{-1}$, the radian being considered as nondimensional.

COMPONENTS OF A POINT ON THE EARTH

The earth rotates, when viewed from above the Northern Hemisphere, in a counter-clockwise manner. At the North Pole the angular-velocity vector would point outward along the axis of rotation and at the South Pole it would point inward. Note that when viewed from above the Southern Hemisphere, the earth rotates in the same

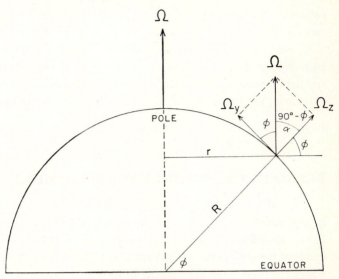

FIGURE 7-1
Components of rotation at latitude φ.

direction as the hands of a clock.   The magnitude of the two vectors at the poles is 2π rad per sidereal day or 7.292×10^{-5} per sec.

At any point on the earth, other than at the poles and on the equator, the rotation can be resolved into two components—a radial component and a tangential component.   At the pole it is all radial and at the equator it is all tangential.   For a spherical earth the radial direction is vertically upward from the surface and the tangential direction is northward.   The local vertical component represents a local turning of the surface of the earth in the tangential plane or, locally speaking, in the horizontal plane of the earth's surface.   The tangential component represents a turning in the radial or vertical plane.

The components at latitude φ of the earth's rotational vector Ω are illustrated in Fig. 7-1.   The vector Ω forms the angle α with the radial line which forms the vertical line at latitude φ.   But $\alpha = 90 - \varphi$, as shown by the construction.   The vertical component at latitude φ is therefore $\Omega \sin \varphi$.   The northward component $\Omega_y = \Omega \cos \varphi$ as shown in the figure.

The vertical component of the rotation, that is, the turning of the surface at all latitudes except the equator (where $\sin \varphi = 0$), has a profound influence on atmospheric motions and has to be referred to in all computations concerning the air in motion.

It should not be confused with the movement of a point on the earth about the axis of rotation; all points on the earth move around the axis with the same angular velocity. As already shown by Eq. (7-2), the linear velocity of a point fixed on the earth would be given by $U = \Omega r$, where r is the distance from the axis of rotation. Since r decreases with increasing latitude, U would be a maximum at the equator and go to zero at the poles where r is zero. If, as in Fig. 7-1, the radius of the earth is R and the distance of the surface from the axis at latitude φ is r, then $r = R\cos\varphi$. At any latitude φ, the linear velocity of a point fixed on the surface of the earth is

$$U = \Omega R \cos \varphi \qquad (7-3)$$

FOUCAULT'S PENDULUM EXPERIMENT

In 1851 the French physicist Foucault, inventor of the gyroscope, demonstrated experimentally the presence and the magnitude of the local vertical component of the earth's rotation. He used a free suspension to attach a long, heavy pendulum to the inside of the high dome of the Pantheon in Paris. The point of suspension was 200 ft above the floor. On the floor he heaped a ring of sand in such a way that near the end of each swing the pointed tip placed on the weighted ball of the pendulum would make a mark in the sand. The pendulum was started in free oscillation in a chosen plane. It was not long before it was noted that either the pendulum was gradually changing its plane of oscillation or the floor was rotating under it. Foucault explained to the spectators that it had to be the latter effect and, after 24 hr, demonstrated that the turning had amounted to that fraction of a complete circle equal to the sine of the latitude of Paris, or $2\pi \sin \varphi$ rad.

Foucault pendulums have been set up in many places throughout the world, and today the student may see them in operation at the Museum of Science and Industry in Chicago, the Franklin Institute in Philadelphia, the United Nations building in New York, the Smithsonian Institution in Washington, D.C., and at exhibits in London, Paris, and a number of other cities, college campuses, etc.

CENTRIPETAL FORCE

According to Newton's first law of motion, a body in motion will continue in the same direction in a straight line and with the same speed unless acted upon by some external force. This means that, in order for a body to move in a curved path, some force must be continually applied. The force restraining bodies to move in a curved path is called the *centripetal force*, and it is always directed toward the center of rotation. When a rock is whirled around on a string, the centripetal force is exerted by the

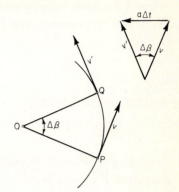

FIGURE 7-2

tension of the string.   In the game of "crack-the-whip," the end man goes off on a straight line when the centripetal force required to keep him moving in a circle can no longer be maintained.

The centripetal force sometimes is regarded in the reverse sense as a *centrifugal force*.   This is not exact, however, since the force applied in curved motion is inward. When one is standing fixed on a rotating platform one has the definite impression of an outward-acting centrifugal force.   Actually, this feeling comes from the action of the body in its tendency toward motion in a straight path.   On the rotating surface of the earth, where it is convenient to refer motions and forces to the rotating system rather than fixed stars, the apparent centrifugal force becomes quite real.   Meteorologists and others in the geophysical sciences often forgo formalities and refer to it directly as a centrifugal force.

It is easiest to derive the expression for centripetal force by graphical means. Consider a mass point moving at constant speed from P to Q around a center at O an angular distance $\Delta\beta$, as in Fig. 7-2.   Although its speed is the same its direction and therefore its velocity has changed in this interval from v to v', represented in the figure by two vectors of equal length.   These linear velocities are tangential and therefore perpendicular to the radius.   If a radius line is rotated $\Delta\beta$ degrees, the perpendicular to it must be rotated the same amount.   Thus the angle between the two velocity vectors shown in the inset of the figure must be $\Delta\beta$.   The mass point has departed from a straight path owing to the centripetal acceleration a, which produces the velocity-vector difference $a\,\Delta t$.   Since v and v' are equal in length, they form an isosceles triangle with angle $\Delta\beta$ at the vertex just as do the two radii and the subtended chord.   From the similarity of the triangles we note the ratios

$$\frac{v}{r} = \frac{a\,\Delta t}{\bar{v}\,\Delta t}$$

where $\bar{v}$ is a mean velocity between the two points on the curve and $\bar{v} \, \Delta t$ is the chord. As Δt and $\Delta \beta$ approach zero, the chord approaches the arc, and $\bar{v} \to v$; so

$$a = \frac{v^2}{r} \qquad (7\text{-}4)$$

directed toward the center. In terms of angular velocity the usual substitution shows

$$a = \omega^2 r$$

According to Newton's second law of motion these accelerations are the force per unit mass, and $F = ma = mv^2/r$.

CONSERVATION OF ANGULAR MOMENTUM

Perhaps everyone as a child has taken a weight on the end of a string and spun it around his upright index finger, letting the string wind up on his finger. As the string becomes shorter the spinning becomes faster. The same effect is produced when a figure skater starts spinning with arms extended, then increases his spinning speed by bringing his arms in close to his body. What is happening is that the angular momentum is conserved. It can be shown that in the absence of friction $vr = \text{const}$ or $\omega r^2 = \text{const}$, where $vr = \omega r^2$ is the angular momentum.

The conservation of angular momentum is an expression of a balance of work and energy. In the case of the string on the finger, work is done to increase the kinetic energy [$dW = d(KE)$], the work being provided by the centripetal force acting through the tension of the string. The work done on the whirling mass is the force times the distance of displacement, or $dW = -F \, dr$. The balance is

$$-F \, dr = d(\tfrac{1}{2}mv^2) \qquad (7\text{-}5)$$

F is the centripetal force mv^2/r; so

$$-\frac{mv^2}{r} \, dr = mv \, dv$$

or

$$r \, dv + v \, dr = 0 \qquad (7\text{-}6)$$

which is the derivative of $vr = \text{const}$, and thus we have $\omega r^2 = \text{const}$.

On the earth, poleward-moving particles are forced toward the axis of rotation. Since the distance from the axis of rotation is $r = R \cos \varphi$, the conservation of angular momentum means $\Omega R^2 \cos^2 \varphi = \text{const}$ or $vR \cos \varphi = \text{const}$. A particle at rest with respect to the earth has the absolute velocity, according to Eq. (7-3), given by $U = \Omega R \cos \varphi$. At the equator $U = \Omega R$. If particles initially at rest are displaced from the equator toward the pole to latitude φ, their absolute west-to-east speed, with angular momentum conserved, would be such that

$$vr = U_0 R$$

$$vR \cos \varphi = U_0 R$$

$$v = \frac{U_0}{\cos \varphi}$$

At latitude 60°, $\cos \varphi = \frac{1}{2}$; so the absolute speed at that latitude would have increased to twice the linear speed of a fixed point at the equator. The implications of this effect on broad-scale motions over the earth will be treated in the chapter on the general circulation.

APPARENT AND TRUE GRAVITY

We wish to investigate the motion of a particle on a rotating earth where we shall assume that the only real force acting is the attraction of gravity. On the rotating earth, however, the gravitational acceleration that we usually measure and use for our calculations includes also the centrifugal effects of the rotation, and as such, is known as the "apparent" gravity. The centrifugal effects are of great importance. The flattening of the earth at the poles results from them.

The inverse-square law of attraction between two bodies should be well known to students of elementary physics. Between two bodies there is a force of attraction proportional to the product of the masses and inversely proportional to the squares of their distances apart, and independent of the nature of the bodies. This law of attraction or gravitation may be stated in the formula

$$F = G \frac{mm'}{r^2}$$

where G is called the *constant of gravitation*. Newton demonstrated that a sphere that is either homogeneous or may be regarded as made up of concentric shells, each of which is homogeneous, will attract as if the mass of the sphere were concentrated at its center. A particle of mass 1 g on the surface of the earth would be attracted toward the center of the earth by the force, which we know is about 980 dyn. Since this is the force acting on a unit mass, it is termed the *acceleration due to true gravity*. The earth, being an imperfect sphere, does not attract exactly toward its center but very nearly so.

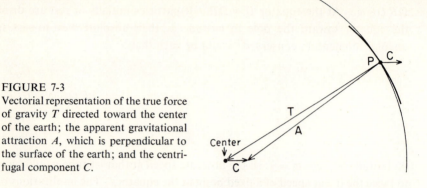

FIGURE 7-3
Vectorial representation of the true force
of gravity T directed toward the center
of the earth; the apparent gravitational
attraction A, which is perpendicular to
the surface of the earth; and the centrifugal component C.

The effects of rotation modify this true gravity considerably. Let us consider
the earth as represented in Fig. 7-3. A particle at the point P on the surface is subjected to a centrifugal action indicated by the vector C, acting outward from the axis
of rotation. The true force of gravity is represented by the vector T. The resultant
force is the vector A, which is the *apparent gravitational attraction*. Note that in the
figure the earth is not represented as a perfect sphere and that the vector A is perpendicular to the surface at P. The earth has adjusted its shape in such a way that
there is no net force along the surface; the southward component of the centrifugal
vector C is exactly balanced by the northward component of the true gravity T. In
the figure, the magnitude of the vector C is exaggerated in comparison with T, thus
making the earth appear less spherical than is actually the case. The exaggeration is
for purposes of clarity in the diagrammatic representation.

A plumb line at P would point in the direction of the vector A. The apparent
gravity is the actually measured value of the acceleration of gravity at any point on the
earth. The vertical component of the centrifugal action C varies with latitude, being
zero at the poles and a maximum at the equator. The true attraction however, increases toward the poles because of the decreasing distance from the center of the
flattened earth. The resulting apparent attraction is greatest at the poles because of
both the nearness to the center and the absence of a vertical centrifugal action. At sea
level at lat 45° the acceleration of gravity is taken as 980.616. From this value it varies
with latitude φ according to the formula

$$g_\varphi = 980.616(1 - 0.0026373 \cos 2\varphi + 0.0000059 \cos^2 2\varphi)$$

Note that for $\varphi = 45°$, $\cos 2\varphi = 0$. The decrease with height is small through the
altitude intervals within the atmosphere. A formula for this variation, given in the
Smithsonian Meteorological Tables, Sixth Edition (1951) and adopted by the WMO is
as follows:

$$g = g_\varphi - (3.085462 \times 10^{-4} + 2.27 \times 10^{-7} \cos 2\varphi)Z$$
$$+ (7.254 \times 10^{-11} + 1.0 \times 10^{-13} \cos 2\varphi)Z^2$$
$$- (1.517 \times 10^{-17} + 6 \times 10^{-20} \cos 2\varphi)Z^3$$

where Z is the height in meters.   At 10 km ($Z = 10^4$ m), the correction for altitude to be applied to g_φ would be about 3 cm sec^{-2}.

MOTIONS IN ACCELERATING COORDINATE SYSTEMS

In considering motions in the atmosphere, it is desirable to use a system of coordinates that is fixed on the surface of the earth, such as the coordinates of latitude and longitude or, more simply, north-south and east-west coordinates measured from any convenient point of origin on the earth.   Since the earth is rotating, such a system of coordinates is also rotating.   In applying Newton's laws of motion to rotation as in Eq. (7-4), we note that at any point not at the center of rotation there must be a centripetal acceleration acting.   So also, in terms of Newtonian mechanics, any system of coordinates that is fixed on a rotating surface is an *accelerating* system of coordinates.

Before examining the rotating case, it may be helpful for an understanding of the problem to consider the simplest case of an accelerating system of coordinates— cartesian axes which are accelerating in a straight line at a uniform rate without rotation.   If you were an observer moving with this system, an object moving through the area would appear to you with respect to your moving system as having an acceleration equal and opposite to your own acceleration.   Perhaps you have had the experience of looking out the window of a train standing in a railroad station and seeing another train alongside appear to accelerate toward your rear, only to discover that your own train is accelerating forward instead.

Another example arises when, while moving along with your accelerating coordinate system, you see an object in that system that is to all appearances stationary. Actually it must be subjected to the same acceleration that you have, but you would not be able to discern this or state it quantitatively unless you were completely aware of your own acceleration.

With this point of view in mind, let us suppose you have established a coordinate system on a turntable.   When the turntable is rotating, objects can remain fixed on it only if a centripetal acceleration or tension prevents them from accelerating outward. If you should place on the table a ball that is free to roll, it would have an apparent acceleration outward and backward with respect to the rotating coordinates.   If the platform is rotating in a counterclockwise direction, as seen from above, the ball

150

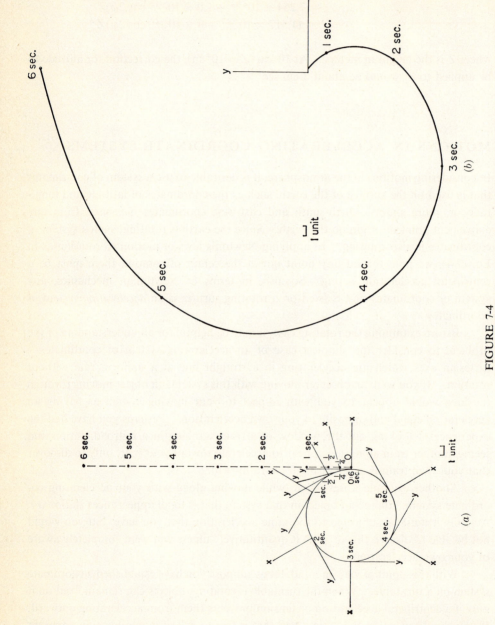

FIGURE 7-4
Motion on a rotating platform in fixed coordinates (*a*) and with respect to the rotating coordinate system (*b*).

would inscribe a path with respect to the turntable in which it would be continuously accelerated in an outward spiral to the right.   Yet, in terms of a fixed coordinate system, the ball would have moved in a straight line; it has moved according to Newton's first law.

In considering the motion of the ball, it is convenient to deal with its absolute motion, that is, its motion in terms of fixed coordinates, and its relative motion. There are, similarly, absolute and relative accelerations. The combined effect of these two produces the apparent acceleration of the ball which would be noted by an observer on the turntable.

In Fig. 7-4 a ball moving outward at uniform absolute velocity is represented in order to show the simplest case.   The motion in fixed coordinates is illustrated in Fig. 7-4a, starting from point O, which is not at the center of rotation.   Each heavy dot represents the position at each 1-sec interval.   Meanwhile the rotating coordinate system (the turntable) makes one complete revolution in 6 sec, as represented on the circle in the diagram.   To simplify the picture, the x axis is radial and the y axis is tangential.   In Fig. 7-4b the coordinates of the ball with respect to the x and y axes are plotted, showing the accelerating outward spiral to the right.   The x and y coordinates measured from the moving origin are here plotted for the ball under the assumption that the axes are fixed.   This is how it would appear on the turntable to an observer unaware of any fixed reference system, and this is the actual path the ball would inscribe on the turntable, although in fixed space it would be traveling in a straight line at constant speed.   Since the absolute acceleration is zero, the relative acceleration is the apparent acceleration.   The acceleration noticeable from the plot by virtue of the increasing separation of the points taken at 1-sec intervals is an apparent acceleration which appears because the coordinate system is accelerating, and it is known that a centripetal acceleration is acting.   Figure 7-4 can, in fact, be used as a form of graphical demonstration that rotational motion is accelerated motion.

A quick demonstration of the relative path can be obtained by placing a cardboard disk on a phonograph turntable and, while it is turning, rapidly drawing a pencil line on the cardboard from near the center straight toward some fixed object, such as a corner of the machine.   It would be most like the ball case just described if the pencil were started from a place where it would be going initially in a tangential rather than a radial direction.

The term *apparent* in dealing with an acceleration or force of this kind should not be taken lightly, because in terms of the rotating surface it can be very real.   For example, one might mount a toy railroad with the track running straight along a radial line upon the turntable.   If the train is sent along the track while the platform is rotating counterclockwise, an added force would be exerted by the train on the right-hand rail.   If the rotation were in a clockwise sense the force would be exerted on the left-hand rail.

INERTIAL, NONINERTIAL FORCES

A better way to denote the apparent forces described above is to call them noninertial forces. We start by defining an inertial force as one which, in accordance with Newton's second law, is equal to the rate of change with time of the momentum, or

$$F = \frac{d}{dt}(mv) = ma$$

the last equality being obtained by considering m as constant and letting $dv/dt = a$, the acceleration. In the case of the ball on the rotating platform there was no change in momentum and therefore no inertial force. In terms of the rotating system, an acceleration and therefore a force appeared which was noninertial.

For motions in the atmosphere and oceans, the most important noninertial force or acceleration is the *coriolis force* or *acceleration*, named after G. G. Coriolis, the French mathematical physicist who demonstrated these effects quantitatively. It will be examined for a rotating plane and for a rotating sphere such as the earth in the next sections.

QUANTITATIVE DETERMINATION IN A ROTATING PLANE

We now wish to obtain an expression for the relative and apparent accelerations, including the coriolis acceleration, acting on a particle moving on a rotating plane. In order to conform with a later application to the spherical earth, a coordinate system different from that of Fig. 7-4 will be used. As illustrated in Fig. 7-5, the y axis points radially inward and the x axis is tangential, positive in the direction of the counterclockwise rotation. Initially, the radial component r, positive outward from the center in a sense opposite to that of y, will be used. The relative velocity in the x direction will be designated as u and in the y direction as v_p, where the subscript refers to a plane as contrasted with a sphere, to be treated later. To simplify the derivation, the relative motion will be considered separately in its two components. We shall treat first the accelerations acting in the x direction.

The relative linear velocity in the x direction is $u = \omega r$, where ω is the relative angular velocity. The relative acceleration in the x direction is made up of the acceleration of the relative motion u at constant r plus the acceleration due to changing r at constant ω. We may write

$$\text{Relative } a_x = \frac{du}{dt} + \omega \frac{dr}{dt}$$

$$= \frac{du}{dt} - \frac{u}{r} v_p \qquad (7\text{-}7)$$

the last term changing sign because v_p is positive inward in the negative r direction.

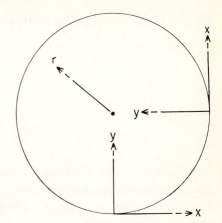

FIGURE 7-5

As stated before, the combined effect of the relative and absolute accelerations produces the apparent acceleration, such that

$$\text{Apparent } a_x = \text{relative } a_x - \text{absolute } a_x$$

It is necessary to have an equation for the absolute motion in addition to that for the relative motion in order to obtain the apparent a_x.   We may take the case where the absolute motion is uniform as in Fig. 7-4.   Since there is no inertial force acting, the absolute momentum will remain constant, and this applies also to the absolute angular momentum.   As the ball starts out from its initial position at a distance r_0 from the center, it has the angular momentum of a fixed point at r_0 which is Ωr_0^2. After it has moved radially to a point at the greater distance r from the center, it has developed an angular velocity ω relative to the turntable, involving at r a linear velocity u; so its total angular velocity is $\Omega + \omega$.   With the conservation of the absolute angular momentum, as expressed by the integral of Eq. (7-6), we have

$$\Omega r_0^2 = (\Omega + \omega)r^2 = \left(\Omega + \frac{u}{r}\right)r^2 = \Omega r^2 + ur \qquad (7\text{-}8)$$

$$u = \frac{\Omega r_0^2}{r} - \Omega r \qquad (7\text{-}9)$$

The acceleration of this velocity is

$$\frac{du}{dt} = -\frac{\Omega r_0^2}{r^2}\frac{dr}{dt} - \Omega\frac{dr}{dt}$$

$$= -\left(\Omega + \frac{u}{r}\right)\frac{dr}{dt} - \Omega\frac{dr}{dt} \qquad (7\text{-}10)$$

the part in parentheses being substituted from Eq. (7-9).   Combining terms, we have

$$\frac{du}{dt} = -2\Omega\frac{dr}{dt} - \frac{u}{r}\frac{dr}{dt}$$

$$= 2\Omega v_p + \frac{u}{r}v_p \qquad (7\text{-}11)$$

$$2\Omega v_p = \frac{du}{dt} - \frac{u}{r}v_p$$

But it is shown in Eq. (7-7) that the expression on the right is the relative acceleration. Since the absolute acceleration is zero, the apparent acceleration is equal to the relative acceleration; therefore, $2\Omega v_p$ is the apparent acceleration.   It is what is referred to as the *coriolis acceleration.*

For counterclockwise rotation of the plane, the coriolis acceleration is positive when v_p is positive; that is, if the particle is moving inward, the acceleration is in the positive x direction, or toward the right.   If the particle is moving outward, v_p is negative and the coriolis acceleration is in the negative x direction, also accelerating the particle toward the right.   If the rotation were clockwise, Ω would be negative, and the acceleration would be toward the left.

The nature of the coriolis acceleration, at least on the y component of the relative motion, is to accelerate the particle toward the right over a counterclockwise-rotating surface and toward the left if the rotation is clockwise, and in both cases, for a given rotation speed, with a magnitude proportional to the relative speed of the particle.   We will next show that this is also true for the acceleration in the y direction acting on the x component of the motion.

The apparent acceleration in the y direction acts on the x component of the motion as an apparent *centrifugal* acceleration.   We have

$$\text{Relative } a_y = \frac{u^2}{r} \qquad \text{Absolute } a_y = \frac{(U + u)^2}{r} \qquad (7\text{-}12)$$

where $U = \Omega r$ is the linear velocity of a point fixed on the turntable and $u = \omega r$ is the relative linear velocity.   Now, since

$$\text{Relative } a_y = \text{absolute } a_y + \text{apparent } a_y$$

we may make the substitution

$$\frac{u^2}{r} = \frac{(U + u)^2}{r} + \text{apparent } a_y$$

$$\frac{u^2}{r} = \frac{U^2}{r} + \frac{2Uu}{r} + \frac{u^2}{r} + \text{apparent } a_y$$

and substitute Ωr for U, to obtain

$$\text{Apparent } a_y = -2\Omega u - \Omega^2 r \qquad (7\text{-}13)$$

The first term is the coriolis acceleration and the second term is an apparent centrifugal acceleration equal and opposite to the centripetal acceleration (tension) required to hold the object at a point on the platform.

Considering our y axis pointing inward, u as positive in a relative sense counterclockwise, and Ω also positive counterclockwise, we see that the y component of the coriolis acceleration would be outward with positive u and inward with negative u, thus always to the right of motion over a counterclockwise turntable. For clockwise rotation of the platform, the coriolis acceleration would be toward the left. The second or centrifugal term in the apparent acceleration always acts outward.

The general conclusion we can draw from this analysis is that on a plane surface rotating counterclockwise a moving object will be accelerated toward the right. If the rotation is clockwise, the acceleration is toward the left. The magnitude of the acceleration is determined by the relative speed of the object at any given rotation speed (angular velocity).

CORIOLIS ACCELERATION ON THE SPHERICAL EARTH

The derivation of the coriolis acceleration in cartesian coordinates on a rotating sphere can be accomplished by the same methods used for the rotating plane. On the earth the x and y axes will be taken in a plane tangential to the spherical surface. We can take into account the curvature of the earth by taking a new tangential plane for every point in question so that we may have as many as we like. The y axis is chosen so as to point toward the North Pole and the x axis is taken as positive toward the east, in the sense of the earth's rotation. A third or z axis, positive upward, will be considered.

The y component is not the same as that in the platform, not being radial now except at the pole itself. The radial distance r is not measured from the North Pole but from the axis of rotation and, as shown on page 144, is given as $r = R \cos \varphi$, where R is the radius of the earth and φ is the latitude. We take the derivative of this r to substitute for dr/dt in the platform equation (7-11), which then gives the equation for the eastward acceleration on a rotating spherical earth as follows:

$$\frac{du}{dt} = (2\Omega + \omega)R \sin \varphi \, \frac{d\varphi}{dt} \qquad (7\text{-}14)$$

But by Eq. (7-2) we note that $R \, d\varphi/dt$ would be the linear velocity on a meridian circle, or $dy/dt = v$. The acceleration then may be written:

$$\frac{du}{dt} = 2\Omega v \sin \varphi + \omega v \sin \varphi \qquad (7\text{-}15)$$

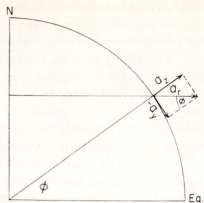

FIGURE 7-6
Acceleration at latitude φ.

It is interesting to note that, as shown in Fig. 7-1, $\Omega \sin \varphi$ is the local vertical component of the earth's rotation at latitude φ. The first term on the right therefore is the velocity multiplied by twice the local vertical component of the rotation. It is defined as the coriolis acceleration. More will be said about the last term in a subsequent paragraph. At the pole, where $\sin \varphi = 1$, Eq. (7-15) reduces to Eq. (7-11).

In considering the effects on motions in the x direction, we recognize that the apparent radial acceleration of Eq. (7-13) has a component toward the equator, that is, in the negative y direction. It also has a vertical or z component, to be treated in a subsequent paragraph. Figure 7-6 illustrates these components. It is seen from the construction of the small right triangle that $-a_y = a_r \sin \varphi$.

Applying this to Eq. (7-13), we obtain

$$-a_y = -\frac{dv}{dt} = a_r \sin \varphi = -\frac{dv_p}{dt} \sin \varphi = 2\Omega u \sin \varphi + \frac{u^2}{r} \sin \varphi$$

$$a_y = \frac{dv}{dt} = -2\Omega u \sin \varphi - \frac{u^2}{r} \sin \varphi \qquad (7\text{-}16)$$

Here again the first term on the right, defined as the coriolis acceleration, consists of the velocity multiplied by twice the local vertical component of rotation.

For the vertical acceleration we have, as seen in Fig. 7-6,

$$a_z = \frac{dw}{dt} = a_r \cos \varphi = 2\Omega u \cos \varphi + \frac{u^2}{r} \cos \varphi \qquad (7\text{-}17)$$

This is an acceleration in the x, z plane and is due to the northward component of the earth's rotation as described on page 143. The coriolis acceleration here is given by the u component of the velocity multiplied by twice the local northward component of the rotation. Note that it can operate only on the u component of velocity.

It is evident that the spherical case is the same as that of the plane except that the local rather than the total rotation is used. The acceleration on a particle is that which it would have if it were moving on a plane rotating at the speed $\Omega \sin \varphi$ at a given latitude φ. For example, at lat 30° the acceleration would be that corresponding to a plane rotating at half the speed of the earth.

When we compare, in each of the Eqs. (7-11), (7-13), (7-15), and (7-16), the last term with the coriolis term, we find that in most cases it can be neglected. In Eq. (7-15) the relation of the first term to the second is

$$\frac{2\Omega v \sin \varphi}{v\omega \sin \varphi} = \frac{2\Omega}{\omega} = \frac{2U}{u}$$

the last ratio being obtained by substituting U/r for Ω and u/r for ω. In Eq. (7-16) the ratio is

$$\frac{2\Omega u \sin \varphi}{\omega^2 r \sin \varphi} = \frac{2\Omega \omega r}{\omega^2 r} = \frac{2\Omega}{\omega} = \frac{2U}{u}$$

In Eq. (7-3) the linear velocity of a fixed point on the earth was found to be $U = \Omega R \cos \varphi$ where R is the radius of the earth, approximately 4000 miles. If we take an extreme case of relative wind u of 100 mph, we find that the ratio of $2U/u$ is 21 at the equator, 18 at 30°, 10 at 60°, and 4.6 at 80°. Therefore, except very near the poles, we are justified in neglecting the second term in meteorological applications. In the polar regions one would not use this type of cartesian-coordinate system anyway. With more nearly normal winds in middle latitudes, such as 10 mph at 45°, the ratio is about 150 to 1. Thus we can conclude that only the first term in Eqs. (7-13) and (7-15) is needed to represent the noninertial acceleration.

Considering force per unit mass, which is the same as the acceleration, we have the noninertial force expressed as the coriolis force in the components:

$$F'_x = 2\Omega v \sin \varphi \qquad (7\text{-}18a)$$
$$F'_y = -2\Omega u \sin \varphi \qquad (7\text{-}18b)$$
$$F'_z = 2\Omega u \cos \varphi \qquad (7\text{-}18c)$$

PRESSURE GRADIENT

As soon as systematic weather charting was practiced, an expected relationship between the winds and the horizontal distribution of pressure was noted. The flow of the air was found to be associated with a horizontal pressure difference somewhat as the flow of fluid through a pipe is governed by a pressure difference between the ends of the pipe. It was at once evident, however, that in the atmosphere some extremely important modifications or complications are involved which ordinarily do not enter into laboratory or engineering flow problems.

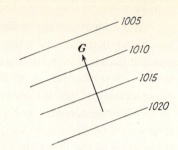

FIGURE 7-7
Pressure gradient.

The pressure effect in wind flow, if comparable to ordinary fluid flow in the laboratory, would be such that the air would move from high to low pressure by the most direct route.   This direct path from high to low pressure is the one along which the pressure is changing most rapidly and is called the *pressure gradient*.   It is customarily measured in the direction of *decreasing* pressure.

Thus it is seen that the gradient has a direction as well as a magnitude; in other words, it is a vector.   This can be illustrated in terms of pressure lines on a map, as in Fig. 7-7.   The lines connect points having the same pressure at a given level (sea level in this case) and are constructed for intervals of 5 mb.   Such lines are called *isobars*. It is obvious that the pressure is decreasing with distance most rapidly along the arrow marked G, and it is to be noted that the gradient $\vec{G}$ will always be in a direction normal to (at right angles to) the isobars.

If the horizontal pressure gradient provided the only horizontal force acting on the air, the wind would blow directly across the isobars in the direction of $\vec{G}$ and, for a given mass, with an acceleration proportional to the magnitude of the gradient $\vec{G}$.

In the form of a derivative, $\vec{G} = -\partial p/\partial n$, or the decrease in pressure with unit distance measured along a line n, normal to the isobars.   We may express $\vec{G}$ in terms of its components along two perpendicular axes, x and y, to obtain $\vec{G} = \vec{G}_x + \vec{G}_y$, the sum of two vectors.   The magnitude of $\vec{G} = -\partial p/\partial n$ is $\sqrt{(\partial p/\partial x)^2 + (\partial p/\partial y)^2}$ as indicated in Fig. 7-8.

To indicate how the pressure gradient exerts a force, we may consider a small rectangular block of the atmosphere having the dimensions δx, δy, and δz as in Fig. 7-9. The pressure acting over the area of the left end may be considered as producing a force F_1 and over the right end F_2.   If $F_1 > F_2$, there will be a net force toward the right along the x axis equal to the difference between the forces at the two ends, or $F_1 - F_2$.   But since pressure is force per unit area, $F_1 = p_1\,\delta y\,\delta z$, $F_2 = p_2\,\delta y\,\delta z$, and $F_1 - F_2 = (p_1 - p_2)\,\delta y\,\delta z$.   In terms of a gradient,

$$p_1 - p_2 = -\frac{\partial p}{\partial x}\,\delta x$$

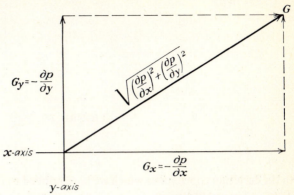

FIGURE 7-8
Components of the gradient.

The net force in the x direction, therefore, is given by

$$F_1 - F_2 = -\frac{\partial p}{\partial x}\,\delta x\,\delta y\,\delta z = -\frac{\partial p}{\partial x}\,V$$

where V is the volume of the block.

Again, we wish to deal with the force acting on a unit mass. In that case the volume would be that occupied by a unit mass, the specific volume α. Per unit mass, the force is

$$F_x = -\alpha\frac{\partial p}{\partial x} \qquad (7\text{-}19a)$$

and, similarly in the other components

$$F_y = -\alpha\frac{\partial p}{\partial y} \qquad (7\text{-}19b)$$

$$F_z = -\alpha\frac{\partial p}{\partial z} \qquad (7\text{-}19c)$$

These are the components of the pressure-gradient force which acts in the direction of decreasing pressure. Its actual direction would be down the gradient, normal to the isobars, so that the total force per unit mass would be

$$F_n = -\alpha\frac{\partial p}{\partial n} \qquad (7\text{-}20)$$

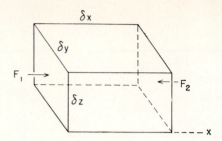

FIGURE 7-9

In the cartesian coordinates which we have applied to the earth, the relations show that if the pressure increases toward the east, F_x will be toward the west, etc. In the z component, $\partial p/\partial z$ is always negative; so F_z is always positive, that is, upward. As will be seen presently, this component is balanced by the acceleration of gravity in hydrostatic equilibrium.

BALANCE OF FORCES

We have outlined two forces acting on a unit mass in the atmosphere—the coriolis force F_x', F_y', F_z' and the pressure-gradient force F_x, F_y, F_z. If we designate all other forces in the x direction by X, in the y direction by Y, and in the z direction by Z, then all forces will be in balance when

$$F_x + F_x' + X = 0$$

or

$$\alpha \frac{\partial p}{\partial x} = 2\Omega v \sin \varphi + X \qquad (7\text{-}21a)$$

and

$$F_y + F_y' + Y = 0$$

or

$$\alpha \frac{\partial p}{\partial y} = -2\Omega u \sin \varphi + Y \qquad (7\text{-}21b)$$

and

$$F_z' + F_z' + Z = 0$$

or

$$\alpha \frac{\partial p}{\partial z} = 2\Omega u \cos \varphi + Z \qquad (7\text{-}21c)$$

Let us dispose of the vertical component first. A very strong force per unit mass, the acceleration of gravity g, acts downward in the vertical component. We substitute it for Z to obtain

$$\alpha \frac{\partial p}{\partial z} = 2\Omega u \cos \varphi - g \qquad (7\text{-}22)$$

Without the first term on the right, this is merely the hydrostatic equation (5-55). If we substitute approximate values occurring in the atmosphere, we have 10^3 cm^3 g^{-1} for α and 1 dyn cm^{-2} cm^{-1} for $\partial p/\partial z$, giving 10^3 dyn g^{-1} for the term on the left. The value of g is approximately 980 dyn. For 2Ω we have 1.4×10^{-4} sec^{-1} and, taking u to be 10^3 cm sec^{-1} and cos φ as 1 ($\varphi = 0$), we get 0.14 cm sec^{-2} for the first term on the right. Each of the other terms is seen to outweigh it by almost 10,000 to 1. Thus we neglect the northward component of the coriolis force and write the vertical balance in the form of the hydrostatic equation.

In the horizontal plane, two forces in addition to the coriolis and pressure-gradient forces are important, namely, (1) the frictional force, including frictional stresses at the surface of the earth and internal stresses within the atmosphere, and (2) a centripetal force which arises when the air is moving in a curved path relative to the earth. We will defer discussion of these two forces until after we have dwelt upon the situation that exists when these two forces are absent or negligible. In this situation the air is said to have *geostrophic motion*.

GEOSTROPHIC BALANCE

For frictionless flow in a straight path on a horizontal plane, the coriolis force and the pressure-gradient force are in balance. This is known as the *geostrophic balance*. It is represented by the equations

$$\alpha \frac{\partial p}{\partial x} = 2\Omega v \sin \varphi \qquad (7\text{-}23a)$$

and

$$\alpha \frac{\partial p}{\partial y} = -2\Omega u \sin \varphi \qquad (7\text{-}23b)$$

The velocities n and v are the components of the *geostrophic wind*. If the gradient is measured directly to determine $-\partial p/\partial n$, where n is the distance measured in a direction normal to the isobars, then we may write

$$-\alpha \frac{\partial p}{\partial n} = 2\Omega c \sin \varphi \qquad (7\text{-}24)$$

where c is the total geostrophic-wind velocity, having the magnitude $\sqrt{u^2 + v^2}$ and a direction always specified as 90° toward the right of the pressure gradient in the

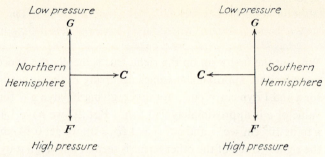

FIGURE 7-10
Geostrophic balance in the Northern and Southern Hemispheres. G is pressure-gradient force, F is coriolis force, and C is geostrophic wind.

Northern Hemisphere and toward the left in the Southern Hemisphere. Solving for c and designating $2\Omega \sin \varphi$ by f, we obtain the simple practical equation

$$c = \frac{\alpha G}{f} \qquad (7\text{-}25)$$

where $G = -\partial p/\partial n$.

Since Ω is 2π rad per sidereal day or $2\pi/86{,}164 = 7.29 \times 10^{-5}$ rad per sec, $f = 1.458 \times 10^{-4} \sin \varphi$ and has the following values at the various latitudes:

$\varphi°$	0	10	20	30	40	50	60	70	80	90
$f \times 10^4$ sec^{-1}	0	0.253	0.499	0.729	0.938	1.117	1.263	1.370	1.436	1.458

It has the value of exactly 10^{-4} at lat $43°18'30''$.

From Eq. (7-25), it is apparent that, in magnitude, the geostrophic wind is greater the steeper the pressure gradient, the lower the latitude, and the higher the specific volume (the lower the density). The relationship is such that a given pressure gradient at a given density or specific volume of the air would be associated with a stronger wind in low latitudes than near the poles. Also, the wind would be stronger for a given gradient at high altitudes (lower density).

The direction of the steady geostrophic wind in relation to the pressure gradient and coriolis forces is shown for the two hemispheres in Fig. 7-10. One may visualize a particle in the Northern Hemisphere starting out down the pressure gradient but deflected toward the right more and more until it reaches a point where further deflection would give it a component against the gradient. This would be at the time it is moving at right angles to the gradient, i.e., parallel to the isobars, with high pressure to the right and low pressure to the left.   During the turning, the coriolis force is always acting to the right; therefore, its direction is finally adjusted 90° to the right of the actual wind, or 180° from the direction of the pressure-gradient force.   In geostrophic balance, the coriolis force is an apparent force directed toward high pressure, and the pressure-gradient force is directed toward low pressure.

These considerations bear testimony to the old rule of meteorology, *Buys-Ballot's law*, which states: *If, in the Northern Hemisphere, you stand with your back to the wind, the high pressure is on your right and the low pressure on your left; in the Southern Hemisphere, the high pressure is on your left and the low pressure on your right.*

When we express the flow in terms of c and $\partial p/\partial n$, we are introducing a system of coordinates called the *natural* orthogonal coordinate system.   An axis is taken along (tangent to) the motion, and the n axis is normal to it, positive toward the left of the velocity vector.   Thus in the Northern Hemisphere n is positive down the gradient toward lower pressure.   For west-to-east flow this system coincides with the standard x, y orthogonal system we have been using, with n coinciding with y, but the natural system can be changed in orientation to fit the direction of flow.   We have only to apply Buys-Ballot's law to see that n points to the left of c, or in the direction of G as in Fig. 7-10 for the Northern Hemisphere.

CYCLONES AND ANTICYCLONES

Centers or areas of low pressure are called *cyclones* and centers or areas of high pressure are called *anticyclones*.[1]   Cyclones are usually a few hundred miles in diameter and anticyclones are generally somewhat larger and often more eccentric, having in some cases elongated axes of 2000 miles or more.

The frictionless wind, in the case of geostrophic motion, flows along the isobars. In cyclones and anticyclones the isobars are curved but the wind follows them. Therefore, the circulation around cyclones is counterclockwise in the Northern Hemisphere and clockwise in the Southern Hemisphere.   Around anticyclones, the air circulates clockwise in the Northern Hemisphere and counterclockwise in the Southern Hemisphere.

In Fig. 7-11 the appearance of cyclones and anticyclones on a surface weather map is shown.   Certain details (fronts) of the cyclones are left out.   It will be noted that the dimensions of the anticyclones are greater and the curvatures less sharp than in the cyclones.

---

[1] In newspapers and popular conversation in the United States the name *cyclone* is often applied to very intense, twisting storms of diameter less than a mile, more properly called *tornadoes.*

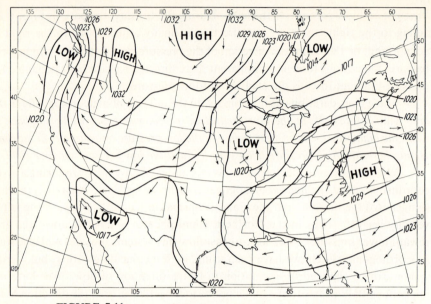

FIGURE 7-11
Surface weather map (sea-level isobars) made from simultaneous observations over the United States, showing high-pressure centers (anticyclones), low-pressure centers (cyclones), and approximate wind-flow arrows.

EFFECTS OF FRICTION

Frictional effects in fluids are due to a molecular property called *viscosity*. The meaning of viscosity can be explained by considering two plane solid surfaces, one above the other, containing between them a continuous fluid, such as glycerin. If we apply a constant force in a horizontal direction to one of the surfaces, the surface will move in the direction of the applied force, dragging adjacent molecules of the fluid with it. Farther from the moving plate the molecules would not be carried along with the speed of the plate but would have a lesser velocity depending on their distance from the moving plate and the horizontal frictional drag between the molecules. A velocity gradient, or *shear*, would be established in the fluid. The force per unit area applied in this case is called the *shearing stress*. The viscosity is defined as a coefficient of proportionality between the shearing stress and the velocity gradient. The velocity profile is assumed to be linear. In its two horizontal components the shearing stress is expressed as

$$\tau_x = \eta \frac{\partial u}{\partial z} \qquad (7\text{-}26a)$$

$$\tau_y = \eta \frac{\partial v}{\partial z} \qquad (7\text{-}26b)$$

where η is the coefficient of viscosity. Since τ is force per unit area, η has the dimensions g cm^{-1} sec^{-1}, to form a unit called the *poise*, after J. L. M. Poiseuille, who did pioneering work on frictional flow. It is seen that the viscosity coefficient may be defined as the shearing force per unit area (shearing stress) between parallel planes for unit velocity gradient. It is sometimes called the dynamic viscosity to distinguish it from the kinematic viscosity, $v = \eta/\rho$ where ρ is the density. The latter has the dimensions cm^2 sec^{-1} in the cgs units.

Since τ is the force per unit area, the force per unit volume (area $\times$ height) would be

$$\frac{\partial \tau_x}{\partial z} = \eta \frac{\partial^2 u}{\partial z^2} \qquad \frac{\partial \tau_y}{\partial z} = \eta \frac{\partial^2 y}{\partial z^2} \qquad (7\text{-}27)$$

Per unit mass (volume $\times$ density) the force would be

$$\frac{1}{\rho} \frac{\partial \tau_x}{\partial z} = v \frac{\partial^2 u}{\partial z^2} \qquad \frac{1}{\rho} \frac{\partial \tau_y}{\partial z} = v \frac{\partial^2 v}{\partial z^2} \qquad (7\text{-}28)$$

In the turbulent flow of the atmosphere a coefficient several orders of magnitude greater than the nominal or molecular viscosity is measured. It is referred to as the *eddy viscosity*. The air flows in eddies of various sizes and shapes which interfere with the direct progress of the motion. Since eddies often carry air parcels vertically from their original positions, the thermal stability of the layer in question affects the growth and dissipation of the eddies; so we find the eddy viscosity varying through a wide range of values from hour to hour, from day to day, and from place to place. The roughness of the underlying surface and the mean wind speed, as well as the wind gradient or shear, affect it.

The eddy viscosity cannot be specified under any conditions except indirectly through special measurements and information not readily available. An entire branch of meteorology is devoted to studies of it and of turbulent flow in general, including the transfer by eddies of energy and properties of the air. In these treatments more meaningful ways of expressing the effects of friction and turbulence are sought with considerable success. Assuming that parameters equivalent to a simple eddy viscosity have been obtained, one may add the friction force to Eqs. (7-23a) and (7-23b), writing

$$\alpha \frac{\partial p}{\partial x} = 2\Omega \sin \varphi \cdot v + v \frac{\partial^2 v}{\partial z^2} \qquad (7\text{-}29a)$$

$$\alpha \frac{\partial p}{\partial y} = -2\Omega \sin \varphi \cdot u + v \frac{\partial^2 u}{\partial z^2} \qquad (7\text{-}29b)$$

Qualitatively, it can be seen that friction reduces the wind speed near the surface of the earth and thus reduces the magnitude of the coriolis force. The wind at the surface is not strong enough for the coriolis force to balance the pressure-gradient force; so the latter dominates and the air moves with a component across the isobars toward lower pressure. In the free atmosphere, above the first 500 to 1000 m, the friction is so slight in relation to the other forces that the flow is essentially parallel to the isobars.

The frictionally retarded surface air exerts a frictional drag on the air above it. Thus the air at 100 m above the surface shows the effects of friction, although it is not in contact with a rough surface. The *gradient-wind level*, that is, the height at which the wind is in equilibrium flow in accordance with the surface pressure gradient, is usually found at 500 to 1000 m above the surface. If the pressure gradient itself is changing rapidly with height through the first few hundred meters, there may be no level at which the surface-gradient wind is noted.

Since friction influences both the speed and direction of the wind, the wind vectors, starting out with low speeds and strong cross-isobar components in the bottom layer, increase in magnitude and take on a direction more in keeping with the isobars as the retarding effect of the lower layers gradually diminishes upward. Thus a balloon rising upward out of the layer of surface friction would curve, in the first 500 m or so, toward the right as it ascends in the Northern Hemisphere, toward the left in the Southern Hemisphere.

At anemometer levels over the oceans and flat grasslands the angle between the wind and the isobars is on the order of 20°. Over rooftops in cities the angle is from 30 to 45°, depending on details of the exposure, and in mountainous areas, especially in the valleys, the flow only approximates the large-scale isobaric pattern. Of course, one could never apply Buys-Ballot's rule in the bottom of the Grand Canyon or on a street corner in downtown New York.

MOTION IN A CURVED PATH

In terms of our earth-bound frame of reference, mass points moving in a curved path with respect to the earth are subjected to a local centripetal force. For a unit mass, this force is given, as usual, as

$$F_c = \frac{c^2}{r}$$

where c is the velocity, having the components u and v (horizontal motion only being considered), and r is the radius of curvature.

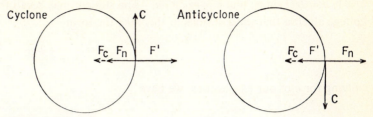

FIGURE 7-12
Forces in a cyclone and anticyclone of the Northern Hemisphere. F_c is centripetal force, F' is coriolis force, and F_n is pressure-gradient force.

In frictionless flow in a curved path the balance of forces is

$$F_n + F' \pm F_c = 0$$

or

$$-\alpha \frac{\partial p}{\partial n} - 2\Omega c \sin \varphi \pm \frac{c^2}{r} = 0 \qquad (7\text{-}30)$$

The two alternative signs in the last term indicate that, with respect to the direction n, F_c will be positive or negative depending on whether the rotation is counterclockwise or clockwise. Since F_c acts inward, it will act along the direction of the coriolis force in an anticyclone and opposite to the coriolis force in a cyclone.

Thus in the Northern Hemisphere we have for an anticyclone

$$\frac{c^2}{r} = 2\Omega c \sin \varphi + \alpha \frac{\partial p}{\partial n} \qquad (7\text{-}31)$$

$$c^2 - (2r\Omega \sin \varphi)c - r\alpha \frac{\partial p}{\partial n} = 0 \qquad (7\text{-}32)$$

and for a cyclone

$$c^2 + (2r\Omega \sin \varphi)c + r\alpha \frac{\partial p}{\partial n} = 0 \qquad (7\text{-}33)$$

Equations (7-32) and (7-33) permit balanced flow to be along the isobars in curved, frictionless motion. The balance of the forces is shown in Fig. 7-12. The Southern Hemisphere case is obtained by either regarding Ω as negative or considering the south latitudes as negative. In the Northern Hemisphere anticyclone the centripetal and coriolis forces both act inward, whereas in the cyclone they act in opposite directions. The reverse is true in the Southern Hemisphere.

For a given pressure gradient and radius of curvature in a given latitude, expressions (7-32) and (7-33) have the form of a simple quadratic $ax^2 + bx + c = 0$, with

$a = 1$, $b = 2r\Omega \sin \varphi$, and $c = r\alpha \, \partial p/\partial n$.   The student may recall from high school or college algebra that the solutions of this take the form

$$x = \frac{-b \pm \sqrt{b^2 - 4ac}}{2a}$$

Solving for c of our expressions, we have

$$c = r\Omega \sin \varphi \pm \sqrt{(r\Omega \sin \varphi)^2 + r\alpha \frac{\partial p}{\partial n}} \qquad (7\text{-}34)$$

for an anticyclone, and

$$c = -r\Omega \sin \varphi \pm \sqrt{(r\Omega \sin \varphi)^2 - r\alpha \frac{\partial p}{\partial n}} \qquad (7\text{-}35)$$

for a cyclone.

To determine whether the positive or the negative root is valid in each case, we introduce the requirement that when $\partial p/\partial n = 0$, $c = 0$; in other words, that there can be no motion without a pressure-gradient force.   In Eq. (7-34) only the negative root and in Eq. (7-35) only the positive root will satisfy this requirement.   Without this requirement, the positive root in Eq. (7-34) with $\partial p/\partial n = 0$ describes what is called *inertial motion* in which

$$c = 2r\Omega \sin \varphi \qquad (7\text{-}36)$$

This type of anticyclonic motion without a pressure gradient can be produced in laboratory models, and is of theoretical interest in the atmosphere.   A mass point following this motion would inscribe a path on the surface of the earth which, if latitude changes along the path were neglected, would form a circle, called the *inertia circle*, having the radius r of Eq. (7-36) or

$$r = \frac{c}{2\Omega \sin \varphi} \qquad (7\text{-}37)$$

With latitude changes during the motion, the inertia path forms an elongated loop.

In the cyclonic case the alternative root (negative) has no physical meaning, since it would produce negative values of c.   Note that in the natural coordinates used here, c is always positive and n is always to the left of it; therefore, $\partial p/\partial n$ is always negative in the Northern Hemisphere.   In the anticyclonic case the quantity under the radical sign could become negative when the second term is greater in absolute value than the first.   When these two are equal, the radical is zero and c has its maximum real value.   If the radical is not to become negative, then the gradient must decrease rapidly as r decreases.   An anticyclone ought to obey the conditions that

$$\left| \frac{\partial p}{\partial n} \right| \gtrless \rho r (\Omega \sin \varphi)^2 \gtrless kr \qquad (7\text{-}38)$$

where the symbolism on the left refers to the absolute value of the gradient.   The constant k is appropriate when considering one latitude and one density.   We find, in fact, that in anticyclones the pressure gradient is very small near the center where r is small and that high-pressure areas are characterized by a large central region with no appreciable pressure gradients and calm winds.

Under certain conditions the wind appears to curve anticyclonically and with a fairly steep gradient so that $\partial p/\partial n > kr$.   This seems to occur over small areas for short periods of time and is regarded as a dynamically unstable type of flow.   Consider an example at 43° lat with air density of 10^{-3} g per cm³ and a radius of curvature of 400 km.   We have $\Omega \sin \varphi$ with a value of 0.5×10^{-4}; so

$$\rho r(\Omega \sin \varphi)^2 = 10^{-3} \times 4 \times 10^7 (0.5 \times 10^{-4})^2 = 10^{-4}$$

If the gradient is to be less than this, it must be less than $10^3/10^7$ dyn per cm³ or 1 mb per 100 km.   These are rare but wholly reasonable values, and it is seen that this type of dynamic instability with $|\partial p/\partial n| > kr$ is possible in the atmosphere.

SUMMARY OF MOTIONS

The principal types of balanced motion in the atmosphere for frictionless flow may now be listed.

Geostrophic wind   The pressure gradient and coriolis forces are in balance and

$$c = -\frac{\alpha \, \partial p/\partial n}{2\Omega \sin \varphi} = \frac{\alpha G}{f} \qquad (7\text{-}39)$$

Gradient wind   The pressure gradient, coriolis forces, and centripetal forces are in balance and

$$c = r\Omega \sin \varphi - \sqrt{(r\Omega \sin \varphi)^2 + r\alpha \frac{\partial p}{\partial n}} \qquad (7\text{-}40)$$

for anticyclones,

$$c = -r\Omega \sin \varphi + \sqrt{(r\Omega \sin \varphi)^2 - r\alpha \frac{\partial p}{\partial n}} \qquad (7\text{-}41)$$

for cyclones.

Inertia motion   The pressure-gradient force is zero and the coriolis force balances the centripetal force in the anticyclonic sense, so that

$$c = 2r\Omega \sin \varphi = rf \qquad (7\text{-}42)$$

Cyclostrophic wind   The pressure-gradient force balances the centripetal force and

$$c = \sqrt{r\alpha G} \qquad (7\text{-}43)$$

All the possible combinations of the three forces in frictionless flow have been defined, but only the first two—the geostrophic wind and the gradient wind—are of importance. The cyclostrophic balance has not been mentioned before. It is important near the centers of tropical cyclones (hurricanes or typhoons) in low latitudes where the centripetal force may outweigh the coriolis force by as much as 25 to 1.

The geostrophic wind is the basic wind relationship used as a starting point in a great variety of meteorological applications. The curvature effect can be added as a correction term to the geostrophic wind to obtain the gradient wind. Writing the gradient-wind equation (7-30) in the form

$$\alpha G = fc \pm \frac{c^2}{r} \qquad (7\text{-}44)$$

we divide through by f to obtain

$$\frac{\alpha G}{f} = c \pm \frac{c^2}{fr}$$

where c is the gradient wind. Then

$$c = \frac{\alpha G}{f} \pm \frac{c^2}{fr} \qquad (7\text{-}45)$$

the plus sign in this last equation being for an anticyclone and the minus sign for a cyclone. The first term on the right is the geostrophic "component" of the wind. From a geostrophic-wind scale or tables, one can obtain $c = \alpha G/f$. Ordinarily the geostrophic instead of the gradient wind is substituted in the second term to arrive at the true value of the gradient wind by steps of approximation from the geostrophic wind.

Equation (7-45) shows that with the positive correction we have the anticyclonic case. This means that *for a given pressure gradient*, winds in an anticyclone are stronger than in a cyclone, and also stronger than geostrophic. In spite of this, winds are nearly always stronger in cyclones than in anticyclones because the pressure gradients are very much stronger.

REPRESENTATION ON CONSTANT-PRESSURE SURFACES

In meteorological practice it has been found that the more important types of computation are simplified if the motion is considered on a constant-pressure surface instead of on a horizontal or level surface. The equivalence of height contours on a constant-pressure surface to isobars on a level surface is seen qualitatively in Figs. 7-11 and

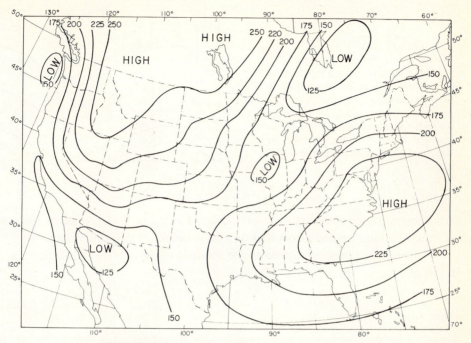

FIGURE 7-13
Same as in Fig. 7-11, represented as contours, meters, of the 1000-mb surface.

7-13.   Figure 7-13 is a contour map of the 1000-mb surface for the same situation as that represented in terms of sea-level isobars in Fig. 7-11.   The contours are labeled according to height in meters above sea level.   Figure 7-14 shows schematically a vertical cross section through the lower atmosphere along the 40th parallel.   The western edge of the map is on the left.   The vertical scale is exaggerated to 10,000 times the horizontal scale in order to make the isobaric slopes noticeable.   It is evident from a comparison of the three figures that where the pressure on a level surface is low, the pressure surfaces have dipped downward, and where high pressure exists, the pressure surfaces have bulged upward.   Just as one represents relief features by contours on a map of a section of the earth's surface, so on a pressure surface, one uses contours to reveal such features as troughs, ridges, domes, and depressions (low centers).   From these representations it should be clear that isobars on a level surface delineate the intersections of the corresponding pressure surfaces with that surface. Also, it can be shown that for a given rate of pressure decrease with height (therefore a given vertical spacing of the pressure surfaces) the horizontal pressure gradient will be given by the slope of the isobaric surfaces.   Finally, as evidenced by the extreme

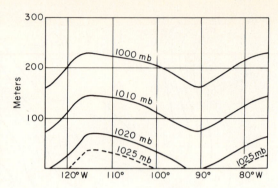

FIGURE 7-14
Vertical cross section along the 40th parallel of latitude showing the slopes of the isobaric surfaces for the situation represented in Figs. 7-11 and 7-13. The problems involved in using pressures near sea level are emphasized when it is realized that much of the land between the 100th and 120th meridians is above 1500 m.

vertical exaggeration necessary to reveal the slopes as in Fig. 7-14, motion on pressure surfaces differs immaterially from that on a level surface at the same height.

The slope of a pressure surface can be obtained by considering first the x, z plane. In this plane

$$dp = \frac{\partial p}{\partial x} dx + \frac{\partial p}{\partial z} dz \qquad (7\text{-}46)$$

Along an isobaric surface such as represented in vertical cross section by AC in Fig. 7-15, $dp = 0$; so

$$\frac{\partial p}{\partial x} dx = -\frac{\partial p}{\partial z} dz \qquad (7\text{-}47)$$

and

$$\left(\frac{dz}{dx}\right)_{AC} = -\frac{\partial p/\partial x}{\partial p/\partial z} \qquad (7\text{-}48)$$

With z representing the height of the isobaric surface, we write

$$\left(\frac{dz}{dx}\right)_{AC} = \left(\frac{\partial z}{\partial x}\right)_p \qquad (7\text{-}49)$$

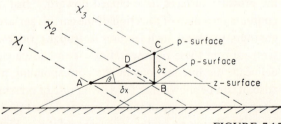

FIGURE 7-15

where the subscript p signifies that the partial derivative is taken with p constant, that is, along an isobar in the x component. Therefore

$$\left(\frac{\partial z}{\partial x}\right)_p = -\frac{\partial p/\partial x}{\partial p/\partial z} \qquad (7\text{-}50)$$

Under hydrostatic equilibrium,

$$-\frac{\partial p}{\partial z} = \frac{g}{\alpha}$$

which may be substituted into Eq. (7-50) to give

$$\left(\frac{\partial z}{\partial x}\right)_p = \frac{\alpha}{g}\frac{\partial p}{\partial x} \qquad (7\text{-}51)$$

We do not need to carry along the g if, remembering the geopotential height relationship of Eq. (5-68) $d\Phi = g\,dz$, we make the substitution

$$\left(\frac{\partial \Phi}{\partial x}\right)_p = \alpha\frac{\partial p}{\partial x} \qquad (7\text{-}52)$$

This is especially convenient because soundings and charts are represented in geopotential heights.

Wherever the term $\alpha\,\partial p/\partial n$ or $\alpha\,\partial p/\partial x$, etc., appears in the equations of motion, $g(\partial z/\partial n)_p$ or $(\partial \Phi/\partial n)_p$ or $(\partial/\partial x)_p$, $(\partial/\partial y)_p$ of these height quantities may be substituted to express the equations of motion on a constant-pressure surface.

One simplification resulting from using constant-pressure surfaces instead of level surfaces is already apparent: the specific volume or its reciprocal, the density, has dropped out. Since α varies with temperature and pressure and increases rapidly with height, it is helpful to be rid of it. Now the same geostrophic-wind scale can be used for all heights or pressures at the same latitude. Getting rid of the acceleration of gravity g is not so important, since its variation with altitude and from place to place is relatively slight. The equations are written either in terms of z or in terms of Φ. One has to remember that the g goes along with z and not with Φ. In most applications, geopotential height is used.

OTHER PROPERTIES ON CONSTANT-PRESSURE SURFACES

The transformation to a constant-pressure surface of any property that is a continuous function of the space coordinates can be obtained from well-known theorems of the calculus. It can be demonstrated in simple terms by reference to Fig. 7-15. The dashed lines are isolines of any property χ which is distributed in the atmosphere.

Since the pressure surface slopes, the distribution along it is made up of a horizontal component and a vertical component.

Consider the small triangle ABC of Fig. 7-15. The change or gradient of χ from A to C on the pressure surface is to be determined. The change in the horizontal is that occurring from A to B in the distance δx. In the same horizontal distance δx projected on AC there is an additional change represented by the difference in value of χ between D and C, which is the same as the difference between B and C. So the change from A to C is the change from A to B plus the change from B to C. In other words:

$$\text{Change on } p \text{ surface} = \text{change on } z \text{ surface} + \text{change through } \delta z$$

or

$$\left(\frac{\partial \chi}{\partial x}\right)_p \delta x = \left(\frac{\partial \chi}{\partial x}\right)_z \delta x + \frac{\partial \chi}{\partial z} \delta z \qquad (7\text{-}53)$$

but

$$\delta z = \delta x \tan \beta = \delta x \left(\frac{\partial z}{\partial x}\right)_p$$

which, when substituted in (7-53), results in a common factor δx throughout, leaving

$$\left(\frac{\partial \chi}{\partial x}\right)_p = \left(\frac{\partial \chi}{\partial x}\right)_z + \frac{\partial \chi}{\partial z}\left(\frac{\partial z}{\partial x}\right)_p \qquad (7\text{-}54)$$

Stated in words, this demonstrates that the increase in value of any property χ in a given direction on a constant-pressure surface is given by the increase in that direction on a horizontal surface plus the change in the quantity with height multiplied by the slope of the pressure surface in the direction in question. Since the slopes of pressure surfaces are slight, the last term will be small in most cases and the gradients of properties on the pressure surface will be essentially the same as on level surfaces. In these expressions geopotential height may be used instead of geometric height.

It is also of interest at this point to introduce a vertical velocity in terms of pressure instead of in terms of z. Vertical velocity has been expressed as $w = dz/dt$, but now we want to define a velocity of air particles downward through the pressure surfaces as $w_p = Dp/Dt$. The capitalized D is used here, in accordance with a common practice in fluid mechanics, to indicate the change that would be measured while moving along with the particle. This distinction is necessary because pressure may also change at a fixed point, causing the height of the pressure surfaces to change.

If the pressure surfaces remain fixed, the capitalized Dp/Dt is the only pressure change with time, and we may transform the differentials in our space-determined field as follows:

$$w_p = \frac{Dp}{Dt} = \frac{Dz}{Dt}\frac{Dp}{Dz} = -\frac{g}{\alpha}w \qquad (7\text{-}55)$$

An advantage in using $w_p = Dp/Dt$ is seen when the change in thermodynamic properties of ascending air is examined.   Consider, for example, Eq. (6-23)

$$C_p \, dT - \frac{R}{m} T \frac{dp}{p} + L \, dq = 0$$

which represents the temperature change during a saturation-adiabatic expansion. (Specific humidity q is used here in place of mixing ratio w in order to avoid the confusion created by also using w for vertical velocity.)   Since this is a temperature change that would occur in ascent following along with the parcel, the derivatives should be written with a capital D.   The time derivative is

$$\frac{DT}{Dt} = \frac{RT}{pmC_p} \frac{Dp}{Dt} - \frac{L}{C_p} \frac{Dq}{Dt}$$

but

$$\frac{Dq}{Dt} = \frac{Dq}{Dp} \frac{Dp}{Dt} \quad \text{and} \quad \frac{RT}{mp} = \alpha$$

so

$$\frac{DT}{Dt} = \frac{\alpha}{C_p} \frac{Dp}{Dt} - \frac{L}{C_p} \frac{Dq}{Dp} \frac{Dp}{Dt} = \frac{w_p}{C_p}\left(\alpha - L \frac{Dq}{Dp}\right) \qquad (7\text{-}56)$$

The first term on the right is the temperature change for the dry-adiabatic process and the second term is the temperature change in the parcel due to evaporation or condensation.   All the variables are in terms of pressure, and it should be pointed out that Dq/Dp can be measured directly on a thermodynamic diagram by counting the intersections of saturation mixing-ratio lines as one traces along a pseudoadiabatic line in the case of pure parcel ascent or along the computed entrainment adiabat when mixing occurs.   It can also be argued that vertical velocities obtained from aircraft flights must always be based on pressure changes.   This point is trivial, however, since through a small temperature range, height changes are a direct function of the change in the logarithm of the pressure which the altimeter or rate-of-climb indicator measures.

A special feature of representation on a pressure surface is that, with allowance for inherent differences in dimensions, the isolines (isopleths) of the various properties of state of the air, such as temperature, potential temperature, specific volume, and density, will coincide.   From the equation of state, it is seen that for constant p,

$$\frac{T}{\alpha} = \rho T = \text{const}$$

and, from Poisson's equation (5-74),

$$\frac{\theta}{T} = \text{const}$$

Thus all these properties have a given numerical ratio to each other on a given pressure surface; their values and their gradients will vary together. A set of lines showing the distribution of one of these properties will give the distribution of the others.

CHANGE OF THE GEOSTROPHIC WIND WITH HEIGHT

Hydrostatic equilibrium requires that the concentration in the vertical of a given set of isobars shall be inversely proportional to the mean temperature through the vertical height in question, in accordance with the relationship

$$\frac{1}{p}\frac{\partial p}{\partial z} = -\frac{mg}{R\overline{T}}$$

so the fractional rate of pressure decrease is proportional to $1/\overline{T}$, where $\overline{T}$ is the mean temperature. If the atmosphere is warm, the pressure changes slowly with height; if it is cold, the pressure changes rapidly with height.

Let us consider the system of isobaric surfaces represented in vertical cross section in Fig. 7-16. In this case it is assumed either that there is no pressure gradient at the ground or that the cross section is taken along a surface isobar or pressure contour. The air to the right is warm, and hence the pressure decreases slowly with height. The air on the left is cold; hence the pressure there decreases rapidly with height. The result is that aloft there is an increasingly steep slope of the isobaric surfaces and increasingly strong pressure gradient from high pressure in the warm region to low pressure in the cold region. This would result in a geostrophic-wind component directed into the page, with the high pressure and warm air on the right. If this is considered to be a north-south cross section with north to the left, then the given temperature distribution would give an increasingly strong westerly-wind component with height. At the surface, since the cross section is taken along an isobar running in a north-south direction, the geostrophic wind would be from either the north or the south. For Southern Hemisphere conditions, all upper winds would be in the reverse sense.

The case of a west wind at the ground with cold air to the north and warm air to the south, as in the Northern Hemisphere, is illustrated in Fig. 7-17a. In this case, the south-to-north pressure gradient, and hence the west wind, increases with height. Figure 7-17b shows the same temperature distribution but with an east wind at the surface. The north-to-south pressure gradient decreases with height, disappears, and finally reverses to a south-to-north gradient and a westerly wind at upper levels. With the normal temperature gradient of the Northern Hemisphere having cold air to the north and the Southern Hemisphere having cold air to the south, the great persistence of westerly components in the upper air in both hemispheres is verified. It should be

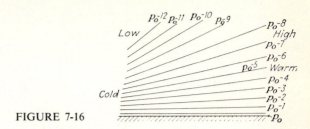

FIGURE 7-16

noted that in the stratosphere and substratosphere levels the temperature gradients are reversed; hence the westerlies increase to a maximum in the upper troposphere and decrease upward through the stratosphere.

One way of studying the details of the change of the wind with height is by means of the *hodograph*.   A hodograph that might be considered as corresponding to the conditions in Fig. 7-16 is shown in Fig. 7-18.   Here the wind vectors are drawn from the common origin O for each 1000 ft, starting with a south-southeast wind at the surface.   In this case the isobars run from south to north at the surface with the pressure decreasing toward the west.   The geostrophic wind is therefore from the south and occurs at the top of the friction layer, or at about 2000 ft.   At 10,000 ft the wind has turned to west-southwest and is quite strong.

The vector difference between the velocity at one level and some other level below is a vector in the general direction of the hodograph curve as shown in the dashed line from P to Q in Fig. 7-18 and as shown in detail in Fig. 7-19 for the 4000- and 5000-ft levels.   In vector notation, $\overline{OB} - \overline{OA} = \overline{AB}$.   The vector $\overline{AB}$ is called the *shear vector* because it measures the shear of the wind with height.   It has a direction that is parallel to the isotherms of mean temperature of the layer, with high temperatures on the right and low temperatures on the left.   Furthermore, it has a magnitude proportional to the gradient of this mean temperature.   Because of this relationship to the temperature field, it is also called the *thermal-wind vector*.   A vector from O normal to $\overline{AB}$ and extending to $\overline{AB}$, designated in Fig. 7-19 by the vector $\overline{OC}$, represents the cross-isotherm component of the wind in the layer from 4000 to 5000 ft. This component of the wind can bring in warm air by advection if no heat is added or

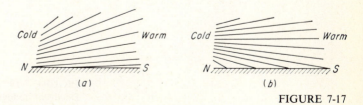

FIGURE 7-17

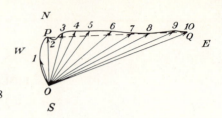

FIGURE 7-18
Hodograph.

removed from air parcels during the process.   Warm-air advection is indicated in this case because it is coming from the right of the vector $\overline{AB}$.   If the situation were reversed, so that $\overline{OB}$ were the 4000-ft wind and $\overline{OA}$ the 5000 ft, the shear or thermal-wind vector would be $\overline{OA} - \overline{OB} = \overline{BA}$, which would point in the opposite direction from vector $\overline{AB}$ of the figure; so the vector $\overline{OC}$ would be coming in from the left of $\overline{AB}$ and would therefore represent cold-air advection.

A wind that changes in a clockwise direction is said to *veer*, and a wind changing in a counterclockwise direction is said to *back*.   These names come from old nautical terminology.   We have the rule: *Winds veering with height indicate warm-air advection; winds backing with height signify cold-air advection.*   The rule is reversed in the Southern Hemisphere.

To obtain the change in the geostrophic wind with height, we follow the present-day practice of using geopotential heights on an isobaric surface.   In these terms the components of the geostrophic wind are

$$v = \frac{1}{f}\left(\frac{\partial \Phi}{\partial x}\right)_p \qquad u = -\frac{1}{f}\left(\frac{\partial \Phi}{\partial y}\right)_p \qquad (7\text{-}57)$$

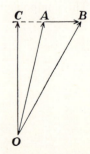

FIGURE 7-19
Details of hodograph.

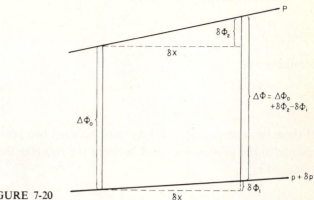

FIGURE 7-20

We want to obtain the change of u and v as we go up or down in terms of pressure. We have

$$\frac{\partial v}{\partial p} = \frac{1}{f} \frac{\partial}{\partial p} \left(\frac{\partial \Phi}{\partial x} \right)_p \qquad (7\text{-}58)$$

which expresses the fact that at a given point the change is proportional to the difference between the slope of the surface p and that of the surface $p + dp$. The differential of geopotential without the y component may be written as

$$d\Phi = \frac{\partial \Phi}{\partial x} dx + \frac{\partial \Phi}{\partial p} dp \qquad (7\text{-}59)$$

This is a total differential, and therefore

$$\frac{\partial}{\partial p} \left(\frac{\partial \Phi}{\partial x} \right) = \frac{\partial}{\partial x} \left(\frac{\partial \Phi}{\partial p} \right) \qquad (7\text{-}60)$$

as can be seen from Fig. 7-20, wherein $\delta\Phi_2 - \delta\Phi_1 = \Delta\Phi - \Delta\Phi_0$ and therefore

$$\frac{1}{\Delta p} \frac{\delta\Phi_2 - \delta\Phi_1}{\delta x} = \frac{1}{\delta x} \frac{\Delta\Phi - \Delta\Phi_0}{\Delta p}$$

In other words, the difference in the slopes in the x direction gives the change along x of the vertical separation between the two surfaces. That is,

$$\left(\frac{\delta\Phi}{\delta x} \right)_p - \left(\frac{\delta\Phi}{\delta x} \right)_{p+dp} = \Delta\Phi_1 - \Delta\Phi_0$$

We have found, then, that

$$\frac{\partial v}{\partial p} = \frac{1}{f} \frac{\partial}{\partial x} \left(\frac{\partial \Phi}{\partial p} \right) \qquad (7\text{-}61)$$

Similarly

$$\frac{\partial u}{\partial p} = -\frac{1}{f} \frac{\partial}{\partial y} \left(\frac{\partial \Phi}{\partial p} \right) \qquad (7\text{-}62)$$

If these two components are integrated between two pressure surfaces p_1 and p_2, the change in the geostrophic wind between the two (the thermal wind for the layer) is obtained as

$$v_T = v_2 - v_1 = \frac{1}{f} \frac{\partial}{\partial x} \int_{p_1}^{p_2} \frac{\partial \Phi}{\partial p} \, dp = \frac{1}{f} \frac{\partial(\Delta\Phi)}{\partial x} \qquad (7\text{-}63)$$

and similarly

$$u_T = u_2 - u_1 = -\frac{1}{f} \frac{\partial(\Delta\Phi)}{\partial y} \qquad (7\text{-}64)$$

where $\Delta\Phi$ is the thickness between p_1 and p_2. In natural coordinates, with n perpendicular and to the left of c_T, we have

$$\tilde{c}_T = -\frac{1}{f} \frac{\partial(\Delta\Phi)}{\partial n} \qquad (7\text{-}65)$$

The direction of the thermal wind is along the thickness lines with high values of $\Delta\Phi$ to the right in the Northern Hemisphere.[1]

In meteorological centers isolines of thickness are plotted on upper-air charts, such as between the 1000- and 500-mb surfaces. The magnitude and direction of the gradient of this quantity at any point on the map is easily seen.

To indicate the thermal relation of the so-called "thermal wind," we note from the hydrostatic equation that

$$\frac{\partial \Phi}{\partial p} = -\alpha$$

and therefore for the two components

$$\frac{1}{f} \frac{\partial}{\partial x} \left(\frac{\partial \Phi}{\partial p} \right) = -\frac{1}{f} \frac{\partial \alpha}{\partial x} \qquad\qquad -\frac{1}{f} \frac{\partial}{\partial y} \left(\frac{\partial \Phi}{\partial p} \right) = \frac{1}{f} \frac{\partial \alpha}{\partial y} \qquad (7\text{-}66)$$

---

[1] Note that the derivation is not carried out in the natural coordinates. That system changes orientation with changing wind direction, and the varying vectors do not adapt to straightforward integration.

and with a substitution for α from the equation of state,

$$\frac{\partial v}{\partial p} = -\frac{R}{fmp}\left(\frac{\partial T}{\partial x}\right)_p \qquad (7\text{-}67)$$

$$\frac{\partial u}{\partial p} = \frac{R}{fmp}\left(\frac{\partial T}{\partial y}\right)_p \qquad (7\text{-}68)$$

Thus on an isobaric surface the trend of the thermal wind can be represented by the temperature field. This field is displayed by drawing isotherms on the surface. The relationships show that with increasing height the contours and isotherms approach coincidence. A common situation seen on synoptic charts constructed for the various levels, including the surface, is one in which the low-level winds and therefore the contours cross the isotherms at a considerable angle, but from the middle troposphere upward to the tropopause the contours and isotherms coincide. In storm conditions nonequilibrium distributions can exist through a large part of the atmosphere, but the tendency toward in-phase isotherm and contour patterns is always present.

SUMMARY

At this point the student is ready to look with some discernment at the array of synoptic maps and charts that issue forth at weather stations several times daily. Contour maps of the surfaces of 1000, 850, 700, 500, 300, 200 mb and higher are prepared at analysis centers and transmitted by facsimile or other wire or wireless systems. On these charts measured winds from rawinsonde stations (see Appendix B) are entered. Thickness lines 1000 to 500 mb or between more proximate surfaces are included on some maps. Contours of the isobaric surfaces are drawn in terms of geopotential meters. Geostrophic motion is along the contours, as would also be the gradient wind in the case of curved contours. One can note how these lines tend to coincide with the thickness lines and the isotherms in the upper troposphere. And one can see how the greater thicknesses in the tropics than in the higher latitudes must be matched by west-to-east winds aloft at nearly all latitudes.

This perfect balance is not always maintained. In the next chapter we shall examine some of the characteristics of fluid motions associated with imbalances.

EXERCISES

1 (*a*) Give the vertical and northward components (sec^{-1}) of the earth's rotational vector at your latitude.

(*b*) Determine the linear speed (and direction) of a fixed point on the earth at your latitude.

(c) How long would a Foucault pendulum take to complete a circuit at your latitude?

(d) What is the angular momentum of a unit mass point on the earth at your latitude?

2   Construct the same details as in Fig. 7-4 but for Southern Hemisphere rotation.

3   Apply Eq. (7-25) to determine the magnitude of the geostrophic wind for every 10° north or south of the equator for a pressure gradient of 1 mb in 100 km and a specific volume of 10^{-3} cm^3 g^{-1}.   Give your answer in meters per second.

4   Draw an x, y coordinate system with x positive toward the east and y positive toward the north.   Superimpose compass points for every 90°.   Plot vectors of wind from NW, NE, SW, SE and from 180, 210, 10, 320, 160, and 270°.   Taking the speed in each case as 10 m sec^{-1}, obtain the x and y components of each, paying attention to sign.

5   At 40°N a pressure gradient of 1 mb in 250 km exists at a level where the air density is 10^{-3} g cm^{-3}.   Calculate (a) the speed of the gradient wind for an anticyclonic curvature of radius 300 km; (b) the speed of the gradient wind in a cyclonic circulation with the same radius of curvature; (c) the geostrophic wind speed.

6   Show that the radius of the inertia circle changes with latitude according to

$$\frac{dr}{d\varphi} = -\frac{c \cos \varphi}{2\Omega \sin^2 \varphi}$$

for a given particle speed c.   Make a graph of the values of r against φ for a particle speed of 3 m sec^{-1}.   Would the particle move in the inertia circle in a clockwise or counterclockwise direction in the Northern Hemisphere?

7   Apply Eq. (7-38) to obtain the critical radii of curvature versus pressure gradients in anticyclonic flow for radii of 100 to 1000 km at latitudes 30 and 60° and an air density of 1.2×10^{-3} g cm^{-3}.   Express the gradients in millibars per 100 km.

8   With respect to constant-pressure surfaces, answer the following:

(a) Show that the difference between the temperature gradient on a constant-pressure surface and the temperature gradient on a level surface is equal to the lapse rate multiplied by the slope of the constant-pressure surface.

(b) If the isobaric surface slopes upward in the direction of increasing temperature, show that the temperature gradient on it will be less than on the constant-level surface, assuming a normal lapse rate.

(c) Taking the slope of the 1000-mb surface in Fig. 7-13 between Saint Louis and Nashville, determine the magnitude of this difference.

9   You will find in Chap. 9, Figs. 9-2, 9-3, 9-4, and 9-5, maps for sea level and 500 mb for mean conditions in January and July.   From these maps sketch relative geostrophic-wind vectors, assuming in each case that the 500-mb wind speed is 5 times the sea-level value, for (a) January at (i) Tokyo, (ii) Seattle, (iii) Halifax; (b) July at same stations.   Construct hodographs from your sketches.   Show the vector difference (thermal-wind vectors), plot perpendiculars to them from the origin, and in each case state whether advection of cold or warm air is indicated.

10   Returning to the temperature map in Chap. 4, Fig. 4-5, sketch in vertical cross section:

(a) The general trend of the isobaric surfaces in January between the Siberian cold center and Japan. The trend of the slopes should be in general agreement with the gradients in Figs. 9-2 and 9-4, according to (*b*) as follows:

(*b*) The hypsometric tables show that the 1000-mb surface would be at 300 m in the Siberian cold spot and 160 m over Japan. Give the 1000- to 500-mb thicknesses by reading the 500-mb chart of Fig. 9-4. Comment on the nearness of coincidence of the thickness patterns and the 500-mb contours.

8

CHARACTERISTICS OF FLUID FLOW
APPLIED TO THE ATMOSPHERE

In the preceding chapter the air flow which would result from a balance of forces was discussed. The tendency for a mutual adjustment of these forces is so marked in the atmosphere that it is difficult to observe or measure any broad-scale imbalance. Yet it is recognized that nothing but constant and steady motion would ever occur in a completely balanced atmosphere; disturbed weather would be unheard of, no new flows would be initiated, and what is more serious, no mechanism would develop for accomplishing the general circulation of the atmosphere which is necessary to prevent the latitudinal and geographic heat differences from getting out of hand. An imbalance, however small and difficult it may be to detect, must exist either continuously or sporadically. Predicting whether nonequilibrium conditions will or will not develop is one of the chief problems of weather forecasting.

Characteristics of fluid flow which are well known from classical hydrodynamics can be applied to the atmosphere to help in solving these problems. It is found that in the accelerating coordinates of the earth the development of some important systems of relative motion (wind systems) can be accounted for in this way; this would never result from a consideration of balanced flow.

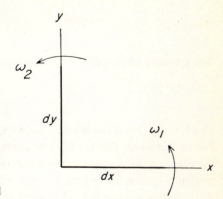

FIGURE 8-1

VORTICITY

When the equation representing the relation between linear velocity and angular velocity $c = \omega r$ is differentiated with respect to r, the result is

$$\frac{\partial c}{\partial r} = \omega$$

For solid rotation, ω at any given instant is constant throughout and the linear speed must increase at a fixed rate ω with radial distance.   Fluids, since they are not constrained to rotate as solids, may turn at varying speeds both in time and in space, and not all the particles of the fluid may have the same center of rotation or the same radius of curvature.

To obtain a quantitative understanding of these fluid motions, we first consider fluid elements which are small enough so that the change of ω in the infinitesimal distances used is zero.   It is convenient to express the motion in the cartesian coordinates x and y.   The rotation is considered for two infinitesimal lines of fluid particles represented initially in dx and dy on the x and y axes as in Fig. 8-1.   The v component represents the linear velocity of the rotation at the x axis and the u component at the y axis.   The angular velocity does not vary in the distance dx or dy but it may be different on the two axes.   If rotation is considered positive in a counterclockwise sense, as in the figure, then

$$v = \omega_1 x \qquad -u = \omega_2 y$$

$$\frac{\partial v}{\partial x} = \omega_1 \qquad -\frac{\partial u}{\partial y} = \omega_2$$

The average of the two angular velocities is

$$\frac{1}{2}(\omega_1 + \omega_2) = \frac{1}{2}\left(\frac{\partial v}{\partial x} - \frac{\partial u}{\partial y}\right) \qquad (8\text{-}1)$$

The quantity enclosed in the second parentheses is called the *vorticity*, or

$$\zeta = \frac{\partial v}{\partial x} - \frac{\partial u}{\partial y} \qquad (8\text{-}2)$$

which is 2ω (twice the local average angular velocity). Since only the x, y plane has been considered, this is only one component of the vorticity, but in the atmosphere the other two components are seldom taken into account because of the overwhelming tendency for the air to flow in laminar fashion. Just as in the case of earth rotation, this is called the vertical component of the vorticity and is represented vectorially by a vertically pointing vector.[1]

We must remember that the earth has its own counterclockwise vertical component of vorticity given by $f = 2\Omega \sin \varphi$, also twice the local angular velocity. We then define the *absolute vorticity* as the sum of the two vorticities; so we have $f + \zeta$. The vorticity ζ is measured relative to the surface of the earth and is referred to as the relative vorticity. The relative vorticity, like the vorticity of the earth's surface, is positive in a counterclockwise sense. A field of motion in the atmosphere can also have relative vorticity in terms of the horizontal wind shear even though the flow is straight. A simple example of this can be seen if Eq. (8-2) is applied with a south wind (northward-pointing wind vectors) increasing in magnitude along the x axis. Then $\partial v/\partial x$ would be positive, and in the absence of a u component the vorticity would be positive. The general rule is that if, along an axis normal to the wind direction, the wind increases to the right, then the relative vorticity is positive. For the Northern Hemisphere a wind field increasing to the right of the flow is said to have a cyclonic wind shear. If the wind increases to the left, the shear is anticyclonic.

The principal use of the combined or absolute vorticity in studying air flow comes from the fact that it is advected and tends to be conserved during atmospheric displacements. The theorem may be stated as

$$f + \zeta = \text{const} \qquad (8\text{-}3)$$

We shall see later how this needs to be modified to take into consideration at least one additional effect. The time derivative of the absolute vorticity is

$$\frac{d}{dt}(f + \zeta) = \frac{df}{dt} + \frac{d\zeta}{dt} = 0 \qquad (8\text{-}4)$$

---

[1] In the notation of vector algebra $\zeta = k \cdot \nabla \times C$, where k is a unit vertical vector. $\nabla \times C$ is called the *curl* of C.

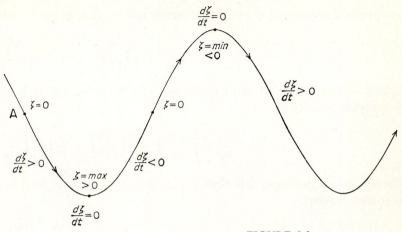

FIGURE 8-2
Vorticity in sinusoidal trajectory.

In air moving over the rotating earth, latitude φ changes and therefore changes in f occur. By writing

$$\frac{d\zeta}{dt} = -\frac{df}{d\varphi}\frac{d\varphi}{dt} = -\frac{df}{dy}\frac{dy}{dt} = -v\frac{df}{dy} \qquad (8\text{-}5)$$

it can be seen that the relative vorticity will be increased in particles carried along in a north wind (negative v) and decreased in south winds. Flow from the north generates cyclonic vorticity or decreases anticyclonic vorticity in the Northern Hemisphere while flow from the south generates anticyclonic vorticity or decreases cyclonic vorticity.

For forecasting purposes, future constant-vorticity trajectories can be computed. The simplest type of trajectory, shown in Fig. 8-2, can develop a sinusoidal wave form. Starting at A the air moves southward with the relative vorticity increasing until the particles curve cyclonically so much that they turn back toward the north. Then the relative vorticity decreases through zero and becomes negative and anticyclonic to the extent that the air is turned back toward the south again to form a ridge.

A number of practical applications are based on the "geostrophic" vorticity in which the curvature of flow is not considered. In most of the region of the upper westerlies the curvature term in the wind-balance equation adds or subtracts less than a tenth. Because of the geostrophic relationship between velocity and slope of the isobaric surfaces, the velocity field as expressed in the vorticity field defines the undulations of the pressure surfaces and vice versa, as will now be shown. The two geostrophic components on an isobaric surface are, as noted in the preceding chapter,

$$u_g = -\frac{1}{f}\left(\frac{\partial \Phi}{\partial y}\right)_p \qquad v_g = \frac{1}{f}\left(\frac{\partial \Phi}{\partial x}\right)_p$$

Taking account of the fact that f varies only with y, $(\partial f/\partial x) = 0$, we may express the geostrophic vorticity as

$$\zeta_g = \frac{\partial v_g}{\partial x} - \frac{\partial u_g}{\partial y} = \frac{1}{f}\left(\frac{\partial^2 \Phi}{\partial x^2}\right)_p + \frac{1}{f}\left(\frac{\partial^2 \Phi}{\partial y^2}\right)_p - \frac{1}{f^2}\left(\frac{\partial \Phi}{\partial y}\right)_p \frac{\partial f}{\partial y} \qquad (8\text{-}6)$$

The last term is ordinarily less than one-tenth of the first two; so it is customarily neglected, making

$$\zeta_g \cong \frac{1}{f}\left(\frac{\partial^2 \Phi}{\partial x^2} + \frac{\partial^2 \Phi}{\partial y^2}\right)_p \qquad (8\text{-}7)$$

This second space derivative of a scalar quantity is called the *Laplacian* of that quantity, designated in vector notation as $(\nabla^2\Phi)_p$. The second derivative in space of the height of an isobaric surface is really the curvature of that surface. A positive value (cyclonic vorticity) is characteristic of a depression of the surface and a negative value (anticyclonic vorticity) is characteristic of a dome or ridge. This is in agreement with the fact that bulges and depressions on isobaric surfaces represent anticyclones and cyclones, respectively. Thus it is seen that the geostrophic vorticity gives the Laplacian which describes the configuration of an isobaric surface.

The absolute geostrophic vorticity much used in meteorological computations is, of course,

$$f + \zeta_g = f + \frac{1}{f}(\nabla^2\Phi)_p \qquad (8\text{-}8)$$

If there is a tendency for geostrophic absolute vorticity patterns to be advected, this quantity has value in prediction. A numerical prediction scheme based on this principle has been widely used.

STREAMLINE, STREAM FUNCTION, AND VELOCITY POTENTIAL

A streamline is defined as a curve whose tangent at any point gives the direction of the velocity vector at that point. In Fig. 8-3, the curve S represents a streamline in which the general sense of the motion is toward the northeastern part of the diagram.

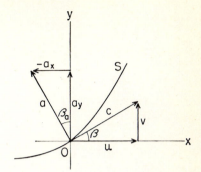

FIGURE 8-3

At the point O, the tangent is in the direction of the velocity vector c which character-izes the flow at that point. The u and v components of c are as shown. The vector c forms an angle β with the x axis, so that

$$\tan \beta = \frac{v}{u} = \frac{dy/dt}{dx/dt} = \frac{dy}{dx} \qquad (8\text{-}9)$$

$$v\,dx = u\,dy \qquad (8\text{-}10)$$

$$v\,dx - u\,dy = 0 \qquad (8\text{-}11)$$

$$\frac{dx}{u} = \frac{dv}{v} \qquad (8\text{-}12)$$

The latter expression is usually given as the mathematical definition (differential equation) of a streamline in two dimensions. When the streamlines retain the same shape and same position at all times, the motion is said to be *steady*. The particles will then have followed a path along a streamline. With changing streamlines, that is, with unsteady flow, the actual paths or *trajectories* have to be determined over small increments of time from the streamlines prevailing at those times.

Streamlines can be drawn on maps in such a way that their horizontal spacing, like the spacing of isobars or contours of a pressure surface, is proportional to the wind speed. A *stream function* ψ is then defined such that

$$u = -\frac{\partial \psi}{\partial y} \qquad v = \frac{\partial \psi}{\partial x} \qquad (8\text{-}13)$$

with streamlines labeled in arbitrary values of ψ, ascendant toward the right of the flow in the Northern Hemisphere, toward the left in the Southern Hemisphere. The streamlines and stream functions can be represented and computed on any desired surface—horizontal, isobaric, or isentropic. It is apparent that isobars are stream-lines for gradient-equilibrium flow on a horizontal surface and that contours serve the same purpose on an isobaric surface.

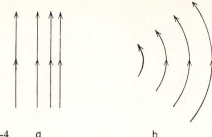

FIGURE 8-4   a   b

Orthogonal to the streamlines is a set of lines represented in Fig. 8-3 by the vector a, perpendicular to and of the same magnitude as the vector c, and having the components a_y and $-a_x$. Note that the two vector triangles are the same, so that in magnitudes $\beta_a = \beta$, $a_y = u$, and $a_x = v$. The slope of a is

$$\tan(90 - \beta) = \cot \beta = \frac{dy}{dx} = -\frac{a_y}{a_x} = -\frac{u}{v} \qquad (8\text{-}14)$$

and

$$u\,dx = -v\,dy \qquad (8\text{-}15)$$

$$u\,dx + v\,dy = 0 \qquad (8\text{-}16)$$

The number of these lines in a unit distance can also be made proportional to the wind speed to produce a *velocity potential* φ defined by

$$u = -\frac{\partial \varphi}{\partial x} \qquad v = -\frac{\partial \varphi}{\partial y} \qquad (8\text{-}17)$$

The three-dimensional case can be considered by equating each of the two terms in Eq. (8-12) to dz/w. In three dimensions, the flow is in streamline tubes bounded by streamlines. The velocity-potential lines become surfaces cutting the streamlines at right angles.

If a velocity potential exists, the flow must be *irrotational;* that is, the vorticity must be zero. This can be shown by obtaining $\partial v/\partial x$ and $\partial u/\partial y$ by differentiating Eq. (8-17).

$$\frac{\partial u}{\partial y} = -\frac{\partial \varphi}{\partial x\,\partial y} \qquad \frac{\partial v}{\partial x} = -\frac{\partial \varphi}{\partial x\,\partial y} \qquad (8\text{-}18)$$

$$\frac{\partial v}{\partial x} - \frac{\partial u}{\partial y} = 0 \qquad (8\text{-}19)$$

This result also is suggested by graphical investigation. Take, for example, the streamlines in Fig. 8-4a which show a wind shear by virtue of increasing crowding to the right, so that the stream function is increasing ($\partial\psi/\partial x > 0$).   It is impossible to construct a set of lines representing velocity potential which would be orthogonal to the streamlines.   If the velocity is to be proportional to $\partial\varphi/\partial x$, the lines of potential must become closer together toward the right, a requirement incompatible with orthogonality.   In Fig. 8-4b the velocity is constant, as shown by equal spacing of the streamlines.   The orthogonal lines are radial lines which separate as they extend outward from the center; thus the gradient of the potential cannot represent the velocity.   These arguments substantiate the theorem that potential flow is irrotational. The concept of potential flow, however, is useful in examining certain properties of vorticity and other flow problems.

CIRCULATION AND VORTICITY

In Fig. 8-5, fluid is considered as flowing along the straight path s, an infinitesimal part of which is represented by ds, with components dx and dy.   It is seen that

$$\frac{dx}{ds} = \cos\beta \qquad \frac{dy}{ds} = \cos\theta \qquad (8\text{-}20)$$

These may be designated as the directional cosines l and m of s.

$$dx = l\,ds \qquad dy = m\,ds$$

Equation (8-16) becomes

$$(ul + vm)\,ds = 0 \qquad (8\text{-}21)$$

Also

$$u = \frac{dx}{dt} = l\frac{ds}{dt} = lc \qquad (8\text{-}22)$$

$$v = \frac{dy}{dt} = m\frac{ds}{dt} = mc \qquad (8\text{-}23)$$

$$ul + vm = (l^2 + m^2)c \qquad (8\text{-}24)$$

$$m = \cos\theta = \sin\beta$$

$$\sin^2\beta + \cos^2\beta = 1$$

Therefore, $lu + vm = c$ but since $l = dx/ds$ and $m = dy/ds$,

$$u\frac{dx}{ds} + v\frac{dy}{ds} = c$$

and

$$u\,dx + v\,dy = c\,ds \qquad (8\text{-}25)$$

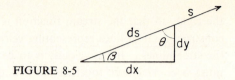

FIGURE 8-5

where c is the actual velocity ds/dt along s.   If the fluid flows at constant speed c_0 a distance s, the expression will be

$$c_0 s = c_0 \int_0^s ds \qquad (8\text{-}26)$$

The value of the integral

$$\int_{s_1}^{s_2} u\, dx + v\, dy = \int_{s_1}^{s_2} \left(u\, \frac{dx}{ds} + v\, \frac{dy}{ds} \right) ds = \int_{s_1}^{s_2} c\, ds \qquad (8\text{-}27)$$

is the *flow* of the fluid from s_1 to s_2.   It has the dimensions cm² per sec or the product of speed and distance.

When the flow is not of the simple form described in Fig. 8-5 and in the last two equations, u and v may each be a function of both x and y in the fluid.   In order to perform the integration, it is necessary to know what these functions are.   They can be specified only along a given line of flow, and the resulting integral along this line is called a *line integral*, since the u and v as functions of x and y are valid only along the chosen line.   This can be a curved line, a line with angular turns in it, or a closed circuit (circular, square, rectangular, elliptical, or in any odd form).

The flow around a closed circuit, that is, the line integral of Eq. (8-27) around a closed path, is called the *circulation*, which we will designate by C.

$$C = \int_c (u\, dx + v\, dy) = \int_c \left(u\, \frac{dx}{ds} + v\, \frac{dy}{ds} \right) ds \qquad (8\text{-}28)$$

where the subscript c of the integral sign indicates a line integral around a closed circuit.

The area A of any surface, such as in Fig. 8-6, may be divided by a double series of lines crossing it into infinitely small elements, one of which is represented as δA in the figure.   The sum of the circulations around the boundaries of these elements, taken all in the same sense, is equal to the circulation around the boundary of the whole area.   This is apparent when, in summing the circulations, the flow along each side common to two elements comes in twice, once for each element, but with opposite signs, and therefore disappears from the result.   There remain only the flows along

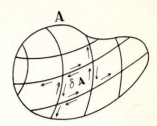

FIGURE 8-6

those sides which are parts of the boundary of A.   If C represents the circulation around A and δC that around δA, we have

$$C = \int_A \delta C \qquad (8\text{-}29)$$

where the summation includes all the elements δA.

Consider the infinitesimal area, now assumed to be a rectangle $\delta A = \delta x \, \delta y$ shown in Fig. 8-7.   On the opposite sides the u and v are not the same, since they are each functions of x and y.   Starting in the lower left corner, having the coordinates x, y, the flow is counterclockwise around the rectangle to y, $(x + \delta x)$, thence up to $(x + \delta x)$, $(y + \delta y)$, back to x, $(y + \delta y)$, and down to x, y again.   The velocities and flows are as follows:

Along bottom, from x, y to $(x + \delta x)$, y, velocity is u, flow is $u \, \delta x$.

Along right, from $(x + \delta x)$, y to $(x + \delta x)$, $(y + \delta y)$, velocity is $v + \dfrac{\partial v}{\partial x} \delta x$, flow is $\left(v + \dfrac{\partial v}{\partial x} \delta x \right) \delta y$.

Along top, from $(x + \delta x)$, $(y + \delta y)$ to x, $(y + \delta y)$, velocity is $-\left(u + \dfrac{\partial u}{\partial y} \delta y \right)$, flow is $-\left(u + \dfrac{\partial u}{\partial y} \delta y \right) \delta x$.

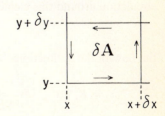

FIGURE 8-7

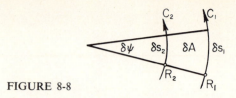

FIGURE 8-8

Along left, from $x, (y + \delta y)$ to x, y, velocity is $-v$, flow is $-v\,\delta y$. The circulation around δA is, then,

$$\delta C = u\,\delta x + \left(v + \frac{\partial v}{\partial x}\,\delta x\right)\delta y - \left(u + \frac{\partial u}{\partial y}\,\delta y\right)\delta x - v\,\delta y$$

$$= \frac{\partial v}{\partial x}\,\delta x\,\delta y - \frac{\partial u}{\partial y}\,\delta x\,\delta y$$

$$= \delta A\left(\frac{\partial v}{\partial x} - \frac{\partial u}{\partial y}\right) \tag{8-30}$$

$$\frac{\delta C}{\delta A} = \frac{\partial v}{\partial x} - \frac{\partial u}{\partial y} = \zeta \tag{8-31}$$

Thus it is seen that the vertical component of the vorticity is equivalent to the circulation around unit horizontal area. Checking dimensions, we see that the flow divided by area is in units of $\sec^{-1}$, appropriate to vorticity or angular velocity.

We can write

$$C = \int_A \zeta\,\delta A \tag{8-32}$$

by substituting (8-31) into (8-29) to eliminate δC. Equation (8-32) is an expression for Stokes' theorem: The circulation around the boundary of any finite horizontal area is equal to the integral of the vertical component of vorticity taken over the area.

It is useful to express the vorticity in coordinates along and normal to a streamline—the natural coordinates. In Fig. 8-8 the two streamlines and the two normals, the latter also lying along radii of curvature, bound an area δA as indicated. The circulation around this element of area is

$$\delta C = c_1\,\delta s_1 - c_2\,\delta s_2 \tag{8-33}$$

the flow in the n directions being zero. But

$$\delta s_1 = R_1\,\delta \psi \qquad \delta s_2 = R_2\,\delta \psi \qquad R_2 = R_1 - \delta n \qquad c_2 = c_1 + \frac{\partial c}{\partial n}\,\delta n \tag{8-34}$$

where n is taken positive to the left of the flow. Thus

$$\delta C = c_1 R_1\, \delta\psi - \left(c_1 + \frac{\partial c}{\partial n}\, \delta n\right)(R_1 - \delta n)\, \delta\psi \qquad (8\text{-}35)$$

Multiplying together the members of the second term, we obtain

$$\delta C = c_1 R_1\, \delta\psi - \left(c_1 R_1 + R_1 \frac{\partial c}{\partial n}\, \delta n - c_1\, \delta n - \frac{\partial c}{\partial n}\, \delta n^2\right)\delta\psi \qquad (8\text{-}36)$$

$$\delta C = c_1\, \delta n\, \delta\psi - \frac{\partial c}{\partial n}\, R_1\, \delta n\, \delta\psi + \frac{\partial c}{\partial n}\, \delta n^2\, \delta\psi \qquad (8\text{-}37)$$

We then divide by $R_1\, \delta n\, \delta\psi = \delta A$, and find

$$\frac{\delta C}{\delta A} = \frac{c_1}{R_1} - \frac{\partial c}{\partial n} + \frac{1}{R_1}\frac{\partial c}{\partial n}\, \delta n \qquad (8\text{-}38)$$

The third term approaches zero as δn approaches zero. We then get the circulation per unit area at the streamline c_1:

$$\frac{\delta C}{\delta A} = \zeta = \frac{c_1}{R_1} - \frac{\partial c}{\partial n} \qquad (8\text{-}39)$$

It is thus evident that the vorticity can be expressed as a shear term $-\partial c/\partial n$ plus a curvature term c/R. It is to be noted that since n increases to the left of c, the shear term takes the positive sign for cyclonic shear. Also, $c/R = \omega$, the angular velocity, which we take as positive in a cyclonic rotation. Thus from the streamlines, or from the contours in gradient flow, the vorticity can be quickly determined. Another practical method is treated in the chapter on synoptic analysis.

For geostrophic flow there is no curvature term and we have the geostrophic vorticity

$$\zeta_g = -\frac{\partial c}{\partial n} \qquad (8\text{-}40)$$

Note that in fluid mechanics rotational motion exists whenever there is a velocity shear whether the path is curved or not. This is not true of a rigid body. As noted before, pure solid rotation is possible only if $\partial v/\partial x = -\partial u/\partial y$, for then $\omega_1 = \omega_2$. For a fluid, rotation as a solid is only one of an infinite variety of forms of rotation. In the atmosphere in the Northern Hemisphere positive or counterclockwise vorticity is cyclonic vorticity. Regardless of wind direction, the relative vorticity is cyclonic in the Northern Hemisphere if the linear speeds of the winds increase toward the right. Anticyclonic vorticity is indicated by linear speeds increasing toward the left.

CIRCULATION AND SOLENOIDS

Circulations in the atmosphere which nearly everyone has observed on a local scale are those arising from local heat differences.   Cold air sinks and flows along at the ground while warm air rises and moves aloft.   Sea breezes, lake breezes, valley breezes, "drainage" winds, and various mountain air circulations are all driven by density and pressure differences associated with temperature differences.   In terms of the basic forces driving it, the general circulation of the atmosphere which transports heat between low and high latitudes is of this type.   The cold air surrounding the poles sinks and spreads toward the equator while the tropical air is forced upward, moving poleward aloft.   The coriolis acceleration with the resulting tendency for geostrophic balance makes the meridional motions hard to detect, but this highly inefficient circulation is nevertheless driven by the heat differences.

In Chap. 5 the buoyancy acceleration arising from density differences was treated in one dimension in the atmosphere on the basis of an individual sounding. It was shown that the energy available for accelerating the motion can be represented on a p, α diagram, a $T \ln p$, or other thermodynamically related diagram.   We are now required to consider a similar problem in the three dimensions of the atmosphere. However, the analysis will be aimed toward investigating the circulation through atmospheric space rather than processes represented on an energy diagram.

Just as pressure lines become surfaces in space, so also do other lines of state, such as specific volume, temperature, and potential temperature.   Any two sets of surfaces may intersect to form a honeycomb-like set of tubes, adjacent ones having common walls and appearing approximately as parallelograms in cross section.   In Fig. 8-9 a schematic cross section in a vertical plane through a portion of the atmosphere shows some of the tubes formed by intersection of isobaric and *isosteric* (constant specific volume) surfaces.   Other planes would intersect the tubes at a different angle, but normally we are interested only in vertical and horizontal planes.   These tubes, formed by intersection of surfaces of pressure and specific volume or of surfaces of properties which bear a similar relationship to each other in a thermodynamic sense, are called *solenoids*.   A *unit p, α solenoid* is bounded on two opposite sides by isobaric surfaces corresponding to values differing from each other by one unit of pressure and on the other pair of opposite sides by isosteric surfaces representing two values of specific volume one unit apart.

Before going into a physical-mathematical discussion, it is useful to consider descriptively the relation of solenoids to the circulation.   In the atmosphere, pressure decreases and specific volume increases with height.   At any given pressure, the specific volume increases with increasing temperature.   The pressure decreases more rapidly with height in cold air than in warm air at the same levels.   These conditions are shown in Fig. 8-9 where the p, α solenoids are represented schematically in a

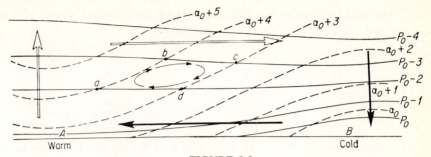

FIGURE 8-9
Solenoidal field as viewed in vertical cross section.

vertical cross section between a cold and warm region.   This might be typical of the solenoidal field between the antarctic and the surrounding oceans.

The cold air is denser (has a smaller α) than the surrounding air and therefore sinks.   The warm air, less dense (greater α) than its environment, rises.   The circulation is completed by flow from cold to warm in the low levels and in the reverse sense aloft.   Note that, since the isobars are expanded upward over the heat source, the isobaric slopes tend to fit the upper branch of the circulation.   The high pressure at the surface, which is often characteristic of cold areas, produces a pressure gradient in keeping with the surface flow.

If a unit solenoid *abcd* is taken, it is noted that the circulation must be in the sense associated with flow down the gradient of pressure and with the ascendant of specific volume, in other words, from *a* to *b* to *c* to *d*.

For a physical-mathematical analysis, let us consider the development of circulation in a vertical or y, z plane intersected by p, α solenoids.   As defined in Eq. (8-28) the circulation is

$$C = \int_c (v\,dy + w\,dz)$$

The acceleration of the circulation is obtained by differentiating C with respect to time, giving

$$\frac{dC}{dt} = \int_c \left(\frac{dv}{dt}\,dy + \frac{dw}{dt}\,dz\right) \qquad (8\text{-}41)$$

The integral can be evaluated by substituting for dv/dt and dw/dt in unbalanced flow.   In terms of the absolute motion, without coriolis acceleration, the only other accelerations would be associated with the pressure-gradient force, the acceleration of gravity, and frictional forces.   The latter can be neglected for the present in order to study the frictionless case and can be injected into the problem later, if desired.

The accelerations in the imbalance under investigation would be expressed, according to Eqs. (7-19b,c) and (7-22), as follows:

$$\frac{dv}{dt} = -\alpha \frac{\partial p}{\partial y} \qquad (8\text{-}42)$$

$$\frac{dw}{dt} = -g - \alpha \frac{\partial p}{\partial z} \qquad (8\text{-}43)$$

Upon substitution into (8-41) the acceleration of the circulation in the vertical y, z plane becomes

$$\frac{dC}{dt} = - \int_c \alpha\left(\frac{\partial p}{\partial y}\,dy + \frac{\partial p}{\partial z}\,dz\right) - \int_c g\,dz \qquad (8\text{-}44)$$

Since, around a closed path, one returns to the same height z, the second integral is zero. We also note that, considering only the y, z plane,

$$\frac{\partial p}{\partial y}\,dy + \frac{\partial p}{\partial z}\,dz = dp$$

Thus the integral becomes

$$\frac{dC}{dt} = - \int_c \alpha\,dp \qquad (8\text{-}45)$$

The acceleration is positive in the expected sense of flow in the direction of the pressure gradient $-\partial p/\partial n$ and the specific-volume ascendant $+\partial \alpha/\partial n$. Its sign depends on these quantities alone and is not determined by the sense of the coordinate axes x, y, z. It can be the same as or the opposite of the sense of circulation associated with the existing vorticity. The sense of the solenoids may be represented by a vector pointing in the direction of advance of a right-handed screw turning in the direction of the circulation accelerated by the solenoids. In Fig. 8-9 the sense would be inward toward the page.

Solenoids may also intersect a horizontal plane in the atmosphere, though at a very small angle. They may have an effect when the conservation of the vertical component of the absolute vorticity is considered. They may operate either to increase or to decrease the vorticity. Ordinarily the effect is small, but it can be eliminated altogether if the vorticity is considered on a constant-pressure surface. A constant-pressure surface forms one wall of a family of solenoids, and therefore solenoids cannot intersect it. This is another advantage of representation in pressure surfaces.

The acceleration in Eq. (8-45) around any closed curve is given by the number of unit solenoids enclosed by the curve. This can be demonstrated by projecting the

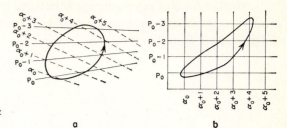

FIGURE 8-10
Projection of solenoidal field on p, α
diagram.

a   b

curve on a $-p$, α diagram as in Fig. 8-10.   The closed curve in Fig. 8-10a encloses
the same number of unit solenoids as that in the $-p$, α diagram of Fig. 8-10b.   Since
the coordinates of the latter are in units of $-p$ and α, the integral measures the actual
area enclosed by the curve, and the unit solenoids each represent a unit of area.   The
area is the number of unit areas, therefore the number of unit solenoids.   This is the
same number of solenoids enclosed by the curve in Fig. 8-10a.   In terms of p and α
the two integrals are the same; so the number of unit solenoids enclosed determines
the value of the integral in each case.   We may write

$$- \int_c \alpha \, dp = N \qquad (8\text{-}46)$$

where N is the number of unit solenoids of $-p$, α contained within the closed curve.
    Two other practical forms of solenoids may be represented by transforming the
integral into expressions involving other thermodynamic properties.
    Since, from the equation of state, $\alpha = RT/mp$, we may write

$$- \int_c \alpha \, dp = - \frac{R}{m} \int_c T \, d(\ln p) \qquad (8\text{-}47)$$

Thus solenoids of temperature[1] and natural logarithm of pressure may be used.
These are indeed practical, since pressure and temperature are in greater use than any
other set of properties in charting the atmosphere.   The pressure surfaces have a
$-\ln p$ distribution with height in the atmosphere, thus making this form quite convenient, and when the plotting of temperature is on constant-pressure surfaces, the
T, $-\ln p$ solenoids are immediately obtained.

[1] More properly, virtual temperature should be used throughout, but we recall that
    this is a trivial matter which becomes even more trivial in affecting the count of
    solenoids.

Another form of representation is in terms of temperature and potential temperature. The differential equation relating potential temperature to the temperature and pressure is, from Eq. (5-77),

$$C_p \frac{dT}{T} = C_p \frac{d\theta}{\theta} + \frac{R}{m} \frac{dp}{p}$$

$$C_p \, dT = C_p T \frac{d\theta}{\theta} + \frac{RT}{mp} \, dp$$

$$= C_p T \frac{d\theta}{\theta} + \alpha \, dp$$

$$\alpha \, dp = C_p \left(dT - T \frac{d\theta}{\theta} \right) \qquad (8\text{-}48)$$

The first term on the right, when integrated around a closed path, becomes zero, since the circuit comes back to the same temperature and no variable other than T is involved. We have, therefore,

$$-\int_c \alpha \, dp = C_p \int_c T \, d(\ln \theta) \qquad (8\text{-}49)$$

showing that temperature and the natural logarithm of potential temperature[1] can be used to construct solenoids. These are useful in working with isentropic charts and vertical cross sections through the atmosphere.

The axes of the three forms of solenoids are always parallel. Along a line of intersection of the isobaric and isosteric surfaces, pressure and specific volume are, by definition, constant. Temperature is determined from these two quantities from the equation of state; so it is also constant along this line. The potential temperature, determined by temperature and pressure, must also be constant along this line. The walls of the first two forms of solenoids have the same orientation.

The relation between the number of solenoids of different kinds in a given cross-sectional area can be stated as follows:

$$N = \frac{R}{m} N' = C_p N'' \qquad (8\text{-}50)$$

where N, N', and N'' refer to the number, respectively, of pressure–specific volume, pressure-temperature, and temperature–potential temperature solenoids.

Before leaving this subject, it should be stressed again that the coriolis acceleration and frictional effects cannot be neglected in applying the solenoidal circulation to actual motions. The coriolis acceleration is so strong in most situations that the solenoidal effect is an essentially undetectable influence on the geostrophic balance. In seeking to outline the main features of the general circulation of the atmosphere,

---

[1] Again, to be proper, based on virtual temperature.

meteorologists find great difficulty owing to the almost complete masking of the meridional circulation which must develop from the strong solenoidal field present in a north-south vertical plane.

In the local solenoidal circulations, such as those of the afternoon sea breezes, the closed circulation is almost never seen because of the widespread prevailing, essentially geostrophic winds which are present above the surface inflow.   The return flow is dissipated, at least in part, by the frictional stress of the upper wind.

It is probable that a steady-state circulation can exist on a local scale in which the acceleration of the circulation is exactly balanced by the surface and internal frictional stresses.   In the general circulation of the atmosphere, such a mean steady state involves a balance of the solenoidal acceleration against all accelerations acting on the air between the equator and the poles.   Actually, as is quite apparent to residents outside the tropics, the exchange of air involved in the general circulation is anything but steady, but in the mean it is sometimes convenient to regard it as so.

It is assumed that even the beginning student of meteorology is acquainted with the concept of fronts as depicted in newspaper and television presentations.   By definition, a front must be a zone in which solenoids are highly concentrated.   Many of the violent features of cold fronts are developed from extreme density gradients producing accelerations measurable through solenoids.   In general, intensifying storms have concentrated solenoidal fields associated with them.   On a localized scale, thunderstorms represent continually changing circulations accompanied by intense solenoidal concentrations.   They change so rapidly that geostrophic or gradient balance is not achieved and the air flows almost directly across the locally distorted isobars.

BAROTROPIC AND BAROCLINIC ATMOSPHERES

A portion of the atmosphere in which the surfaces of pressure, specific volume (or density), temperature, or potential temperature are all parallel is called *barotropic*. In other words, a barotropic atmosphere is one without solenoids.   The atmosphere is approximately barotropic in large regions in the tropics.   As a first approximation for the solution of certain problems in meteorology, the atmosphere sometimes is assumed to be barotropic when it really is not.   Such simplifications are common in theoretical studies in order to handle otherwise intractable problems and thus to gain incomplete yet important knowledge about intricate processes.

The opposite of barotropic is *baroclinic*, i.e., characterized by the presence of a solenoidal field.   The natural atmosphere is mainly baroclinic, since horizontal temperature gradients are the rule and often have a direction opposite to that of the pressure gradients.

EULERIAN EXPANSION

Fluid motion may be regarded in two different ways. One may investigate the motion and properties at all points in the fluid at various times or one may determine the changes in velocity and properties of a single particle. In the first case, the measurements are made at fixed points, while in the latter the measuring system must be able to follow the course of the particle. The equations of motion have different forms in the two methods, the first called the *Eulerian* (pronounced oil-air'-ian) form and the second the *Lagrangian* form.

Euler introduced the complete expression for the changes in velocity or properties involving both the moving particles and the distributions through the fluid space. To arrive at the so-called "Eulerian expansion," consider any quantity Q which varies in space and in time in a moving particle. If the particle is moving with velocity u in the x axis starting at time t from the point x, then at the time $t + \delta t$ (Fig. 8-11), its position will be at

$$x + \delta x = x + u\,\delta t$$

If the quantity Q changes both in time and in space, the new value of Q is

$$Q_{x+\delta x,\, t+\delta t} = Q_{x,\, t} + \delta Q$$

$$= Q_{x,\, t} + \frac{\partial Q}{\partial x}\,\delta x + \frac{\partial Q}{\partial t}\,\delta t = Q_{x,\, t} + u\,\delta t\,\frac{\partial Q}{\partial x} + \frac{\partial Q}{\partial t}\,\delta t \qquad (8\text{-}51)$$

If, as previously, we introduce the symbol D/Dt to denote differentiation following the motion of the particle, the new value of Q is

$$Q_{x+\delta x,\, t+\delta t} = Q_{x,\, t} + \frac{DQ}{Dt}\,\delta t \qquad (8\text{-}52)$$

and, from (8-51), the change of Q in the x direction with time would be

$$\left(\frac{DQ}{Dt}\right)_x = \frac{\partial Q}{\partial t} + u\,\frac{\partial Q}{\partial x} \qquad (8\text{-}53)$$

or, considering motion with all three components,

$$\frac{DQ}{Dt} = \frac{\partial Q}{\partial t} + u\,\frac{\partial Q}{\partial x} + v\,\frac{\partial Q}{\partial y} + w\,\frac{\partial Q}{\partial z} \qquad (8\text{-}54)$$

The derivative on the left in capital letters we will call the "individual change" and the first term on the right will be designated as the "local change," since it is the change that will be noted by an observer at a fixed point past which many particles are moving.

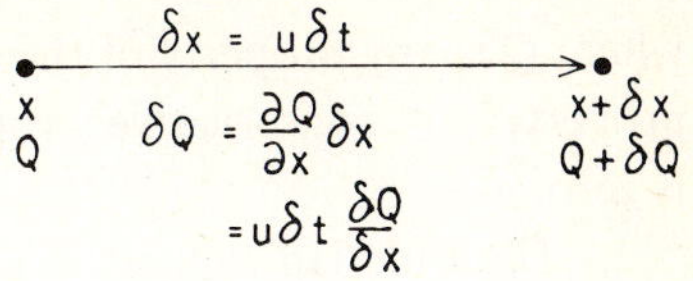

$$\delta x = u\,\delta t$$

$$\begin{array}{ccc} \bullet & & \bullet \\ x & \delta Q = \dfrac{\partial Q}{\partial x}\,\delta x & x+\delta x \\ Q & & Q+\delta Q \\ & = u\,\delta t\,\dfrac{\delta Q}{\delta x} & \end{array}$$

FIGURE 8-11

This distinction between individual and local change may best be illustrated by considering some property of the air such as temperature. If we carried a thermograph along with an air parcel, the temperature change recorded would be the individual change of temperature DT/Dt. If we had a thermograph at a fixed station, it would record the local change of temperature $\partial T/\partial t$. We should have

$$\frac{DT}{Dt} = \frac{\partial T}{\partial t} + u\frac{\partial T}{\partial x} + v\frac{\partial T}{\partial y} + w\frac{\partial T}{\partial z} \qquad (8\text{-}55)$$

If an axis is taken normal to the isotherms, and C_n is the wind component in this axis,

$$\frac{DT}{Dt} = \frac{\partial T}{\partial t} + C_n\frac{\partial T}{\partial n} \qquad (8\text{-}56)$$

where n refers to the normal direction.

The second term on the right is the *advective change* due to the movement of air particles of differing temperatures. The change observed at a given location, the local change, would be

$$\frac{\partial T}{\partial t} = \frac{DT}{Dt} - C_n\frac{\partial T}{\partial n} \qquad (8\text{-}57)$$

In other words, it would be the change due to the advection of air of different temperature (movement of isotherms) corrected for the change of temperature within the air particles as they move along—the fact that the isotherms are not displaced with the same speed as the wind normal to them. If we had no individual change, the temperature would be conservative; that is, the isotherms would move with the speed of the normal wind. Then the local change would be entirely due to advection and the first term on the right would be zero. If C_n is toward increasing temperature, advection would cause the temperature to decrease locally. If the air parcels moved faster than the isotherms, DT/Dt (within the parcels) would be positive. This would counteract to a certain extent the advective cooling.

The expression

$$\frac{DQ}{Dt} = \frac{\partial Q}{\partial t} + C_n\frac{\partial Q}{\partial n} = 0 \qquad (8\text{-}58)$$

where Q is any property of the air, expresses the condition of conservativeness of that property. A conservative property is one of which the individual change DQ/Dt is zero.

The Eulerian expansion applied to the three components of velocity becomes

$$\frac{Du}{Dt} = \frac{\partial u}{\partial t} + u\frac{\partial u}{\partial x} + v\frac{\partial u}{\partial y} + w\frac{\partial u}{\partial z} \qquad (8\text{-}59a)$$

$$\frac{Dv}{Dt} = \frac{\partial v}{\partial t} + u\frac{\partial v}{\partial x} + v\frac{\partial v}{\partial y} + w\frac{\partial v}{\partial z} \qquad (8\text{-}59b)$$

$$\frac{Dw}{Dt} = \frac{\partial w}{\partial t} + u\frac{\partial w}{\partial x} + v\frac{\partial w}{\partial y} + w\frac{\partial w}{\partial z} \qquad (8\text{-}59c)$$

In using isobaric surfaces for studying motions and properties, the p coordinate normal to the pressure surface is substituted in the above equations for z, and $w_p = Dp/Dt$ is substituted for w. In all other respects the Eulerian expanded equations are the same.

CONTINUITY AND DIVERGENCE

A statement of the law of conservation of matter in the form of the "equation of continuity" is useful in meteorology. This equation indicates that, in a continuous fluid or gaseous medium, the mass of fluid material passing into a given volume must be equal to that coming out unless a density change has occurred in the volume.

Let us consider a small volume in the form of a box with sides δx, δy, and δz as in Fig. 8-12. Initially the mass of air in the box is $\delta M = \rho\,\delta V = \rho(\delta x\,\delta y\,\delta z)$. We examine first the motion in and out of the pair of faces $\delta y\,\delta z$ as indicated by the arrow. The contribution of the flow in this component to the net accumulation of mass would be

Inflow at x_0 − outflow at x_1

$$\delta y\,\delta z\left[\rho u - \left(\rho u + \frac{\partial \rho u}{\partial x}\,\delta x\right)\right] = -\frac{\partial \rho u}{\partial x}\,\delta x\,\delta y\,\delta z \qquad (8\text{-}60)$$

If we go through the same reasoning for the other pairs of sides, we obtain the total net inflow of mass in the time dt as

$$-\left(\frac{\partial \rho u}{\partial x} + \frac{\partial \rho v}{\partial y} + \frac{\partial \rho w}{\partial z}\right)\delta V\,dt \qquad (8\text{-}61)$$

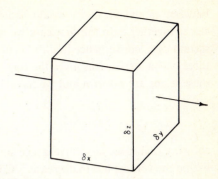

FIGURE 8-12

This net inflow, or accumulation, must give rise to a local increase in the mass, and if δV is to remain constant, the increase in mass in time dt is therefore

$$\left(\rho + \frac{\partial \rho}{\partial t}\, dt\right)\delta V - \rho\,\delta V = \frac{\partial \rho}{\partial t}\,\delta V\,dt \qquad (8\text{-}62)$$

The principle of conservation of mass requires that the two expressions (8-61) and (8-62) be the same; so we have

$$\frac{\partial \rho}{\partial t} = -\left(\frac{\partial \rho u}{\partial x} + \frac{\partial \rho v}{\partial y} + \frac{\partial \rho w}{\partial z}\right) \qquad (8\text{-}63)$$

This may be expanded into

$$\frac{\partial \rho}{\partial t} + u\,\frac{\partial \rho}{\partial x} + v\,\frac{\partial \rho}{\partial y} + w\,\frac{\partial \rho}{\partial z} + \rho\,\frac{\partial u}{\partial x} + \rho\,\frac{\partial v}{\partial y} + \rho\,\frac{\partial w}{\partial z} = 0 \qquad (8\text{-}64)$$

The first four terms give the expression for the individual change of the density. Designating this individual change as $D\rho/Dt$, we may write

$$\frac{D\rho}{Dt} + \rho\left(\frac{\partial u}{\partial x} + \frac{\partial v}{\partial y} + \frac{\partial w}{\partial z}\right) = 0 \qquad (8\text{-}65)$$

or

$$\frac{1}{\rho}\frac{D\rho}{Dt} + \frac{\partial u}{\partial x} + \frac{\partial v}{\partial y} + \frac{\partial w}{\partial z} = 0 \qquad (8\text{-}66)$$

This last expression is the equation of continuity as it is most often used in meteorology.

The partial derivatives $\partial u/\partial x$, $\partial v/\partial y$, $\partial w/\partial z$ express the divergence of the air. This may be seen by considering the speed u to be greater at the x_1 face than at the x_0 face of the box in Fig. 8-12.   Under this condition, $\partial u/\partial x$ would be positive.   The air particles would be pulling or stretching farther apart.   If the speed decreased between x_0 and x_1, $\partial u/\partial x$ would be negative, and the air particles would be pushing closer together. In the first case we would have divergence $[(\partial u/\partial x) > 0]$ and in the second case, convergence $[(\partial u/\partial x) < 0]$ in the direction of the x axis. The sum of the three partial derivatives gives the total divergence in the box. If there is divergence, these terms are positive and we have

$$\frac{\partial u}{\partial x} + \frac{\partial v}{\partial y} + \frac{\partial w}{\partial z} = -\frac{1}{\rho}\frac{D\rho}{Dt} \qquad (8\text{-}67)$$

showing that if we have divergence so that more comes out than goes in, the density within the volume has to decrease. Conversely, if convergence occurs, the term on the left of (8-67) is negative and $(D\rho/Dt) > 0$.

For the case of incompressibility, or when compression does not occur, the divergence is zero, or

$$\frac{\partial u}{\partial x} + \frac{\partial v}{\partial y} + \frac{\partial w}{\partial z} = 0 \qquad (8\text{-}68)$$

In meteorology we are often concerned with the horizontal divergence; so in applying this last expression we write

$$\frac{\partial u}{\partial x} + \frac{\partial v}{\partial y} = -\frac{\partial w}{\partial z} \qquad (8\text{-}69)$$

which emphasizes that horizontal divergence must be compensated by vertical shrinking or vertical convergence, while horizontal convergence must be accompanied by vertical stretching or vertical divergence $[(\partial w/\partial z) > 0]$. Any wind component that is increasing downstream is divergent in that component but may be convergent in one or both of the other two components.

In isobaric coordinates the equation of continuity can be developed for a rectangular parallelepiped with sides δx, δy, and δp. If hydrostatic equilibrium prevails,

$$\left(\frac{\partial u}{\partial x} + \frac{\partial v}{\partial y}\right)_p = -\frac{\partial w_p}{\partial p} \qquad (8\text{-}70)$$

where w_p is positive for downward motion as p increases.

The divergence is often written as div c, where c is the total velocity, or, in vector notation, as $\nabla \cdot c$. In the x and y dimensions it is written as $\text{div}_2\, c$ or $\nabla_2 \cdot c$, meaning the horizontal divergence. In meteorology it is desirable to consider the horizontal component separately in order to arrive at the important vertical component. It is not revealing to compute the three-dimensional divergence to find only that it is zero, as is usually the case in restricted volumes of the atmosphere. By computing the lateral divergence in the same situation, significant positive or negative values may be found that have a profound effect on the vertical motions, already known from Chap.

5 to be important for development of thermal instability.   As will be shown later, the vertical motions are of great importance in causing pressure changes over large areas in the atmosphere.

From the divergence, one cannot determine the vertical velocity directly; only the change of that quantity with height or pressure is given.   Conversely, one cannot infer that the existence of vertical velocity at any point in the atmosphere means divergence or convergence at that point.   Since the atmosphere has a fixed lower boundary at the surface of the earth and also some indefinite upper layer, the presence of vertical motions always indicates horizontal divergence and convergence above and below to complete the continuity.   Thus a region of steady upward motion would be maintained by convergence below and divergence above.   Downward motion would require divergence below and convergence above.

At the surface of the earth the vertical velocity must be zero.   With zero as a starting point, the direction of the vertical motion in convergent or divergent flow immediately above the surface is at once determined from the equation of continuity. Physically this means that if air converges along the ground there is no escape for the accumulating air except upward, and if divergence occurs the deficit can only be supplied by air coming down from above.   In convection, especially in the building stage of thunderstorms, convergence is observed in the low levels and divergence in the upper parts.

The different possibilities in the free atmosphere and at the surface are represented schematically in Fig. 8-13.   In the top row horizontal divergence is represented for the three types of vertical motion—upward, downward, and zero.   The second row shows the same for nondivergent flow.   Continuous vertical flow could occur between this row and the ones above and below it, considered as layers in the atmosphere, except in the last column of the second row which is boxed to indicate that zero vertical velocity would have to exist for some distance above and below in the box.   The third row represents the convergent conditions.   At the ground, indicated in the lowest row, there is only one possibility for each of the three different values of divergence.   In this part of the diagram, the vertical velocities given for the various columns should be regarded as occurring immediately above the ground while at the air-earth boundary itself, w is zero.   These comments, of course, must be modified when one is considering a sloping terrain.   For all the situations it is to be noted that horizontal divergence is accompanied by vertical convergence and that horizontal convergence goes with vertical divergence.

A practical way of considering continuity is to state the condition of zero total divergence as one in which the mass of any given volume of air does not change. This is another way of saying that the density does not change.   If neither the mass nor the density varies, the volume must also remain constant.   While remaining constant, the volume may change its shape, for example, stretching horizontally

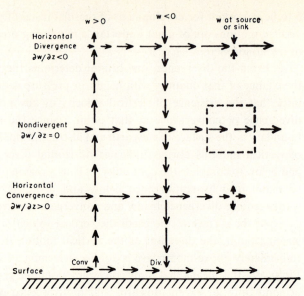

FIGURE 8-13

Divergence-convergence patterns and vertical velocities in the atmosphere.

while shrinking vertically by the same amount.   In this case there would be horizontal divergence and vertical convergence of the air particles.   It is not difficult to obtain a mathematical expression for these effects.

The mass M of a given volume V, such as a cylinder or rectangular parallel-epiped, is given by

$$M = \rho V = \rho A h = \text{const}$$

where A and h are the area and height (depth) of the volume, respectively.   The constant indicates that there is no total divergence.   In the atmosphere the air is not restricted to fixed volumes as in the box example of Fig. 8-12.   If there is horizontal divergence, the area covered by a mass of air will enlarge at the expense of the height between identifiable levels, and if horizontal convergence occurs the area shrinks accompanied by vertical stretching.   In Fig. 5-7, p. 101, the effects of vertical shrinking and stretching on the lapse rate of a layer were demonstrated graphically.   As long as the mass does not change, the volume can be distorted or even increased or decreased. Between the top and bottom of a mass of air, identified by two isentropic surfaces, the pressure difference gives the weight and, for uniform g, the mass per unit area; so the total mass between height h_1 and h_2 and over the area A is given by

$$M = A \int_{h_1}^{h_2} \rho \, dz = -\frac{A}{g} \int_{h_1}^{h_2} \frac{\partial p}{\partial z} \, dz \qquad (8\text{-}71)$$

The right-hand expression is obtained from Eq. (5-56). In hydrostatic equilibrium and with the direction of integration reversed,

$$M = \frac{A}{g}\int_{p_2}^{p_1} dp = \frac{(p_1 - p_2)A}{g} \qquad (8\text{-}72)$$

With conservation of mass, $dM/dt = 0$; so we have

$$\frac{dM}{dt} = \frac{p_1 - p_2}{g}\frac{dA}{dt} + \frac{A}{g}\frac{d}{dt}(p_1 - p_2) = 0 \qquad (8\text{-}73)$$

or

$$\frac{1}{A}\frac{dA}{dt} + \frac{1}{p_1 - p_2}\frac{d}{dt}(p_1 - p_2) = 0 \qquad (8\text{-}74)$$

We represent $p_1 - p_2$ by D_p, the pressure height between the two levels, and write

$$\frac{1}{A}\frac{dA}{dt} = -\frac{1}{D_p}\frac{dD_p}{dt} \qquad (8\text{-}75)$$

On the left is the horizontal stretching and on the right the vertical shrinking of the volume.

To show that $(1/A)(dA/dt)$ does indeed represent the divergence as we have defined it in Eq. (8-69), let us consider a small rectangular area $\delta A = \delta x\,\delta y$ at initial time t as represented by the base plane of Fig. 8-12. At time $t + dt$ the rectangle will be formed into a parallelogram whose angles differ infinitesimally from right angles. To the first order of dt its edges are

$$\delta x + \frac{\partial u}{\partial x}dt\,\delta x \quad \text{and} \quad \delta y + \frac{\partial v}{\partial y}dt\,\delta y \qquad (8\text{-}76)$$

We obtain the area from the product of the two sides. Applying only the first-order terms in dt, we have

$$\delta A + \frac{d(\delta A)}{dt}dt = \left[1 + \left(\frac{\partial u}{\partial x} + \frac{\partial v}{\partial y}\right)dt\right]\delta x\,\delta y \qquad (8\text{-}77)$$

We divide by $\delta A = \delta x\,\delta y$, to obtain

$$\frac{1}{\delta A}\frac{d(\delta A)}{dt} = \frac{\partial u}{\partial x} + \frac{\partial v}{\partial y} \qquad (8\text{-}78)$$

or, for the fractional change with time of any area A,

$$\frac{1}{A}\frac{dA}{dt} = \frac{\partial u}{\partial x} + \frac{\partial v}{\partial y} = \text{div}_2\,\mathbf{c} \qquad (8\text{-}79)$$

as was postulated.

The relationship (8-79) states that the measurement of the fractional change per unit time of the area occupied by an identifiable body of air gives the horizontal divergence. It is also given in the negative sense by the fractional change per unit time of the depth. For the total divergence to be zero, horizontal stretching (divergence) must be compensated by vertical shrinking (convergence) and vice versa.

The area method of measuring horizontal divergence in the atmosphere can be accomplished by the use of carefully tracked balloons. Three or more balloons are released simultaneously from different nearby points so that, having the same free lift, they reach the different levels of the atmosphere at the same time. Three balloons will form the vertices of a triangular area and four or more balloons will form the corners of a figure of four or more sides. The area enclosed by the balloons is measured at frequent time intervals. If the area enlarges as the balloons pass through a layer of finite thickness, say from 300 to 500 m, horizontal divergence is given for that layer by the fractional change of the area. The fractional shrinkage of the area per unit of time as the balloons pass through a layer would measure the horizontal convergence in that layer.

The depth-change method of determining divergence or convergence can be applied in connection with isentropic surfaces. If vertical displacements are accomplished adiabatically (isentropically), then an isentropic surface is a *substantial* surface, that is, one which must move exactly up or down with the displacement of the air particles, or a surface through which air particles cannot move. The fractional variation with time of the pressure-depth contained between two given isentropic surfaces as the air is channeled between them gives a measure of the horizontal divergence and convergence.

A standard method of computing the horizontal divergence is based on an analysis of the wind field to obtain $\partial u/\partial x$ and $\partial v/\partial y$. A map is made of a region, such as the United States or Western Europe, containing isolines of values of u and a similar map of the values of v. The number of unit isolines in a unit x distance at a given location on the first map is $\partial u/\partial x$ and the number in a unit y distance on the second map is $\partial v/\partial y$. The sum of these two values is plotted on a third map, and the isolines constructed therefrom show the pattern of horizontal divergence and convergence.

It is also possible to evaluate the horizontal divergence by means of a streamline chart using the total velocity c. Two sets of lines are needed: the streamlines and the isolines of equal total wind speed, called *isotachs*. Over restricted areas, or if the velocity is irrotational, the isotachs, like lines of velocity potential, are orthogonal to the streamlines. However, this is not a requirement for application of the method. In Fig. 8-14 the two streamlines and the two isotachs form an area *abcd*. The isotachs correspond to the magnitudes of the wind vectors c_1 and c_2 drawn in the middle of the streamline channel. These vectors represent the distance the air particles would be displaced in unit time; so if the streamlines remain fixed, the change in area in unit

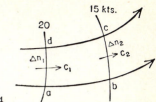

FIGURE 8-14

time would be given by $c_2 \, \Delta n_2 - c_1 \, \Delta n_1$.   Dividing this by the area itself, we would have the horizontal divergence

$$\text{div}_2 \, c = \frac{c_2 \, \Delta n_2 - c_1 \, \Delta n_1}{A} \qquad (8\text{-}80)$$

This method is difficult to apply in the free air because of the difficulty of obtaining true streamlines from the sparse network of slightly inaccurate balloon wind measurements.

It should be noted that convergence of the streamlines does not in itself indicate convergence.   As the separation Δn becomes smaller, c increases and often does so in the same proportion such that the divergence is zero.   This situation is sometimes referred to as *confluence* and the reverse as *diffluence*.

CONTINUITY, ANGULAR MOMENTUM, AND VORTICITY

Earlier in this chapter some consequences of the tendency for conservation of absolute vorticity in the atmosphere were suggested.   Considerations of conservation of mass (continuity) and, indirectly, of the conservation of total angular momentum enter into the problem, as will now be shown.

For a simple approach to the problem, a cylindrical mass of air with its base on the ground as in Fig. 8-15 is considered.   It has the basal area $A = \pi r^2$ and the height D_p, both of which may be permitted to vary together in such a way that the mass is preserved.   It is assumed that the cylindrical column has vertical components of relative vorticity ζ and of angular velocity ω' superimposed on the local vertical components of the earth f and ω, respectively.   The vertical component of the absolute vorticity is $f + \zeta$ and the vertical component of the absolute angular velocity is $\omega + \omega'$.

As stated in Eq. (7-6), the angular momentum is given by

$$M\omega r^2$$

where M is the mass and ω is any angular velocity.   According to the physical law of the conservation of momentum, rotating masses, in the absence of torques, conserve

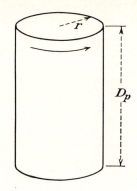

FIGURE 8–15

their angular momentum.   In our example, we are dealing with the absolute angular momentum of the air column; so

$$M(\omega + \omega')r^2 = \text{const} \qquad (8\text{-}81)$$

The vertical component of the absolute vorticity is, according to Eq. (8-1),

$$f + \zeta = \omega + \omega'$$

so we may write Eq. (8-81) in the form

$$M(f + \zeta)r^2 = \text{const} \qquad (8\text{-}82)$$

Substituting the area of the cylinder A for πr^2 and absorbing $1/\pi M$ in the constant, we have, for constant mass,

$$(f + \zeta)A = \text{const} \qquad (8\text{-}83)$$

as a statement of the conservation of absolute angular momentum.   The derivative of this expression with respect to time is

$$(f + \zeta)\frac{dA}{dt} + A\frac{d}{dt}(f + \zeta) = 0 \qquad (8\text{-}84)$$

We divide through this equation by $(f + \zeta)A$ obtaining

$$\frac{1}{A}\frac{dA}{dt} + \frac{1}{f + \zeta}\frac{d}{dt}(f + \zeta) = 0 \qquad (8\text{-}85)$$

The first term is recognized as the horizontal divergence.   We substitute the pressure-depth for the area and obtain the meteorologically important equation

$$\frac{1}{f + \zeta}\frac{d}{dt}(f + \zeta) - \frac{1}{D_p}\frac{dD_p}{dt} = 0 \qquad (8\text{-}86)$$

or
$$\frac{1}{f+\zeta}\frac{d}{dt}(f+\zeta) = \frac{1}{D_p}\frac{dD_p}{dt} = -\operatorname{div}_2 c \qquad (8\text{-}87)$$

This shows that the change in absolute vorticity and the change in depth of an air column must go together and that the fractional change in the absolute vorticity is equal to the horizontal convergence.

To integrate this expression, we multiply through by $D_p(f+\zeta)/D_p{}^2$ and obtain

$$\frac{D_p(d/dt)(f+\zeta) - (f+\zeta)\,dD_p/dt}{D_p{}^2} = 0 \qquad (8\text{-}88)$$

which has the form

$$\frac{d}{dt}\left(\frac{x}{y}\right) = \frac{y(dx/dt) - x(dy/dt)}{y^2}$$

So, the integral of Eq. (8-88) is

$$\frac{f+\zeta}{D_p} = \text{const} \qquad (8\text{-}89)$$

This equation and its time derivative represent one of a few relations that have real value in predicting future atmospheric flow patterns.   As air particles move from one latitude to another, f changes.   Compensating changes must occur in ζ and D_p in order to meet the requirements expressed by these physical equations.   Keeping in mind the fact that vorticity is positive in the cyclonic sense and that divergence is symbolized by decreasing depth of a layer, we can recognize the following qualitative results:

1 In poleward displacements, f increases; so either ζ decreases or D_p increases (convergence) or both adjustments occur.

2 In equatorward displacements, f decreases; so either ζ increases or D_p decreases (divergence) or both adjustments occur.

3 In straight easterly or westerly flows, f does not change; so divergence decreases the relative vorticity ζ and convergence increases it.   The same is true in small circulations, such as thunderstorms, where the motions take place at essentially one latitude and there is no appreciable change in f.

THE TENDENCY EQUATION

At this point it is of interest to use concepts discussed in this chapter to illustrate some of the factors that determine the changes in atmospheric flows, that is, the departures from the steady state.   One application of these concepts, namely, the computation of trajectories with constant absolute vorticity, neglecting the divergence term, has already been mentioned.   Another application of these concepts is in the development of the barometric-tendency equation.   It shows how divergence and vertical motion contribute to the change in the pressure with time either at the surface or in the free atmosphere.

The pressure at any point at height z is given by the total weight of a column of unit cross-sectional area extending from z to the outer limit of the atmosphere.   Or, integrating the hydrostatic equation from z to ∞ where the pressure goes from p to 0, we have

$$-\int_p^0 dp = \int_0^p dp = p = g \int_z^\infty \rho \, dz \qquad (8\text{-}90)$$

The pressure tendency recorded on a barograph at a fixed point at height z would simply be the derivative of the above with respect to time, or

$$\left(\frac{\partial p}{\partial t}\right)_z = g \int_z^\infty \frac{\partial \rho}{\partial t} \, dz \qquad (8\text{-}91)$$

From the equation of continuity we have

$$\frac{\partial \rho}{\partial t} = -\frac{\partial(\rho u)}{\partial x} - \frac{\partial(\rho v)}{\partial y} - \frac{\partial(\rho w)}{\partial z}$$

This may be written as

$$\frac{\partial \rho}{\partial t} = -\frac{\partial(\rho c)}{\partial s} - \frac{\partial(\rho w)}{\partial z} = -\mathrm{div}_2(\rho c) - \frac{\partial(\rho w)}{\partial z} \qquad (8\text{-}92)$$

where c is the total horizontal velocity and s is distance, positive downwind, on a streamline of that velocity.   When this is substituted in Eq. (8-91), the result is

$$\left(\frac{\partial p}{\partial t}\right)_z = -g \int_z^\infty \mathrm{div}_2(\rho c) \, dz - g \int_z^\infty \frac{\partial(\rho w)}{\partial z} \, dz$$

$$= -g \int_z^\infty \mathrm{div}_2(\rho c) \, dz + (g\rho w)_z \qquad (8\text{-}93)$$

the last term being zero at $z = \infty$ where $\rho = 0$.

This equation, which is an explicit expression of the change in weight of the air column above z, shows that this change is contributed to in part by horizontal mass divergence above z and in part by transport of air into or out of the column through the level z.   If more air goes out of the column than into it, by either horizontal divergence (first term) or downflow through the bottom (second term), the pressure at z will fall.

By itself the tendency equation is not very useful because standard synoptic-aerological observations cannot supply the values required for solution.   In practice another type of tendency equation is used.   It involves a determination of the change with time of the geopotential resulting from the advection of vorticity and from the thickness advection.   It is related to numerical prediction methods discussed in Chap. 15.

COMMENT

In this chapter the student has been led through simplified equations of motion that are derived from fluid mechanics and applied to an atmosphere over a rotating, unequally heated planet.   Since the laws of fluid flow are hardly touched upon in elementary college physics courses, the student has no doubt encountered much that is new and, perhaps, difficult.

At this point it must be apparent that meteorology is basically an exact science. Exact methods cannot be realized fully because all the information necessary to apply the hydrodynamical and thermodynamical equations is not available.   It is necessary to deal only with those data derived from the simple surface and aerological observations which can be made routinely and at a price the world economy can afford.

The development of high-speed electronic computing machines has made it possible to apply the equations to the atmosphere for purposes of prediction of the motion.

EXERCISES

1   State whether the relative vorticity ζ is positive or negative in the following cases: (a) south wind increasing in positive x direction; (b) east wind increasing in speed toward the north; (c) north wind increasing in speed toward the west; u positive, increasing in negative y direction; (d) cyclone with winds constant along n.

2   Show the relation between stream function ψ and wind c in natural coordinates.

3   Show that in natural coordinates irrotational motion exists when $c/R = \partial c/\partial n = 0$.

4   Compute the anticyclonic shear in straight flow required to balance the vertical component of the earth's vorticity at lat 45°.   In an anticyclone rotating with a shear of 10^{-10} sec^{-1}, at what radius of curvature (km) would a wind of 10 m sec^{-1} balance the vertical component of the earth's vorticity at lat 45°?

5   Give the units of circulation C, of vorticity, and of divergence.

6   Transpose, in general appearance, the solenoidal field in Fig. 8-9 in terms of (a) T, log p; (b) T, log θ.

7   Pressure is falling at your station at a rate of 15 mb in 3 hr while a low-pressure center of 990 mb is approaching directly toward you at 15 m sec^{-1}.   Your pressure is 994 mb. Is the low-pressure center deepening or filling?   At what rate (mb per 3 hr)?

8   The horizontal convergence under an average thunderstorm is 10^{-3} sec^{-1}. Theoretically, what would be the vertical velocity at 10 m above the ground?

9   A mass of polar air pushes southward from a center between 60°N and 45°N to 30 to 45°N. Assume that it is bounded on the west by the 90th meridian and on the east by the 75th meridian. Find (a) the change in area that occurs in this displacement; (b) the depth change in millibars between two isentropic surfaces initially 110 mb apart, taking the bottom of the layer unchanging at 900 mb; and (c) give the value of the vorticity, paying attention to sign. Determine graphically on a $T \log p$ or similar diagram the change in lapse rate that would occur in this layer which lies between the isentropic surfaces of 290 and 295.

10 Along a streamline the relation in Eq. (8-86) can be expressed as $d[\ln(f + \zeta)] = d(\ln D_p)$. From the indefinite integral of this last expression show that $(f + \zeta)/D_p = $ const.

11 What would the net mass divergence be in the moving air column between the ground and the top of the atmosphere above the cyclone in Exercise 7 to cause the calculated pressure change at the surface? Assume $g = 980$ dyn g^{-1}.

9

THE GENERAL CIRCULATION

In Chap. 3 it is shown that to balance the low-latitude energy accumulation and high-latitude energy deficit, a poleward flux of heat is required. In the Northern Hemisphere the maximum flux is at about 35° latitude where the curves of incoming and outgoing radiation shown in Fig. 3-6 cross. The mean atmosphere is strongly baroclinic in this zone. The energy is transported partly as sensible heat and partly as latent heat, the former far outweighing the latter. A small amount of heat is transported by the ocean currents, of which the Gulf Stream in the North Atlantic and Kuroshio (Japan Current) in the North Pacific are the most important.

In the absence of rotation and surface inhomogeneities the solenoidal field of the troposphere should produce a single meridional circulation cell in each hemisphere, with poleward flow at upper levels and equatorward flow at the surface. This simple, single-cell circulation is not possible on a rotating planet. From the discussion of conservation of angular momentum in Chap. 7 it is clear that any tendency toward such conservation would result in upper westerlies and lower easterlies. Deflection to the right in the Northern Hemisphere and to the left in the Southern Hemisphere is an expression of the same principle.

An insight into the effects of conservation of absolute angular momentum can be gained by taking an example. The west-to-east speed of a fixed point at latitude

φ is $U_\varphi = \Omega r = \Omega R \cos \varphi$, where r is the radius of a latitude circle and R is the radius of the earth. The relative west-to-east speed, which would be the u component of the wind, is $u = U_A - U_\varphi$, where U_A is the absolute west-to-east component. Conservation of the absolute angular momentum means that $U_A R \cos \varphi = \text{const}$, or

$$U_{A_2} R \cos \varphi_2 = U_{A_1} R \cos \varphi_1$$

$$U_{A_2} = U_{A_1} \frac{\cos \varphi_1}{\cos \varphi_2} \qquad (9\text{-}1)$$

Consider particles displaced from the equator to $30°$. Take u_0 as -5 m sec^{-1} (east wind at equator). Then

$$U_{A_0} = U_{\varphi_0} - 5$$

$$U_{A_{30}} = (U_{\varphi_0} - 5)\frac{\cos 0}{\cos 30} = 1.155(U_{\varphi_0} - 5)$$

$$\begin{aligned} u_{30} = U_{A_{30}} - U_{\varphi_{30}} &= 1.155(U_{\varphi_0} - 5) - \Omega R \cos 30 \\ &= 1.155(464.45 - 5) - (464.45 \times 0.866) \\ &= 530.7 - 402.2 \\ &= 128.5 \text{ m sec}^{-1} \end{aligned}$$

West winds with speeds of this magnitude are more than twice the speeds observed in extreme cases in the strongest high-level jet streams. If the above example were carried to latitude $60°$, the wind would be at 686.7 m sec^{-1}. The speeds that would result from conservation of angular momentum are so far from the real speeds that one immediately realizes that something must happen to the wind system. The reduction of the winds to speeds that we normally observe cannot be accounted for by friction, which at least in the upper air, probably could reduce them by only a few percent.

What happens is that the meridional circulation breaks down into smaller circulations. Bergeron[1] depicted a system containing a meridional circulation of three cells. This system, considerably modified, is shown in Fig. 9-1. The net meridional circulation is represented in vertical cross section around the profile of the earth, and the actual wind directions at the surface of the earth are shown within the circumference. Throughout the atmosphere the meridional components are on the average less than one-tenth of the zonal components. Except for a belt of deep easterlies around the equator, the upper winds are prevailingly from the west. At the surface, three wind belts are noted in each hemisphere: the trade winds of low latitudes, the middle-latitude westerlies, and the polar easterlies.

[1] T. Bergeron, Über die Dreidimensional Verknüpfende Wetteranalyse, *Geofysiske Publikasjoner*, vol. 5, no. 6, 1928.

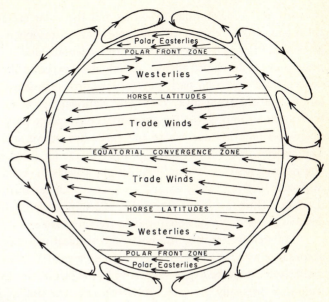

FIGURE 9-1
Schematic representation of the general circulation, modified after Bergeron.

The pressure-wind relationship applied to the surface circulation requires a belt of high pressure at about 30°N and S of the equator. This belt is called the "horse latitudes."[1] Low pressure is required between the westerlies and the polar easterlies.

Except for the easterlies of the trade-wind belt, the surface wind systems are quite variable. The continents develop thermally induced wind and pressure systems, with relatively high pressures over the land areas in winter and lower pressures in summer. The wind systems pictured in Fig. 9-1 are best marked over the oceans, particularly in the Southern Hemisphere where the continental areas are small. The low-pressure area around the latitude of 60° is most inconstant. It is an average made up of the effects of all the moving cyclones of middle and high latitudes. The polar highs are also quite variable, especially in the Northern Hemisphere.

It is apparent from Fig. 9-1 that low-level convergence occurs in the meridional components at the equator between the trade winds of the two hemispheres and also

---

[1] It is thought that the name is derived from the numerous bodies of horses seen floating on the sea in these latitudes in the seventeenth century. It is related that sailing vessels transporting horses to the New World were becalmed in the high-pressure center and, running low on feed and water, reduced their cargoes in mid-ocean.

between the middle-latitude westerlies and the polar easterlies. This latter convergence zone is in a strongly baroclinic part of the atmosphere. The meridional temperature advection tends to concentrate the solenoids and an average front, called the *polar front*, is developed there. As might be expected, however, the convergence is concentrated in certain parts of it where vorticity is generated. The cyclonic circulations pull the front away from any stationary position it may have and severe distortions are imposed. The equatorial convergence zone is in a barotropic atmosphere and in a latitude where the earth has no vertical component of vorticity. The absence of a coriolis force means that the air flows directly across the isobars and no cyclones can develop. At certain seasons the equatorial convergence zone is displaced 10 to 15° away from the equator. Under this circumstance cyclones can develop. Hurricanes and their Far Eastern counterparts, typhoons, sometimes develop in the displaced equatorial convergence zone.

As seen in Fig. 9-1, the polar and tropical meridional circulation cells are in accordance with the expected circulation on a rotating planet with cold poles, but the middle cell cannot be explained easily; so the picture has to be modified considerably from that originally presented by Bergeron. The sinking in the horse latitudes has been explained as resulting from net heat losses by radiation at the upper levels. Although the mean radiation data also show cooling aloft in the tropics, pictures and observations from satellites indicate enough release of heat of condensation in the convergence zone by large convective storms to counteract the radiation effect.

In reality, a *mean* picture of the middle-latitude circulation tends to obscure the real nature of the meridional exchange. The mean consists of a great variety of patterns continually changing in time and space. When the earth is viewed from space, it reveals cloud patterns which obviously trace out major eddies recognized as large cyclones and anticyclones. These account for most of the exchange.

Mean charts of isobars at sea level or contours of the 1000-mb surface show considerable departures from the idealized flow pattern of the general circulation. The charts of Figs. 9-2 and 9-3 represent these distributions in midwinter and midsummer, respectively. The following departures from the idealized pattern should be noted in the Northern Hemisphere:

Winter

Polar anticyclone   Missing.

Polar-front low-pressure belt   Icelandic low and Aleutian low well developed; continents show high pressure, especially the Siberian high.

Horse-latitude high-pressure belt   Broken into cells over oceans, the Pacific anticyclone being the best defined; tendency for merging with continental highs in North America, Europe, and North Africa.

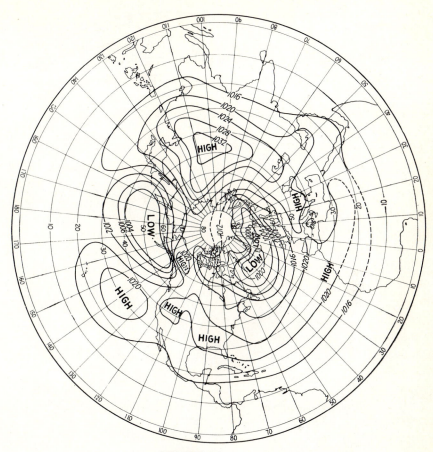

FIGURE 9-2
January normal sea-level pressure, Northern Hemisphere.

Trade-wind and equatorial regions Pushed southward but otherwise little modified.

Summer

Polar anticyclone Fairly well pronounced over Arctic Ocean.

Polar-front low-pressure belt Icelandic low present but weak; Aleutian low not discernible, but trough extends along Arctic Circle into Siberia.

Horse-latitude high-pressure belt Atlantic and Pacific high-pressure cells very well marked; continents in this belt are occupied by heat lows, especially in North Africa and Asia and the Arizona low.

Trade-wind and equatorial regions Trades interrupted by monsoon circulations in Indian Ocean, southern Asia, southwestern North Pacific Ocean areas, Gulf of Panama, and southern part of the bulge of Africa.

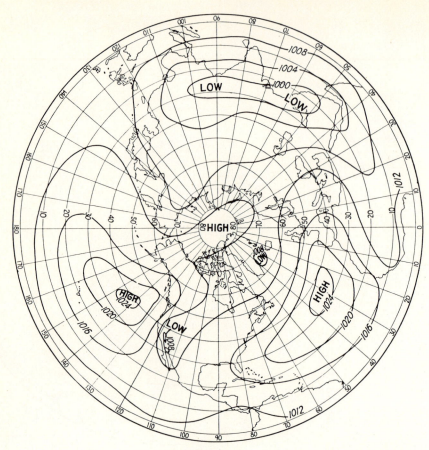

FIGURE 9-3
July normal sea-level pressure, Northern Hemisphere.

In the Southern Hemisphere the circulation follows the idealized pattern more closely than in the Northern Hemisphere. Nevertheless the horse-latitude high-pressure belt is broken up into cells and noticeable thermal effects of the continents are present. For example, over Australia there is a summer (January) heat low and a winter (July) continental high pressure.

The upper-air mean charts shown in Figs. 9-4 and 9-5 exhibit fewer departures from the idealized flow. Note the concentration of the westerlies in a relatively narrow band, to be discussed in a later section, and the asymmetrical character of the polar low.

On the charts the winds may be inferred from the isolines, which, according to the pressure-wind relationship, are approximate streamlines.

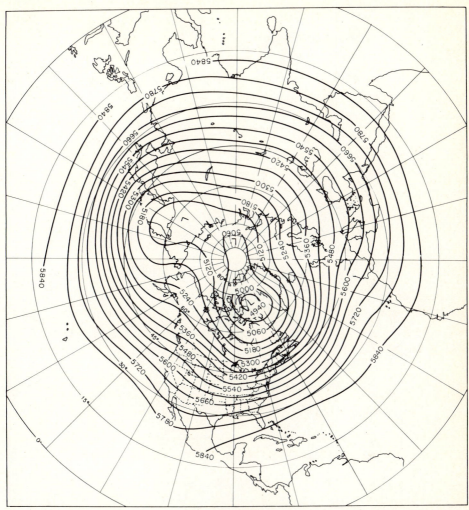

FIGURE 9-4
January normal 500-mb chart, Northern Hemisphere.

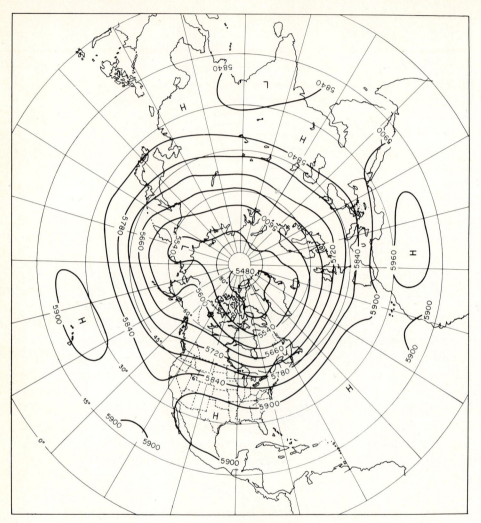

FIGURE 9-5
July normal 500-mb chart, Northern Hemisphere.

MOVING CYCLONES AND ANTICYCLONES

The mean circulations pictured on these pages do not account for the meridional exchange required by the terrestrial balance of heat, any more than the mean position of a swinging pendulum will indicate the motions it goes through. The exchange in the atmosphere is accomplished by the secondary circulations, that is, by the moving cyclones and anticyclones of middle latitudes. Bjerknes and Solberg[1] in 1922 pointed out that this mechanism is the most important in the heat exchange between low and high latitudes. Today, the familiar photographs of the earth from space illustrate the process. Visible cloud bands trace out the motions in storm areas. The large areas of release of heat of condensation to the middle and upper troposphere in heavy tropical cloud masses are also evident from these pictures.

Secondary circulations of this type also appear in laboratory models of rotating fluids with heat sources.[2] As the speed of rotation passes some critical value, the flow becomes asymmetrical and large waves and cyclonic-anticyclonic circulations appear. The transition is designated as going from stable to unstable above the critical speed of rotation, thus suggesting that the atmospheric flow represents the unstable phase. At very high rotation speeds the flow becomes very chaotic. These transitions suggest that planets with a speed of rotation (and heat distribution) different from that of the earth might show quite different types of circulations in their atmospheres.

TRANSFER OF ANGULAR MOMENTUM

A circulation that transfers heat poleward is required in order to preserve the heat balance of the earth. The circulation also transports angular momentum poleward. Such a transport is not obvious in the mean models presented in the charts on these pages. They show an increase with latitude of the speed of the upper westerlies from the equator to about 35°N or S, producing the well-known "jet stream" near the tropopause in the two hemispheres, but the speeds decrease again in middle to high latitudes. One might consider that the exchange of both heat and momentum is accomplished in the large perturbations which do not show up in the mean picture, but it is of interest to see if any steady-state circulation, such as the continuous wavy band of upper westerlies, can produce an average net transfer of momentum poleward. Do such waves have the net effect of transferring momentum poleward around this belt? Jeffreys[3] examined this question, and refinements of his statement were added by Starr.[4] A simple mathematical treatment is presented in the following paragraphs.

[1] J. Bjerknes and H. Solberg, Life Cycle of Cyclones and the Polar Front Theory of Atmospheric Circulation, *Geofysiske Publikasjoner*, vol. 3, no. 1, 1922.

[2] D. Fultz, R. R. Long, G. V. Owens, W. Bohan, R. Kaylor, and J. Weil, Studies of Convection in a Rotating Cylinder, *American Meteorological Society Monographs*, vol. 4, no. 21, 1959.

[3] H. Jeffreys, On the Dynamics of Geostrophic Winds, *Quarterly Journal of the Royal Meteorological Society*, vol. 52, pp. 85–104, 1926.

[4] V. P. Starr, An Essay on the General Circulation of the Earth's Atmosphere, *Journal of Meteorology*, vol. 5, pp. 39–43, 1948.

Consider a ring of unit width extending from the surface to the top of the atmosphere (from p_0 to $p = 0$) around a middle-latitude circle.   The mass for a unit width around the 2π rad of longitude λ is

$$M_\varphi = \frac{\text{weight}}{g} = \frac{1}{g} \int_0^{po} \int_0^{2\pi} r_\varphi \, d\lambda \, dp \qquad (9\text{-}2)$$

where r_φ is the distance from the earth's axis, that is, the radius of the latitude circle. Since the atmosphere is extemely thin in relation to this radius at middle to low latitudes, we may consider r_φ to include the entire depth of the atmosphere.   In the integral the latitudinal-width term $R \, d\varphi$ does not appear because it is taken as unity.

The total angular momentum is the angular momentum of the earth plus the relative momentum of the air having the west-to-east component of velocity u with respect to the earth.   Thus, the total momentum is given by

$$\mu_\varphi = M_\varphi \left(\Omega + \frac{u}{r_\varphi} \right) r_\varphi{}^2 = M_\varphi (\Omega r_\varphi + u) r_\varphi \qquad (9\text{-}3)$$

Substituting for M_φ from Eq. (9-2) we have

$$\mu_\varphi = \frac{r_\varphi{}^2}{g} \int_0^{po} \int_0^{2\pi} (\Omega r_\varphi + u) \, d\lambda \, dp \qquad (9\text{-}4)$$

As we integrate through the 2π rad of longitude, let us also write $r_\varphi = R \cos \varphi$, where R is the radius of the earth.   We then obtain

$$\mu_\varphi = \frac{R^2 \cos^2 \varphi}{g} 2\pi \int_0^{po} (\Omega R \cos \varphi + u) \, dp \qquad (9\text{-}5)$$

In the general circulation the momentum in the ring may be considered to be transported northward by the mean value around the ring of the velocity component v, positive northward, which we shall designate as $\bar{v}$; so the net transport is the combined mean around the ring of the u and v components, or

$$\frac{2\pi R^2 \cos^2 \varphi}{g} \int_0^{po} (\Omega R \cos \varphi \cdot \bar{v} + \overline{uv}) \, dp \qquad (9\text{-}6)$$

Examining the first term under the integral sign, we recognize that if $\bar{v}$ is positive there will be a net accumulation of air to the north and a net deficit to the south, and vice versa for a negative (southward) transport.   Leaving out of consideration the slow seasonal changes in mass distribution due to seasonal thermal effects, we recognize that such accumulations and deficits cannot occur.   Therefore, only the second term in the integral needs to be considered. It should be emphasized, however, that if a limited depth rather than the whole atmospheric depth is considered, the first term of the integral can be very large.

It is seen that northward transfer of positive (west-to-east) relative momentum occurs when the u and v components taken in the mean around the ring are both positive or both negative. If u is negative (component from the east), v must also be negative to transfer a deficit of momentum (negative, east-to-west momentum) southward, which has the effect of transferring momentum (positive) northward. Thus the net meridional transfer of angular momentum depends on a correlation between the u and v components. It is interesting to note that in a sinusoidal wave pattern in the westerlies the northwest-wind negative would cancel the southwest-wind positive.

In the waves in the upper westerlies the correlation occurs when the wind west of the troughs has an east-to-west component (u and v both negative) and the wind east of the troughs has a westerly component (southwest wind, u and v both positive). The southwest correlation is commonly observed, but the northeast correlation is not so frequent. Such a system of "tilted" waves was found by Starr and others[1] to be insufficiently predominant to account for all the momentum transfer. A more important mechanism for exchange is found in the disturbances of cyclonic and anticyclonic scale.

In several treatments of the subject of the role of cyclones and anticyclones, meteorologists[2] have found it convenient to split the applicable second part of the integral in Eq. (9-6) into $\overline{uv}$ representing the *circulation flux* resulting from the mean meridional motion $\bar{v}$, and $\overline{u'v'}$ representing an *eddy flux*. The latter is accomplished by disturbances in the mean flow. The combined terms produce the maximum northward flux of angular momentum in the Northern Hemisphere at approximately 30° latitude. Between that latitude and the tropics there is a net $\bar{v}$ as a characteristic of a net direct, meridional component in the tropical circulation cell. Furthermore, at 30° very strong winds prevail in the upper troposphere in the form of the subtropical jet stream, giving a maximum value of $\bar{u}$ there. Farther north, in middle latitudes, the eddy-flux term dominates. Although half the surface of the earth lies at latitudes lower than 30°, the northward flux of angular momentum in winter, and especially in summer, is slightly greater north of 30° than south of that latitude in the Northern Hemisphere.

---

[1] V. P. Starr and R. M. White, A Hemispherical Study of the Atmospheric Angular-Momentum Balance, *Quarterly Journal of the Royal Meteorological Society*, vol. 78, pp. 215–225, 1951. E. Palmén, The Role of Atmospheric Disturbances in the General Circulation, *Quarterly Journal of the Royal Meteorological Society*, vol. 77, pp. 337–354, 1951.

[2] For a summary of treatments and findings see, for example, E. Palmén and C. W. Newton, "Atmospheric Circulations," Academic Press, Inc., New York, 1969.

ENERGY EXCHANGE

The baroclinic nature of the total atmosphere is associated with a field of potential energy. The general circulation operates to transform this potential energy into kinetic energy. The *mean* circulation data do not show a total kinetic energy of sufficient magnitude to accomplish the transformation. Part of the kinetic energy is found in the cyclonic-anticyclonic systems which are "painted out" in the mean picture. The disturbances, by distorting the field of temperature and solenoids, create potential energy and thus solenoidal fields of their own. While the mean solenoidal field is meridional, that of the disturbances is characterized by a zonal component. Under the effects of the coriolis force the horizontal air motions resulting from these zonal solenoidal fields become meridional in direction.

The mean motions are so lacking in meridional flow that they appear to preserve the global potential-energy distribution more than to transform it into the kinetic energy which would be required to balance the buildup of potential energy. Unstable flow of the disturbance type is necessary to provide the balance of meridional energy exchange. The predominance of zonal motion which results from the earth's rotation greatly limits the efficiency of the atmospheric engine.

JET STREAMS

Fluid-model experiments agree with the observed motions in the atmosphere and in the oceans in showing a pronounced concentration of west-to-east motion in a narrow band. In the atmosphere there are at least two such concentrated currents circling the earth in the vicinity of the tropopause in middle and subtropical latitudes. They are called *jet streams*. On mean charts and vertical cross sections, especially for the winter season, only the subtropical jet stream appears, as shown in Fig. 9-6. This cross section was constructed by Petterssen[1] from temperature and zonal-wind distributions averaged for all the meridians of the Northern Hemisphere in winter. The jet maximum of 40 m per sec is located at about latitude 24° and at a height around 13 km.

At any given observation time along any chosen meridian one usually finds a much stronger jet in winter at some middle latitude. The reason that this second jet does not show up on mean cross sections is that it meanders and fluctuates through various latitudes with the major perturbations of the upper westerlies. In parts of these perturbations the jet may produce extreme meridional flows which would not appear strong in a cross section as in Fig. 9-6, which depicts only the zonal component.

[1] S. Petterssen, *Royal Meteorological Society Centenary Proceedings*, 1950, pp. 120–155.

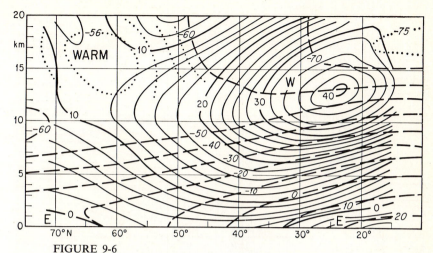

FIGURE 9-6

Meridional cross section of zonal component of the geostrophic wind in meters per second (solid lines) and mean isotherms in degrees Celsius (broken and dotted lines), both for winter season. Data are means for all longitudes in the Northern Hemisphere. (*After Petterssen.*)

This meandering jet is associated with the main low-troposphere fronts and is typically part of the break in the tropopause which so frequently characterizes the transition from subtropical to polar conditions as outlined in Chap. 4. In its equatorward sweeps this jet stream, sometimes called the *polar-front jet*, often merges with the subtropical jet. In daily charts a pattern of confluence and diffluence[1] is common. It is not unusual to find winds of over 100 m per sec (225 mph) in these jet streams. The subtropical jet is generally centered at a higher altitude than the polar-front jet.

Although on mean charts only the subtropical jet appears, on any given day both are apparent, even when average positions are taken around the entire hemisphere. A representation of this type was constructed by Defant and Taba[2] showing in cross section the average meridional distributions in the Northern Hemisphere on January 1, 1956, as represented in Fig. 9-7. The two jet streams are located near the northern and southern termini of a middle-latitude tropopause. The relation of the polar-front jet to the front itself is clearly seen, and the three tropopause segments—polar, middle, and tropical—are sharply delineated by discontinuities in the temperature field.

[1] Confluence means the joining of two flows, as at the confluence of two rivers; diffluence means flowing apart. See discussion of Eq. (8-80).

[2] F. Defant and H. Taba, The Threefold Structure of the Atmosphere and the Characteristics of the Tropopause, *Tellus*, vol. 9, pp. 259–274, 1957.

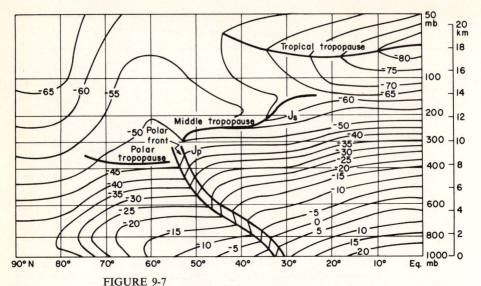

FIGURE 9-7

Meridional cross section of mean conditions on a single day (January 1, 1956) around the Northern Hemisphere. The polar and subtropical jet streams are designated by J_p and J_s, respectively. Isolines are temperatures, degrees Celsius. The double line between 30 and 55° is the mean position of the polar front. Note the three identifiable tropopauses. (*After Defant and Taba.*)

A vertical cross section along approximately the 75th meridian, from Hudson Bay to the mid-Caribbean, on January 15, 1959, is shown in Fig. 9-8 simplified from a representation by Newton and Persson.[1] Here the two jet streams are quite pronounced. In this case the temperature field suggests two tropical tropopauses. Although there are great differences between one situation and another, this model is now recognized as characteristic. It is also noted that the subtropical jet is spread out over a greater range of latitude and is more confined in altitude than the polar-front jet. The middle tropopause under the subtropical jet is often seen as an upper-level front sloping downward toward the tropics and separating warm upper-troposphere air of the tropics from somewhat cooler subtropical air. It is interesting to note that in the simple depiction of the meridional circulation cells in Fig. 9-1, convergence is indicated between the tropical cell and the middle cell at upper levels, which would be favorable for generating a front there.

A tropical east-wind jet stream is observed in summer mainly in Asia at about latitude 15°N. Flowing in a relatively fixed position at a height of about 16 or 17 km,

[1] C. W. Newton and A. V. Persson, Structural Characteristics of the Subtropical Jet Stream and Certain Lower-stratospheric Wind Systems, *Tellus*, vol. 14, pp. 221–241, 1962.

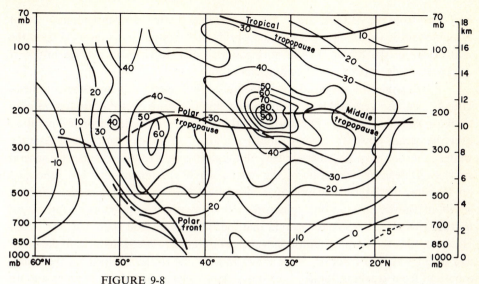

FIGURE 9-8
Vertical cross section along approximately the 80th meridian west on January 15, 1959. Isotachs (meters per second), fronts, and tropopauses are represented. (*After Newton and Persson.*)

it exhibits speeds up to 35 m per sec in the core. Bands of strong winds are also detected in the upper stratosphere, the mesosphere, and beyond. One that has received attention is a jet stream moving just south of the Arctic Circle during the polar night with very high speeds at heights of 25 to 30 km.

MONSOON CIRCULATIONS

We return now to further consideration of the low levels and a direct thermal circulation of a seasonal nature that develops in those regions where thermally produced continental highs in winter or lows in summer dominate the circulation around a continent. Such circulations are called *monsoons*. An ideal monsoon would be characterized by air flowing inward toward the thermal low of a continent in summer and outward from a continental high in winter. This landward flow produces heavy rains in the damp sea air, called *monsoon rains*. Some of the rain is caused by direct convection as the air comes in over the summer-heated land; but where the monsoon rains are most pronounced, for example, in India, there is a convergence zone between the monsoon current and another wind system. In India this convergence takes place between the south and southwest monsoon current and a current from the northwest from the Arabian Sea. The effects are accentuated by mountains. What is reputed to be the rainiest place in the world, Cherrapunji, India, at 4455 ft on the south slope of the Khasi Hills in northeastern India, receives an average of $35\frac{1}{2}$ ft of rain annually, mostly in the summer monsoon.

In the winter monsoon, dry continental air is brought out from the continent. Usually, it is also very stable in the beginning and therefore does not produce rainfall until it travels out to sea. The warmth and high moisture content from evaporation over the ocean, however, quickly modify the air. On the eastern coast of Asia the winter monsoon is quite generally felt, although it is interrupted at times by general storms. It deposits heavy rain on the Asiatic side of the Japanese islands, especially on Honshu, the largest and principal island, which is very mountainous. Farther south, in Taiwan and the Philippines, the winter monsoon appears as a pronounced northeast wind. It is sometimes said that the Indian monsoon is an ideal monsoon. However, only the summer monsoon is present in India. The Himalayas to the north and the mountains of Burma, Yunan, and Indo-China to the east prevent the winter monsoon from reaching India. The only region where both kinds of monsoon are felt in a pronounced way is in the vicinity of Taiwan and the East China Sea, with southerly winds in summer and northerly in winter.

No continents other than Asia show appreciable monsoon effects. In the Southern United States southerly winds predominate in summer, bringing air from the Gulf of Mexico, while northerly components prevail in winter. However, in both summer and winter the circulation is highly variable, and monsoon effects are noticeable only to the practiced eye. In India and southern Asia even the laborer in the rice fields is aware of the monsoon and depends on its regularity to save him from famine.

THE OCEANIC ANTICYCLONES

The high-pressure cells over the oceans in the horse latitudes are the most permanent features of the general circulation. They are sometimes referred to as "centers of action," although they are characterized by inaction. The name comes from their importance as pivot areas around which storms circulate; that is, when the center is displaced abnormally far to the north, storminess prevails farther north than usual, and when it is displaced to the south, the storms enter the continents in more southerly latitudes. These anticyclones are large and permanent and exist over the ocean where the greatest store of moisture is available.

The Pacific and Azores anticyclones are great areas of subsiding air. The circulation is such that the subsidence effects are most noticeable in the eastern parts of these highs. In the western regions of the oceanic highs, convergence and ascent appear to be more prevalent. These differences produce differences in the thermal structure of the air. In the eastern portion, subsidence produces a strong inversion that usually lies at about 500 to 1500 m above sea level. The air above this inversion is very dry, and as a consequence of the dryness and thermal stability, the continents on the eastern sides of the oceans (west coasts) near latitude 30°N and S are markedly deficient in rainfall. Some of the most arid regions of the world are found immediately at the ocean in these places, for example, the Sahara Desert in North Africa, the Kalahari Desert in South Africa, the deserts of Western Australia, and the coastal deserts of northern Chile and Lower California. The western side of the oceanic highs is characterized by moist, conditionally unstable air and quite ample rains. Through the eastern part of the trade-wind belt the inversion is gradually carried to greater heights, often disappearing by the time the western part of the anticyclone is reached.

EXERCISES

1 Assume conservation of angular momentum for a particle from 80°N, wind calm, displaced to 60°N. Compute the resulting wind speed and direction.

2 In Eq. (9-6) compare the preintegral term at 30° latitude with that at 60°. If u has a maximum near 30° in the subtropical jet stream, how might that factor also affect the total expression?

3 Determine a reasonably representative value of the vorticity from the horizontal shear to the north and south of the centers of the two jet streams in Fig. 9-8. Also determine an average value of vorticity through the polar-front zone at 500 mb.

10

VERTICAL STRUCTURE OF CYCLONES AND ANTICYCLONES

In the preceding chapters we have put together the building blocks of atmospheric structure and dynamics. We have seen that, as in all natural systems, everything fits together. Each process is related to others either directly or indirectly, giving us a choice of ways of describing what goes on in the atmosphere, all resulting from the unequal heating of our rotating planet. A picture of the general circulation has been developed from principles of fluid dynamics, thermodynamics, and statics. We are now ready to examine the more important secondary circulations, starting first with building blocks of extratropical cyclones and anticyclones.

Rotating-pan experiments and aerological observations show the polar-front jet stream to be in a strongly baroclinic region. When wave-like perturbations develop in this meandering current, the frontal boundary is likewise displaced. At least in the upper and middle troposphere, the polar air is drawn equatorward in the troughs and the subtropical air is carried northward in the ridges. The tropopause is similarly displaced so that low tropopause conditions prevail in the troughs while the ridges have a high tropopause. Graphically speaking, one can say that the jet stream in its meanderings carries the isotherms with it. The strongest temperature gradients are observed across the jet, irrespective of the wave-like displacements.

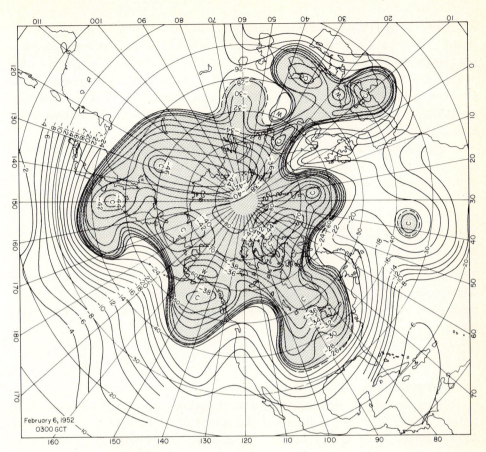

FIGURE 10-1
Isotherms at 500 mb, February 6, 1952, 0300 GCT for the Northern Hemisphere, showing front, *heavy line*, in vicinity of strongest temperature gradient. Shaded areas represent polar air.   Note cutoff cold air mass in middle Atlantic.

In Fig. 10-1 a chart used by Bradbury and Palmén[1] is reproduced to illustrate the consistent pattern of the polar front extending all around the Northern Hemisphere at middle-troposphere levels (500 mb).   Although the level represented here is considerably below the height of the core of the jet stream, the front is nevertheless associated with it and follows its meanderings.

By integrating the hydrostatic equation upward or downward from north and south of the core of the jet stream, one would find that the pressure in the cold air

[1] D. L. Bradbury and E. Palmén, On the Existence of a Polar-Front Zone at the 500-mb Level, *Bulletin of the American Meteorological Society*, vol. 34, pp. 56–62, 1953.

occupying the troughs would increase rapidly downward. Upward it would decrease rapidly until the warm subpolar stratosphere is reached, after which it would decrease less rapidly than in the cool subtropical stratosphere. In the warm ridges, the rate of change with height would be relatively smaller, except in the stratosphere. At a level considerably below or above, the troughs and ridges could become completely smoothed, reversed, displaced, or otherwise modified because of the differences in hydrostatic weight between cold and warm air.

Since the air motion is approximately geostrophic, we can arrive at these changes in pressure pattern with height by applying the expressions for the change of the geostrophic wind with height.

VERTICAL STRUCTURE AS SHOWN BY THE THERMAL WIND

The thermal wind between any two pressure levels is defined from Eq. (7-65) as the vector difference of the geostrophic wind at the upper level minus that at the lower level, given by the horizontal gradient of the thicknesses between the two pressure surfaces.

$$\vec{c}_T = -\frac{1}{f}\frac{\partial(\Delta\Phi)}{\partial n}$$

where $\Delta\Phi$ is the thickness in geopotential units.

It has already been indicated that lines of thickness are likely to coincide with contour lines of the pressure surfaces in the vicinity of the jet stream, such as at the 500-mb surface. One also finds that when the entire layer between 500 mb and 1000 mb is considered, the gradient of the thicknesses is often about the same as the slope of the 500-mb contours. This would mean that the thermal wind for this very thick layer would not be very different from the 500-mb wind. In magnitude, the wind in the jet core may be several times the geostrophic wind at 1000 mb, in agreement with approximate similarity in magnitude between the thermal wind and upper wind. The thickness lines often have a slightly different orientation from the 500-mb contours, so that, even though it is of about the same magnitude as the 500-mb wind, the thermal wind has a slightly different direction. The vector difference between the two results from a 1000-mb wind vector with a direction quite different from that of the 500-mb wind. A common case is that in which polar easterlies are observed at the ground under a westerly jet. Under these circumstances the thermal wind vector would exceed in magnitude the wind in the jet core.

The two charts in Figs. 10-2 and 10-3 illustrate the normal situation in which the 500-mb contours and the thickness lines between 1000 and 500 mb almost coincide. The difference between the pressure-contour patterns of the 1000- and 500-mb

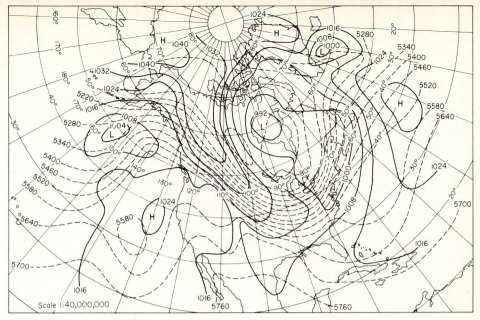

FIGURE 10-2
Sea-level pressure, millibars (solid lines), and isopachs for the 1000- to 500-mb thickness, meters (dashed lines), 0000 GCT January. 31, 1971, over North America.

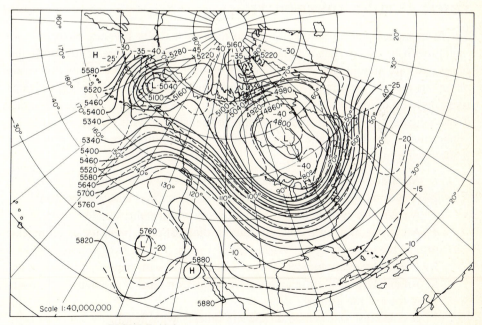

FIGURE 10-3
Contours and isotherms of the 500-mb surface corresponding to the chart of Fig. 10-2.

237

surfaces represented separately in the two figures is characteristic. It may be of interest to the student to use the three sets of lines (500-mb and 1000-mb contours and thickness lines) to construct the corresponding vector triangles at a few points of the maps.

The change in the geostrophic wind going upward in the stratosphere smooths out the perturbations. This should be apparent when one considers that warm stratospheric conditions overlie the cold troughs of the troposphere and that the warm tropospheric ridges have a high, cold stratosphere above. The pressure decreases slowly with height in the polar stratosphere and falls off rapidly with height in the cold subtropical stratosphere. As this tendency continues upward, the pressure slopes can approach zero and finally reverse. The increase of temperature with height in the subtropical stratosphere results in a diminution of the meridional temperature gradient at a certain height; so reversals of the flow pattern are not commonly observed. One cannot carry this line of thought upward too far, for reasons that become apparent in studies of the physics of the outer atmosphere.

REFLECTIONS ON PRESSURE AND TEMPERATURE DISTRIBUTIONS

The hydrostatic equation points directly to the fact that the pressure changes more rapidly with height in cold air than in warm air. Since pressure centers, troughs, ridges, etc., are only identifiable relative to their surroundings, that is, only by virtue of the fact that the pressure is higher or lower than at other nearby places at the same level, they can disappear, intensify, or change position relative to their surroundings if the weights $p_0 - p_z$ or thicknesses vary horizontally. With a little reflection or application of the hydrostatic equation to the various situations, one can arrive at the following rules:

1 A surface warm-core cyclone disappears quickly with height.

2 A cyclone aloft in a warm-air column increases in intensity downward.

3 A cold-core cyclone increases in intensity with height.

4 A cyclone aloft in a cold-air column diminishes in intensity or disappears toward the ground.

5 A warm-core anticyclone increases in intensity with height.

6 An anticyclone aloft in a warm-air column diminishes in intensity or disappears toward the ground.

7 A cold-core anticyclone disappears quickly with height.

8 An anticyclone aloft in a cold-air column increases in intensity downward.

9 Low-pressure centers are displaced with height toward the colder air.

10 A low-pressure center aloft is found in lower levels in the direction of the warmer air.

11 High-pressure centers are displaced with height toward the warmer air.

12 A high-pressure center aloft is found in lower levels in the direction of the colder air.

13 At the surface of the earth, high pressure is favored in cold regions and low pressure in warm regions, but dynamic effects may outweigh the thermal ones, e.g., in the oceanic anticyclones.

DYNAMIC EFFECTS

The static relationships just described determine the structure of the atmosphere and its perturbations but tell us nothing about the mechanics of change. From the static relationships we can expect that surface cyclones will be found near the positions of the mid-troposphere troughs but displaced toward the warmer air. They are generally found to the east of the troughs because that is where warm-air advection is most pronounced, as a thermal-wind hodograph will show, while advection of cold air is common to the west. All factors operate to form an asymmetrical dome of cold air under the trough sloping downward steeply toward the warm air to the east, more gradually to the west. An idealized situation consists of a cold front to the east and a warm front at a considerable distance to the west.

Only the dynamic effects can explain why and where in relation to the trough a surface cyclone will develop. In the first place, since a surface cyclone must produce convergence as soon as its cyclonic vorticity is generated, as shown in Eq. (8-86), it can deepen and intensify only if the divergence of mass aloft exceeds this low-level convergence. The tendency equation (8-93) applied to the surface requires that to produce a local pressure decrease the integral

$$\int_0^\infty \operatorname{div}_2 (\rho c) \, dz \qquad (10\text{-}1)$$

must be positive or that there must be a net horizontal mass divergence in the atmospheric column above the place in question.

From a consideration of the vorticity relationships treated in Chap. 8 and as illustrated in Fig. 8-2 one notes that normally the absolute vorticity should be at its maximum in the trough lines of the mid-troposphere westerlies, decreasing as the air returns northward on the eastern side.

With the absolute vorticity $(f + \zeta)$ designated as ζ_a, Eq. (8-87) may be written as

$$\nabla_2 \cdot \mathbf{c} = -\frac{1}{\zeta_a} \frac{d\zeta_a}{dt} \qquad (10\text{-}2)$$

This time rate of change of the vorticity is made up of (1) the vorticity advected by the air particles moving along the streamlines with wind speed c, and (2) the moving along of the vorticity field characteristic of the upper wave pattern. The first is given by $c\, \partial \zeta_a/\partial s$, where s is distance along the streamline, positive downwind. The second is $-c_s\, \partial \zeta_a/\partial s$, where c_s is the s component of velocity of the isolines representing the vorticity field. The negative sign indicates that the change at a point is positive if ζ_a decreases downstream. We may then write

$$\nabla_2 \cdot \mathbf{c} = -\frac{1}{\zeta_a}(c - c_s)\frac{\partial \zeta_a}{\partial s} \qquad (10\text{-}3)$$

Commonly in the mid-troposphere waves the wind blows through the troughs and ridges such that $c \gg c_s$. This means that to the east of the trough, where ζ_a is decreasing and $c\, \partial \zeta_a/\partial s$ is negative, the first term dominates to produce a positive $\nabla_2 \cdot \mathbf{c}$, that is, divergence. If the trough is part of the jet stream, the first term can be very large. Near the surface the wind is weak; so the second term becomes important, tending to produce negative $\nabla_2 \cdot \mathbf{c}$ or convergence. But near the jet stream or any strong flow through a mid-troposphere trough the divergence aloft easily exceeds the surface convergence. Add to this the fact that this is a strong baroclinic region and you find conditions favorable for the development of a cyclone. Thus, for example, the development of a cyclone in Fig. 10-2 in the Mississippi Valley could be predicted.

Taking the case to the west of the trough where the vorticity is increasing, one finds that mid-troposphere convergence is to be expected. A cold anticyclone is formed at the surface somewhere in this region, depending on the temperature and wind distribution.

In Fig. 10-1, the cutoff center of cold air in the middle Atlantic represents a cutoff low with a cyclonic circulation around it more or less following the isotherms. This type of disturbance is not uncommon in the atmosphere, and is noted for its production of bad weather without frontal or cyclonic manifestations at the surface. In these cases the mid-troposphere vorticity is associated with convergence at those levels. The cold air subsides from these levels and draws the tropopause down to 400 mb or lower. The low pressure is associated with a warm, low stratosphere. The main outflow which balances the convergence occurs in the stratosphere, where static equilibrium would require the presence of an anticyclonic cap.

THE NATURE OF FRONTS

Fronts were first studied from the point of view of an observer at the ground. It is easy to make measurements and observations concerning fronts at the surface of the earth for obvious reasons of accessibility; hence a great deal of detail is known about them as they occur at the bottom of the atmosphere. This is fortunate, because their structure and behavior appear to be more complicated there than in the free atmosphere.

The picture of the continuous polar front at 500 mb as represented in Fig. 10-1 is quite different from that found at the ground. From the remarks on the preceding pages about the changes in wind and pressure distributions between the 300- or 500-mb and 1000-mb surfaces, one would expect to find a difference in frontal distributions. The change in frontal patterns between the two surfaces is greater, however, than can be accounted for in terms of the change in pressure and wind distributions. Several reasons can be given for this situation. One condition that appears frequently in the low levels, particularly over and near continents in winter, is the tendency for solenoidal concentrations to develop or to disappear somewhat independently of the upper jet-stream concentration. Layers of temperature inversion, which are quite common in the low levels, can become frontal zones under some circumstances.

In the previous chapter the representations in vertical cross sections show a single frontal zone, associated with the polar-front jet stream, extending from the tropopause to the surface of the earth. When viewed on a smaller scale, the atmosphere exhibits a number of shallow fronts in addition to the main polar front.

In the simplest sense fronts may be defined as the boundary surfaces separating masses of air of differing densities. While the air masses may exhibit changes in their properties measured over considerable distances, these changes are negligible when compared with those occurring across a front. To maintain this situation the winds cannot be divergent in the vicinity of the front. If they were, they would cause the density or specific-volume field to attenuate rather than to concentrate as required. Fronts in the upper air are located in regions of maximum cyclonic shear $(-\partial p/\partial n)$, whereas at the ground they are most frequently found in troughs, although they can be maintained in zones of strong cyclonic shear. In either case the vorticity is positive—that is, cyclonic.

In the idealized case it is convenient, although wrong, to think of a front as a mathematical discontinuity in property. Actually there is always a transition zone of a certain width, perhaps from 10 to 20 km, where the air is of neither one kind nor another. However, in the scale of the usual synoptic map, 1 : 5,000,000 to 1 : 20,000,000, we are justified in marking the front as a broad pencil line, as is customary.

The front represents a temperature discontinuity, for it is only by differences in temperature that differences in density at any given level (or pressure) can be preserved, except for the slight effect on densities caused by varying amounts of water vapor. Since the air follows the ordinary laws of fluid statics and dynamics, it might be expected that the surface of separation would be horizontal, as in the case of two fluids of different density in a tank, such as oil and water. The air masses can have a sloping boundary separating them and still be in equilibrium with respect to each other. This is because there are other forces besides that due to gravity which come

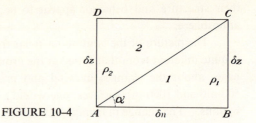

FIGURE 10–4

into play where atmospheric motion is concerned.   Most important of these forces is the coriolis force.

Under atmospheric conditions the denser air forces itself underneath the warmer, lighter air, just as in the case of laboratory fluid experiments.   But the colder air underlies the warm in a wedge-like fashion, the colder current moving on the ground with the warm air above its sloping upper surface.   The representation of fronts on any surface amounts simply to a charting of the line of intersection of the fronts with that surface.

From given wind and temperature conditions in two air masses at a given latitude, it is possible to compute the slope of the front between them.   This was first done by Margules[1] from methods developed previously by Helmholtz.   A simple derivation of the equation for the equilibrium slope can be obtained from a consideration of the horizontal pressure gradient and static relationship above and below the front shown in vertical cross section in Fig. 10-4.   In this figure, we may consider the rectangle $ABCD$ to be of infinitesimal size, having vertical and horizontal sides δz and δn.   Using the natural coordinate δn, positive to the left of the wind, it is seen that the wind must be moving through this cross section in a direction outward from the page.   The slope of the front from A to C would be $\tan \alpha$ or $\delta z / \delta n$.

Our problem first is to find the pressure gradient along AB and along CD in terms of the density of the two air masses and the slope of the surface, using the fact that the pressure gradient along AC is common to both air masses.   The lower, colder air mass is designated by the subscript 1 and the upper, warmer mass by the subscript 2.   We use p with appropriate subscripts to denote the pressure at points in question.

Applying the hydrostatic equations, we describe the pressures at the corners of the rectangle as

$$p_B = p_C + \rho_1 g \, \delta z \qquad (10\text{-}4)$$

$$p_D = p_A - \rho_2 g \, \delta z \qquad (10\text{-}5)$$

[1] M. Margules, Über Temperaturschichtung in stationär bewegter und ruhender Luft, *Meteorologische Zeitschrift*, Hann-Band, p. 243, 1906.

Then, introducing the horizontal pressure gradients and substituting the above two expressions for p_B and p_D, we have

$$\frac{p_B - p_A}{\overline{AB}} = \left(\frac{\delta p}{\delta n}\right)_1 = \frac{p_C - p_A}{\delta n} + \rho_1 g \frac{\delta z}{\delta n} \qquad (10\text{-}6)$$

$$\frac{p_C - p_D}{\overline{DC}} = \left(\frac{\delta p}{\delta n}\right)_2 = \frac{p_C - p_A}{\delta n} + \rho_2 g \frac{\delta z}{\delta n} \qquad (10\text{-}7)$$

We subtract Eq. (10-7) from Eq. (10-6) to obtain

$$\left(\frac{\delta p}{\delta n}\right)_1 - \left(\frac{\delta p}{\delta n}\right)_2 = \rho_1 g \frac{\delta z}{\delta n} - \rho_2 g \frac{\delta z}{\delta n} = g(\rho_1 - \rho_2)\frac{\delta z}{\delta n} \qquad (10\text{-}8)$$

in which $(p_C - p_A)/\delta n$, which is a function of the pressure gradient along the frontal surface, has been eliminated. We then write

$$\frac{\delta z}{\delta n} = \tan \alpha = \frac{(\delta p/\delta n)_1 - (\delta p/\delta n)_2}{g(\rho_1 - \rho_2)} \qquad (10\text{-}9)$$

according to the gradient-wind equation,

$$\left(\frac{\delta p}{\delta n}\right)_1 = \rho_1 \left(\pm \frac{c_1}{r_1} - f c_1\right) \qquad (10\text{-}10)$$

with a similar expression for $(\delta p/\delta n)_2$, in which r is the radius of curvature and c is the component of the wind in a direction along the front with n positive to the left. By substitution, Eq. (10-9) becomes

$$\tan \alpha = \frac{\rho_1(\pm c_1{}^2/r_1 - f c_1) - \rho_2(\pm c_2{}^2/r_2 - f c_2)}{g(\rho_1 - \rho_2)} \qquad (10\text{-}11)$$

Experience has shown that the centripetal term may be neglected in most determinations and that the wind may be considered as geostrophic. It is noteworthy, however, that in a cyclonic circulation c^2/r is positive and thus may contribute to a steeper slope of the front, while an anticyclonic circulation would make this term negative and the slope would be less than for straight-line motion. For geostrophic conditions, we have

$$\tan \alpha = f \frac{\rho_2 c_2 - \rho_1 c_1}{g(\rho_1 - \rho_2)} \qquad (10\text{-}12)$$

from the equation of state, $\rho_1 = p_1 m/RT_1$ and $\rho_2 = p_2 m/RT_2$, but at any point on the frontal surface, $p_1 = p_2$; so Eq. (10-12) becomes

$$\tan \alpha = f \frac{T_1 c_2 - T_2 c_1}{g(T_2 - T_1)} \qquad (10\text{-}13)$$

If the temperature difference $T_2 - T_1$ is not too large and the wind difference is appreciable, we may take the mean of the two temperatures in the numerator and write

$$\tan \alpha = \frac{fT_m}{g} \frac{c_2 - c_1}{T_2 - T_1} \qquad (10\text{-}14)$$

Considering that the pressures are the same at any point on the two sides of the front, we may substitute potential temperature directly for the temperature to obtain

$$\tan \alpha = \frac{f\theta_m}{g} \frac{c_2 - c_1}{\theta_2 - \theta_1} \qquad (10\text{-}15)$$

This last form is useful, especially in making computations from vertical soundings through the front. An inversion or stable layer usually separates the lower, colder air from the upper, warmer air. The potential temperature at the bottom of the inversion characterizes the colder air and the potential temperature at the top of the inversion represents the warm air.

In the natural coordinates used here we have taken n as positive in the direction of the cold air. Positive c is 90° clockwise from n, therefore parallel to the front with n pointing to the left. The requirement that the warm air must lie above the cold air can be met only if $\delta z/\delta n$ is positive. The sign of $\delta z/\delta n$ is determined by the wind difference alone, because with T_2 representing the warmer air, the denominator is always positive. With these conditions, we find the following general rule:

The wind change through a front must be such that the component parallel to the front, taken positive with cold air to the left, increases to the right.

This type of shear is cyclonic ($\partial c/\partial n < 0$), representing positive vorticity. In most cases at the surface, the wind components on the two sides of the front are opposed, with one negative and the other positive. In the usual situation, with n pointing toward the cold air, c_2 in the warm air will have a positive component and c_1 in the cold air a negative one. A corollary of the general rule just stated is the following:

The vorticity is positive (cyclonic) in the frontal zone.

The vorticity can be recognized by the cyclonic curvature or the cyclonic shear or both. Two practical rules for locating fronts are:

At the surface, fronts occur mostly along low-pressure troughs, although they are possible anywhere if there is a cyclonic wind shear, and not possible along a ridge line or in an anticyclonic shear zone.

In the westerlies of the middle and upper troposphere, the main polar front is to the left (generally north) of the maximum wind or polar-front jet stream.

These rules may be applied to the Southern Hemisphere by recognizing the opposite sense of f and ζ, or by some other suitable adjustment.

MOVING FRONTS

If a front has a component of the wind normal to it, then it should, in the ideal case, move with the speed of and in the same sense as that component. If gradient-wind conditions prevailed, a front would move whenever and wherever it was crossed by isobars, and the spacing of the isobars along the front would give the rate of movement in accordance with the gradient-wind equation. Owing to the effects of friction, fronts move with a speed slightly less than that of the normal gradient-wind component. Since pressure is continuous across a front, the normal component of the wind immediately ahead of the front must be the same as that immediately behind. Fronts in the ideal sense may be considered as separating walls in the moving air streams through which the air particles cannot move but which must move along at the same speed as the normal component of the air particles.

When a front moves toward the warmer air, so that the cold air is occupying territory formerly covered by warm air, it is called a *cold front*. If it is moving toward the colder air, with warm air occupying territory formerly covered by cold air, it is called a *warm front*. A front may be moving in one direction at one portion of its length and another direction at another sector; therefore, it may be partly a warm front and partly a cold front.

In a cold front, the wedge of cold air is moving actively forward, and the effect of surface friction is to hold back the part near the ground so that it tends to steepen the front. Sometimes this frictional steepening produces an action like that produced on a wave reaching a beach, causing the upper part to tumble over the lower part in an unstable, top-heavy interaction and mixing of the air masses. The normal condition is shown in Fig. 10-5. The average slope of cold fronts is between $\frac{1}{50}$ and $\frac{1}{150}$. In the illustration, the vertical dimension is exaggerated some five hundred times with respect to the horizontal.

In a warm front, the cold-air wedge is receding and the effect of surface friction is to hold back the front near the ground so that it trails with a small slope, as shown in Fig. 10-6, also exaggerated some five hundred times in the vertical. The average slope of warm fronts is between $\frac{1}{100}$ and $\frac{1}{300}$ with occasional fronts with lesser slopes still discernible at the ground. Sometimes the portion of a warm front near the ground is drawn out such a great distance that some cold air may remain in a shallow layer near the surface after the effective warm front has passed a station. This often makes the exact location of the front difficult to trace on the surface chart. At times this shallow layer is heated by solar radiation or mixes by mechanical turbulence with the upper air in such a way as to cause an apparent movement of the surface position of the warm front that is much faster than the actual movement of the air masses. This effect is especially noticeable in the lee of mountains where the warm air does not readily come down to sweep out the cool shielding layer.

Since at least a kink in the isobars in the form of a cyclonic trough occurs at a

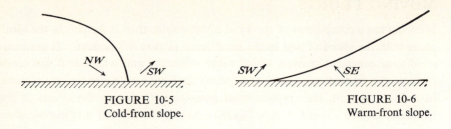

FIGURE 10-5
Cold-front slope.

FIGURE 10-6
Warm-front slope.

front, one observes at a station with a front passage one of the following pressure changes:

1 Falling, then rising
2 Falling, then falling less rapidly
3 Rising, then rising more rapidly

The pressure change before or after a front may be zero, in which case the change on the other side is different in relatively the same sense as the above.

The windshift observed at a station in the Northern Hemisphere with the passage of a front is always such that the wind should veer (turn clockwise) with time. This must always be the case in a trough that is moving with the wind. The fronts must move around the lows in a counterclockwise rotation, and a few graphical considerations on a synoptic chart will prove that as the fronts pass a station the wind must turn in a clockwise direction (veer) with time. A veering wind does not always mean a frontal passage. The wind also veers with the passage of a high-pressure wedge from the west with the anticyclonic center to the north. In an anticyclonic condition, a long period of calm occurs with the change in wind, while in fronts the wind may change almost instantaneously with a very short period of calm or no calm at all.

CYCLONES AND THE UPPER WAVES

It is apparent from the discussions in the preceding pages that the waves in the middle and upper troposphere are associated with surface cyclones. Except near the ground, the troughs are occupied by cold air and the ridges by warm air. It has been explained that the surface troughs or cyclones do not lie directly under the upper troughs. They tend to occur below the forward or eastern portion of these eastward-moving upper troughs. This characteristic is found in the hydrodynamic-model experiments as well as in the natural atmosphere.

Because the mid-troposphere waves are a striking feature of upper-air charts, it is reasonable to associate surface cyclones with energy transformations initiated by these waves.   In some cases, however, there is clear evidence, both in the atmosphere and in models, of disturbances building upward from the surface of the earth and apparently causing the development of the upper waves or centers.   Historically, the lower disturbances were studied first.   After many years of observation and experience a greal deal is known about them; so it is convenient to study fronts and cyclones as they have been identified near the surface.

FRONTS AND THE LIFE CYCLE OF CYCLONES

The existence of approximate discontinuities in the atmosphere and to some extent the association of these with extratropical cyclones were vaguely known to meteorologists of the nineteenth century.   The systematic relationship between fronts and cyclones and the application of these concepts to daily synoptic analysis were developed by meteorologists working at the Geophysical Institute in Bergen, Norway, during and shortly after World War I.   The names of V. Bjerknes, J. Bjerknes, H. Solberg, and T. Bergeron[1] are associated with these early discoveries.   During the war years 1914–1918 the Norwegians were cut off from weather reports from surrounding areas, especially the oceans.   In order to make up for this deficiency, their government established in Norway itself a dense network of observing stations, located every few miles.   Through the preparation of synoptic charts for this region the details of extratropical cyclones and fronts were obtained.   The so-called "polar-front theory" of cyclones was evolved.   This theory recognizes that extratropical cyclones develop on the "polar front" separating the westerlies from the polar winds. Later it was discovered that several fronts may exist, each with cyclone-forming potential.   Several other complications soon became apparent, but the main substance of the theory is today an accepted part of meteorological thought and practice.

WAVE THEORY OF CYCLONES

The frontal theory recognizes that extratropical cyclones form along fronts, a fact that has been verified by countless synoptic studies.   They form at a wave-like twist or perturbation on the front.   They go through a cycle in which either (1) the ampli-

---

[1] J. Bjerknes, On the Structure of Moving Cyclones, *Geofysiske Publikasjoner*, vol. 1, no. 2, 1918.   J. Bjerknes and H. Solberg, Meteorological Conditions for the Formation of Rain, *ibid.*, vol. 2, no. 3, 1921.   J. Bjerknes and H. Solberg, The Life Cycle of Cyclones and the Polar Front Theory, *ibid.*, vol. 3, no. 1, 1922.   T. Bergeron, Über die dreidimensional verknüpfende Wetteranalyse, *ibid.*, vol. 5, no. 6, 1928.

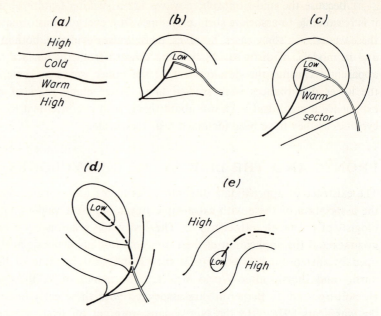

FIGURE 10-7
Life cycle of extratropical cyclone of Northern Hemisphere.

tude of the wave increases until great sweeps of arctic, polar, or tropical air are carried away from the source regions, eventually to become modified and mixed together; or (2) the wave may remain about the same without noticeable further development, eventually dying out without having participated in any great meridional mass exchange. Waves of the first type are called *unstable* waves, because they grow in amplitude until they appear to "break" like waves in a confused sea. The second type is the *stable* wave, having a tendency to be damped out.

The life cycle of an unstable wave is shown in map plan in Fig. 10-7. On the front separating westerlies from the easterlies, a small wave-like indentation is formed. As this develops more of a wave character, a cyclone of increasing intensity forms around it. The air, in adjusting itself toward equilibrium, pushes in toward this weak point in the front and, under the combined gradient and frictional conditions, develops a cyclonic circulation, centered at the wave crest. In the figure, a Northern Hemisphere development is depicted, and it is noted that the westerlies turn into a southwest gradient wind which pushes the eastern part of the front northward as a warm front and the western part southward as a cold front. Each of these fronts is convex in the direction toward which it is moving, like a sail on a ship.

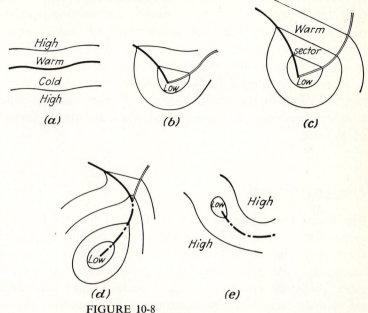

FIGURE 10-8
Life cycle of extratropical cyclone of Southern Hemisphere.

The cold front moves faster than the warm front and closes the warm sector to form a combined front, indicated by the heavy broken line in the figure. This process is called *occlusion*, and the front thus formed is called an *occluded front*. It represents a folding of the frontal surface by the action of the warm and cold fronts in such a way that the warm sector is shut off from the surface and occurs only aloft. As the occlusion process continues, the warm sector is displaced more and more aloft and the cyclone becomes completely surrounded by cold air in the low levels. It is then that the air masses are either completely modified or mixed and the cyclone decreases in intensity until it dies out completely. The cyclone generally reaches its greatest intensity just as occlusion is beginning, or just after reaching " maturity."

Similar conditions for the Southern Hemisphere are depicted in Fig. 10-8. In both hemispheres the cyclones are carried along in the westerlies, with the cold air in high latitudes and the warm air in the direction of the equator. Over the large Northern Hemisphere continents in winter, the warmer air may be to the west over the oceans, producing wave cyclones that move from northwest to southeast.

In terms of the life cycle, it may be said that an unstable wave is one that develops and goes to occlusion, whereas a stable wave is one that does not develop or go to occlusion.

A number of surface effects can cause a wave to form. It is likely that mountains, land-and-sea temperature contrasts, and ocean-current contrasts can affect the temperature-wind relationships sufficiently to create a wave. It is also found that many waves are started by some nearby disturbance. Thus in North America in winter, waves are frequently formed on the arctic front by occluded polar-front cyclones coming in from the Pacific and disturbing the westerlies in such a way that they make inroads against the arctic air. Perhaps the most important effect comes from the appearance of an upper trough which disturbs the circulation in the vicinity of the front.

CYCLOGENESIS

The Norwegian concept is based on the idea that the formation of a cyclone—*cyclogenesis*—occurs in a frontal wave. The picture of the life cycle in Fig. 10-7 shows increasing amplitude of the wave accompanied by a circulation of the air that accentuates the cyclonic nature of the disturbance. Since there is a mutual adjustment between the wind and the isobars, a low-pressure center occurs at the center of the cyclonic circulation. By constructing the streamlines and their corresponding isobars, one can show that the center must be at the crest of the wave, that is, at the pivot point between the warm and cold air, at least until occlusion starts. After occlusion, the lowest pressure follows the northern end of the occluded front. In some cases the center of low pressure remains at the wave crest after an occluded front is formed, causing the latter to be swung counterclockwise around the center, perhaps folding back toward the cold front and disappearing, or forming a secondary cold front.

Another way to describe the action is to relate the increasing cyclonic vorticity to convergence as in Eqs. (8-69) and (10-2) in the form

$$\nabla_2 \cdot \mathbf{c} = -\frac{1}{\zeta}\frac{d\zeta}{dt} = -\frac{\partial w}{\partial z} \qquad (10\text{-}16)$$

Near the surface with horizontal convergence w can only be positive, that is, upward, and must increase with z from zero at the precise surface. The upward motion in a cyclone of this type occurs mainly along the fronts. In convergence the first term in (10-16) is negative; so the vorticity increases and a cyclone is formed.

Sometimes extratropical cyclogenesis occurs in a region where there is no front. If the formation is large enough and if it moves, it will have the same characteristics as a frontal cyclone and may eventually draw fronts into its circulation.

A persistent type of cyclone is the so-called "heat low." Over continents in summer, particularly over the deserts, there is intense heating which causes the lower layers to expand and increase their depth.   The vertical expansion is compensated by a lateral outflow or divergence aloft.   The outflow exceeds the surface inflow to lower the pressure until a balance is reached.   These thermally produced lows usually do not move and are usually found in areas that are almost entirely free of clouds and fronts in summer.   In the United States there is a heat low centered more or less permanently during the warm part of the year in southwestern Arizona and southeastern California.   The most pronounced low of this type is found in the region of the Persian Gulf, affecting the winds over a large part of Asia and the Middle East.

Cyclones sometimes develop in the lee of mountain ranges.   In the United States a region of frequent cyclogenesis is just east of the Rockies in eastern Colorado and the Texas Panhandle.   These cyclones often become important traveling extratropical disturbances, drawing fronts into their circulations after they have been in existence for some time.   To explain the formation of this lee-of-the-mountains trough, one may consider the vorticity equation in the form

$$\frac{1}{D_p}\frac{dD_p}{dt} = \frac{1}{f+\zeta}\frac{d(f+\zeta)}{dt} \qquad (10\text{-}17)$$

and recognize that as the air moves from the west up the slope of the Rockies, the depths of the layers D_p must decrease as they are crowded upward.   The vorticity decreases, making an anticyclonic curvature of the west wind and causing it to become northwest.   As the northwest wind crosses the eastern slope, two things happen: f decreases and D_p increases.   These changes require the local vorticity to increase, producing a trough or cyclone.

As already pointed out, mid-troposphere troughs have an important bearing on cyclogenesis in the low levels.   As a recognition of this fact, 500-mb contour and vorticity charts are a regular feature of the facsimile transmissions from the United States analysis and prediction center.

Cyclogenesis of tropical cyclones is discussed in a subsequent chapter.

SIMPLIFIED VERTICAL STRUCTURE

If a vertical cross section is made through a "mature" idealized cyclone at the stage indicated in Fig. 10-7c, the structure south of the cyclone center through both the cold and warm front would appear in the ideal case in accordance with Fig. 10-9. Here the fronts slope in such a manner as to give a wedge-like shape to the cold air. The smaller slope of the warm front in comparison with the cold front is to be noted, also the greater tendency for ascending motion in the air above the warm-front surface.   The prevailing westerlies aloft above the frontal surfaces are downslope

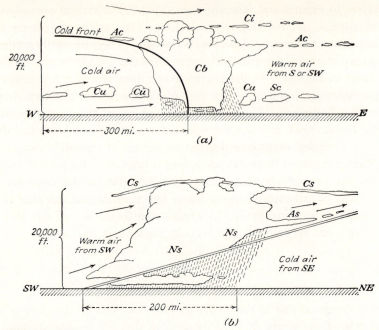

FIGURE 10-9
Idealized cross section through a cold front (*a*) and a warm front (*b*) in a mature cyclone.

winds above the cold front, the upward movement being confined mainly to an upward and outward pushing of the lower warm-sector air by the advancing cold-air wedge.

The main cloudiness and precipitation occurs in the lifted warm-sector air. Clouds may form in the lower cold-air wedge, but these are generally not rain-producing and often form by reevaporation of the water that has fallen from the overrunning air.   The precipitation is to be explained as resulting from the decreasing temperature with lifting due to the adiabatic expansion against the decreasing pressures with height as the warm-sector air is lifted at the fronts.

The Warm Front

In view of the prevailing west-to-east direction of movement of the extratropical cyclone and the arrangement of the fronts about it, it is evident that the warm-front effects should be the first events heralding the approach of one of these storms. These usually make their appearance just as the clear skies and calm winds associated with the preceding anticyclone are being replaced by the return current of polar air moving from the southeast or east on the southwestern side of that anticyclone.   The front is a sloping surface of discontinuity, or temperature and moisture inversion, between this returning mass of polar air and the warm moist current which overruns it from the west.   The upward decrease of pressure as this air slides up over the cold wedge ahead of it causes the cooling by adiabatic expansion necessary to produce clouds and precipitation.   (For cloud definitions and descriptions, see Appendix B.)

The first sign of a warm front coming toward any particular place is the appearance of the clouds that have been caused by the air forced well up the slope to where the cold air is from 15,000 to 20,000 ft deep.   These clouds, of course, are the types characteristic of these and greater heights.   First the cirrus appear, becoming cirrostratus in a continually denser sheet.   If the overrunning air is unstable and turbulent, cirrocumulus will be noted, forming the so-called "mackerel sky" recognized for its import by the old-time sailors.   The nearer approach of the front brings the warm air lower, and the intermediately high clouds are noted—altostratus and altocumulus—then finally stratocumulus, nimbostratus, and stratus, and occasionally cumulonimbus. Rain or snow usually begins from the altostratus clouds as they approach their greatest density and continues until the front has passed at the surface.   In many cases, particularly in maritime locations, lower clouds are present in the cold air so that the sequence of clouds as given above cannot be noted by an observer at the surface.   Although the precipitation is initiated in the overrunning clouds, the low clouds in the cold air under the front contribute much of the water that reaches the ground.   Collision-coalescence of falling drops sweeps out the lower cloud which, in most cyclones, is continually replenished by low-level convergence.

The character of the activity in the overrunning air current depends in a large measure upon the conditions existing in the warm air mass before it was lifted. Furthermore, in view of the slow rate of ascent of the air up the relatively gradually sloping surface, much of the heavy rain that we observe in warm-front conditions cannot be accounted for unless strong convection is present within the overrunning current itself on account of its inherent instability.

The sequence of events at a particular station with the approach of a warm front is the usual condition to be expected with the coming of a cyclone; for the warm front is the first unit of the advancing low.   Obviously, then, a fall in pressure, which becomes more rapid as the front nears, is observed.   The increase in cloudiness and precipitation as well as in the humidity is a characteristic feature.   Generally, the temperature is constant or slowly rising until the surface front reaches the station, at which time there is a sharp increase, depending for its sharpness on the degree of contrast existing between the two air masses on either side of the front.   With the passage of the front, there is a decrease in cloudiness, or a complete clearing, as the station now is in the warm sector.   The cloudiness then depends on the general properties of the air mass that occupies the warm sector, in accordance with temperature, moisture, and lapse-rate conditions for air masses.   In the warm sector, the temperature remains relatively high, and the barometer is steady or shows only a slight falling tendency, unless another front, the cold front, is approaching.

The Cold Front

The greater steepness of the cold front makes it act more violently in producing clouds and precipitation when displacing warm moist air.   The squall line, with its sudden showers and vigorous windshift, occurs when the cold front interacts with moist unstable air.   It is generally characterized by heavy clouds, usually of the cumulonimbus type, gusty turbulent winds, heavy rain, and sometimes thunderstorms. Since it produces within a very short distance the same amount of lifting as occurs over a much broader zone in advance of the warm front, and since its direction of motion is such that the warm air generally retreats from it instead of sliding actively up over it, the cold front is accompanied by a much narrower band of cloudiness and precipitation than is found in the case of the warm front.   In other words, the precipitation and cloud phenomena of a cold front are usually brief and violent.

The sequence of events in a pronounced cold-front passage begins with the observance of a general increase in the southerly winds of the warm sector and the appearance of high cumuliform clouds, such as altocumulus, darkening on the horizon to the west or north.   A fall in the barometer is often noted at about the same time, but this is usually not so pronounced as the marked rise in pressure which occurs immediately following the passage of the front.   The lowering of clouds to cumulonimbus with rain of increasing intensity marks the approach of the front.   The most vigorous squall condition occurs with the actual passage of the front and the shifting of the wind to a westerly or northerly direction.   Normally this is followed by a fairly rapid clearing, unless in a mountainous or moist region, where cumulus or stratocumulus in the following cold air mass may linger for a long time.   As in the case of the warm sector, the cloud conditions in the cold air mass depend on the stability and moisture relationships inherent in the mass.   The presence of the dry superior air mass above the front has a marked effect in limiting the vigor of the frontal interaction with regard to clouds and precipitation, and the existence of subsidence inversions within the cold air mass has a prominent influence on conditions there.

The Occluded Front

The occluded front has more varied characteristics than do cold and warm fronts. Consequently, there are several different kinds of conditions to be expected.   From a consideration of the general process of occlusion as outlined previously in this chapter, it appears that the occluded front is formed near the center of the disturbance in the cold air mass, thus separating the latter into two sections.   Theoretically, these represent two parts of the identical air mass.   In actuality, however, the circulation about the front brings in air from opposing directions; and although originally these oppositely moving currents consisted of the same air mass, their difference in path, or trajectory, has given them different modifications.   Therefore, contrasts develop between them.   Thus, after a certain length of time following its initial formation, every occluded front shows not only a contrast in wind direction but also in most cases temperature differences.

Regardless of the temperature and other differences occurring in the cold air, the definition of the occluded front requires that it consist of a trough of warm air pushed aloft from the warm sector, as illustrated in Fig. 10-10.   In other words, an occluded front means that there is a warm sector at upper levels.   Clouds and precipitation will occur in this warm air above as the cold air squeezes it upward.

The temperature contrasts within the cold air below the warm sector and on either side of the windshift line determine whether the front is (1) a cold occluded front or (2) a warm occluded front.   If the colder of the two cold currents is advancing, then it is a cold type of occluded front.   If the less cold of the two is gaining ground, it is a warm occluded front.   The warm type is represented in the eastern section in Fig. 10-10, and the cold type in the western low. The movement of the front is from west to east.   It will be noted that the intersection of the front with the ground, which is the part represented on the weather map, is ahead of the upper warm sector in the case of the cold type and lags far behind in the warm type.   On many occasions the lower cold-acting or warm-acting front will itself produce clouds and precipitation and otherwise behave as a regular cold or warm front.

The type of weather associated with the occluded front cannot be described definitely because of the variability of its structure and action.   In general, it would be expected that the warm type of occluded front would be preceded by the same sequence of events that comes with the approach of a warm front, except that in some cases the clouds and precipitation may cease before the passage of the surface front. Under such circumstances, one might justifiably conclude that the warm sector aloft was the only part of the system capable of producing rain or snow and that the front at the ground, which for the warm type of occlusion trails behind the lifted warm trough, represented an interaction between air masses too dry or too stable to produce significant weather phenomena.   The cold type of occluded front should have no forerunning clouds and precipitation.   If these were confined solely to the lifted warm sector, then rain or snow would not be observed until after the passage of the surface front, indicated by the windshift.   In many cases, however, the interaction between the two cold surface air masses produces clouds and precipitation in the lower levels accompanying the windshift line.   It is conceivable, and is occasionally sup-

256

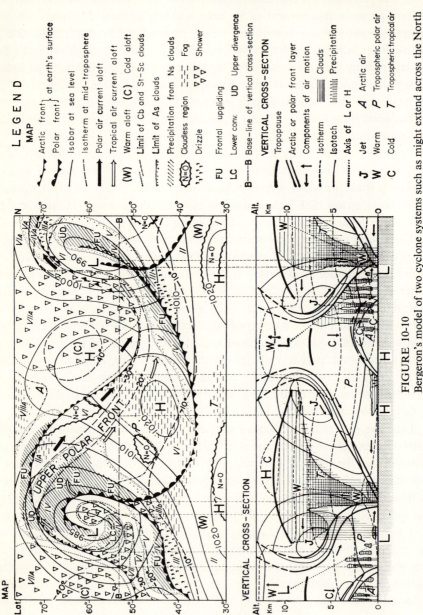

FIGURE 10-10
Bergeron's model of two cyclone systems such as might extend across the North Atlantic.

ported by actual observation, that two definite zones of clouds and rain or snow will follow each other, the one representing the lifted air of the warm sector and the other the interaction of the colder surface air masses.

As the process of occlusion continues, the cold air deepens as it crowds the warm sector upward until the latter has reached a height sufficient for it to spread out over the domes of cold air that have occluded it. It then ceases to be lifted farther and becomes inactive as a cloud- and rain-producing air mass. It is then that the contrasting properties of the two parts of the lower cold air on either side of the occluded front become more important than the interactions with the lifted warm sector. When such a stage is reached, it is good practice to regard only the lower masses of air and to consider the lower discontinuity as being either a warm or a cold front depending on the type represented by the occluded front. Such old occluded fronts often will act as "secondary" cold fronts or as separate warm fronts.

BERGERON'S MODEL OF OCCLUDED CYCLONES

From a study of weather charts of Western Europe and the North Atlantic, Bergeron[1] has concluded that the most frequently observed cyclonic structure is that of the warm-front-type occlusion. His model is shown in Fig. 10-10. The first impression one obtains is that the structure is very complicated. As the details are studied in relation to the legend, a systematic pattern emerges. Bergeron designates the model as a three-dimensional model of main disturbances (two separate cyclone series) in the system of polar front and jet, including the weather systems. The conditions depicted are most typical of autumn. It is noted that in the vertical cross section the positions of the jet are shown, but since it is a zonal cross section, the meridional components of the meandering jet are the ones shown, with northerly winds (toward the viewer) over the cold fronts and southerly winds (toward the page) above the warm front.

In addition to the symbols given in the legend, Roman numerals appear on the map to represent the weather regions related to the main fronts and air masses. Their meanings are as follows:

I   Cloudless in tropical air

II   Stratocumulus-stratus fog in tropical air

III   Drizzle (or rain) in tropical air

IIIb   Showers or convective systems in tropical air, which are especially common in North America

IV   Nimbostratus in polar air

[1] T. Bergeron, Paper presented at the Rome Assembly of the International Union of Geodesy and Geophysics, 1954.

V   Altostratus in polar air
VI   Subsidence in polar air
VII   Showers in polar air
VIII   Stratocumulus-stratus fog in polar air

The suffix *A* refers to weather regions in relation to the arctic front, such that near that front the *A* refers to polar air for the numbers *I* to *III* and arctic air for *IV* to *VIII*.

According to Bergeron, the chief aim of the model is to show the average connection between the weather regions and the three-dimensional front and air-mass structure, including temperature and flow pattern aloft.   The model comprises two whole cyclone series or cyclone "families" and therefore two whole long waves in the upper westerlies.   Thus it may cover, for instance, all the area from Eastern Europe to North America.   For the sake of simplicity, each cyclone series is represented essentially by one major occlusion.   The two occlusions are purposely shown in different stages of development.

The difference between the deep excursions of the polar air, with its vigorous occlusions, and the relatively flat, nonoccluded waves of the arctic front is strongly emphasized.   The two arctic domes have been shown with different structures.   In the eastern or European one, where the arctic air has traveled partly over the sea, the arctic front is penetrated and partly dissolved by the cumulonimbus forming from below.   Over the North American continent the conditions in the western arctic dome are typified.

Bergeron points out that the arctic front may also have a secondary jet associated with it, and mentions the possibility of multiple arctic-front and jet structures. Typically the arctic front overtakes the polar front near the rear of the low-pressure centers, but the two fronts are kept separate in the model to show their different characters.   The broken nature of the polar front in its southwestern part between the two cyclones indicates the "venting" of polar air into the trade-wind region.

FLOW OVER SHALLOW COLD DOMES

As explained in the case of air flowing over mountains, the currents tend to curve anticyclonically as they ascend the warm front and become cyclonic in vorticity as they descend the cold front.

Fronts are sometimes masked at the surface, appearing as upper-air fronts. In the warm type of occluded-front system there is a cold front aloft.   Cold air acts against warm air, but the action occurs above a third still colder air mass which lies in wedge fashion at the ground underneath the occluded front.   Synoptic practice has shown that such activity of cold fronts can occur aloft above an essentially horizontal, nonfrontal inversion.   In the United States and Canada, such fronts, not directly connected with a warm-type occlusion, move for considerable distances across the continent. This type of structure is of the greatest frequency in winter when a shallow layer of continental-arctic air covers much of the land. Cold fronts from the Pacific will move across the continent above this cold-air cushion and will have pronounced effects on the weather. The direction of motion is sometimes quite inconsistent with the surface currents; for example, in the case of east winds in the arctic air mass we find west winds carrying the front along in the opposite direction above.

Upper-air cold fronts are frequently found extending northward from the centers of cyclones, where easterly winds prevail at the ground. Many of them are so near the ground that they produce changes that may be mistaken for the passage of a surface front. However, any front that appears to move from west to east north of a low must be an upper-air front, because the gradient winds north of a cyclonic center would oppose such movement of a surface front. Occasionally, when strong frictional and other nongradient influences are at work, the actual surface air may be moving in such a direction as to be favorable for west-to-east motion despite unfavorable gradient indications. In such cases, the gradient simply does not represent the actual air flow.

Upper-air warm fronts are also important. They are noticeable where a nearly horizontal warm front or other discontinuity surface becomes abruptly steeper. The line along which this change in slope occurs shows many signs of an actual warm-front passage and is called an upper-air warm front. Owing to the steepness of the slope, the advection of warm air is more rapid than in the region of lesser slope. Therefore, the pressure falls rapidly in advance of an upper-air warm front and tends to level off underneath the nearly horizontal portion of the front. Also precipitation may develop just ahead of the upper-air warm fronts, owing to the rapid ascent of the damp air along this portion of the frontal surface. Occasionally these fronts work down to the surface. As a warm front crosses a mountain range, it may encounter colder air to the east and therefore move along as an upper warm front above the shielding layer of cold air. This is a common observation in warm fronts that cross the Appalachians in winter.

ENERGY OF CYCLONES

The important part played by extratropical cyclones in transforming energy in the atmosphere has already been mentioned in the chapter on the general circulation. Cyclones are areas of concentration of kinetic energy, which is equivalent to saying that cyclones are regions of strong winds. This is a very pronounced characteristic of cyclones in the low levels. It is appropriate at this point to consider the transformations of energy somewhat along the lines taken up in Chap. 9, but with particular reference to the concentration of kinetic energy in cyclones.

From the point of view of the fronts and frontal waves, cyclones may be considered as deriving their kinetic energy from the potential energy of distribution of mass—the existence in nonequilibrium juxtaposition of air masses of contrasting densities. The concentrated solenoidal field is distorted in the region of the developing cyclone so that the solenoidal gradients have a marked zonal component. The solenoids are tilted so as to accelerate the circulation both in the vertical and in the horizontal. The acceleration of the circulation must continue as long as the favorable solenoidal field exists; thus the cyclone increases in energy until it reaches its maximum intensity at occlusion.

From the larger point of view, the mid-troposphere waves may be considered as the sources of the cyclone developments. As shown by experiments in rotating fluids, these waves must occur in the atmosphere merely because it is unequally heated and is rotating. These waves distort the jet-stream polar front. Because of the connection, either direct or indirect, between the polar front in the middle troposphere and surface fronts, the latter are also distorted, giving rise to the process described in the preceding paragraph. In certain situations it has been observed that frontal-wave cyclones have developed near the ground and have built upward to become intense cyclones without the presence of a preexisting upper-troposphere wave in the region. Apparently in these cases, strong low-level fronts become distorted without any related disturbance of the upper jet-stream front. Although experience shows that the low-level disturbances are much more frequently associated with upper troughs than not, it suggests that the principal source of energy can come from any level in the atmosphere.

Much of the energy seen in cyclones may have been concentrated there by what might be considered random coincidences. It is normal for the atmosphere to be in a highly disturbed state. The waves and related systems at upper levels normally move at a speed different from that of the low-level disturbances, thereby providing an opportunity for superpositions of upper and lower systems. Thus when two systems of cyclonic vorticity too weak to be noticed as cyclones are superimposed in such a way as to combine their kinetic energies, a vigorous cyclone may develop.

Heat of condensation is irreversibly added to the atmosphere in a cyclone where rain falls to the earth. This source of energy should also be considered. In parts of the cyclones of middle latitudes the lapse rate is favorable for pseudoadiabatic ascent of moist air. Ascending motion induces low-level convergence and the heat of condensation increases the temperature of the air. Both these effects are favorable for cyclonic development. The latent heat has a small effect in most extratropical cyclones. The thermal stratification is stable in most of the cyclonic area, and such instabilities as occur as a result of saturation are not enough to overcome the net stability effect in the cyclone as a whole. What has been said in this paragraph does not apply to tropical cyclones (hurricanes and typhoons), which derive their principal energy from the buoyancy gained in pseudoadiabatic ascent.

FRONTOGENESIS

As stated before, it is possible for fronts to develop near the surface of the earth somewhat independently of the upper-troposphere polar front. The term "frontogenesis" has been applied by Bergeron[1] to the formation of new fronts. The definition may be broadened to include the regeneration of old fronts. Conversely, the degeneration of fronts is called *frontolysis*.

Bergeron showed that wind streams forming what is known in hydrodynamics as a deformation field can be responsible for frontogenesis. The simple type of deformation field which he studied is shown in Fig. 10-11. Petterssen[2] elaborated on this system in giving a quantitative theory of frontogenesis. The streamlines in Fig. 10-11 are rectangular hyperbolas representing a possible flow around two highs and two lows of the Northern Hemisphere, with the origin of coordinates taken at the "neutral point" between them. Such a neutral point and wind system could develop, for example, between a Canadian high to the northwest, the Azores-Bermuda high to the southeast, the Icelandic low to the northeast, and a Mexican low to the southwest. It could also develop on a much smaller scale between two small highs and two small lows.

In the figure, the x axis along which the particles are carried away from the neutral point is called the *axis of dilatation*, and that along which they are transported toward the neutral point, the y axis, is known as the *axis of contraction*. It is easier to visualize how a velocity distribution of this nature can bring particles together by separating it into its two components. This has been done in Fig. 10-12, which gives the x and y components of the motion separately. In each case the length of the arrow is proportional to the wind velocity. The two components, added together, give the hyperbolic streamlines of Fig. 10-11. The deceleration as the particles near the x axis in the y component and the speeding up in the x component as they leave the y axis mean that there must be a crowding as the x axis is approached, because there will always be particles coming in from the rear at higher speeds—contraction. In the x direction, the particles speed up as they move outward from the center; hence they are carried far from each other—dilatation.

[1] T. Bergeron, Über die dreidimensionale verknüpfende Wetteranalyse, *Geofysiske Publikasjoner*, vol. 5, no. 6, 1928.
[2] S. Petterssen, Contribution to the Theory of Frontogenesis, *Geofysiske Publikasjoner*, vol. 9, no. 6, 1936.

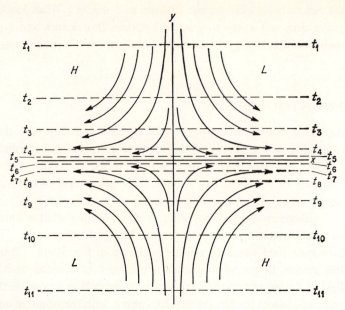

FIGURE 10-11
Simplified deformation field, showing how isotherms (broken lines) become concentrated along the axis of dilatation.

If the elements of air carried along in this circulation preserve their temperature, which is often approximately the case, then the isotherms, unless they parallel the streamlines, will also be carried along. If the distribution of isotherms is such that they are perpendicular to the axis of dilatation, frontolysis will result, for then the isotherms will be more and more widely separated. On the other hand, isotherms forming a high angle with the axis of contraction will be carried toward each other, and the necessary conditions for the formation of a front—the concentration of isotherms and therefore also of solenoids—will be fulfilled. In Fig. 10-11 the isotherms are drawn parallel to the axis of dilatation, a perfect condition for frontogenesis, especially along that axis.

For a neutral point in which the axes are at right angles, Petterssen has shown that an angle of less than 45° between the isotherms and the axis of dilatation is necessary for frontogenesis; if they cut this axis at a greater angle, frontolysis occurs. For nonperpendicular axes, the critical angle is altered correspondingly. In most cases, the line of frontogenesis will fall along the axis of dilatation.

It is not necessary to have a deformation field like that in Fig. 10-11 to have frontogenesis. Any wind and temperature distribution that brings isotherms together and keeps them there is frontogenetic. The only requirement is that the winds shall

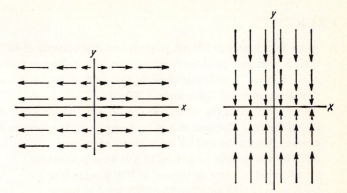

FIGURE 10-12
Components of deformation field.

have a cross-isotherm component and that this component shall decrease downwind. This assumes, of course, that temperatures are at least partially conserved and that the temperature gradient maintains the same approximate direction throughout the region considered; for example, frontogenesis could not occur along a cold zone surrounded by warm air or along a warm zone surrounded by cold air. If the component of the wind normal to the isotherms changes sign (reverses itself) downstream, it will be in many cases the same as a diminution. This is sometimes the case in fronts along windshift lines.

It is possible to express these statements in the form of a mathematical equation, as has been done by Petterssen. A numerical measure of the intensity of the fronto-genetic or frontolytic effect can then be obtained. Designating the frontogenetic function by F, positive for frontogenesis and negative for frontolysis, we define

$$F = -\left|\frac{\partial T}{\partial s}\right|\frac{\partial c}{\partial s}$$

where $|\partial T/\partial s|$ is the magnitude of the temperature gradient, s is normal to the isotherms, and c is the wind. This is a mathematical statement and a measurement of the rule: *If the wind has a component across the isotherms that decreases in its downstream direction, frontogenesis is to be expected.* If any of these conditions fail, that is, if there is no temperature gradient or no cross-isotherm wind component or if this component increases in its downstream direction, no frontogenesis will occur. The intensity of frontogenesis will be greatest with a large temperature gradient, a strong component of wind across the isotherms, and a sharp decrease in this wind downstream.

EXERCISES

1   At lat 38° a trough at 500 mb exhibits a relative vorticity of 10^{-5} sec^{-1}.   The wind speed there is 50 m sec^{-1}, and the isolines of the field of absolute vorticity are moving at a speed c_s of 15 m sec^{-1}, with a gradient in that field of 10^{-5} sec^{-1} in 100 km.   Compute the divergence ahead of this trough at 500 mb.

2   Apply two tests to the validity of the "fronts" in the wind and temperature fields described below.   The tests should be based on (i) the formula for the slope of the front given in this chapter and (ii) the vorticity equation.

   (*a*) NW wind in cold air to west of SW wind in warm air

   (*b*) SW wind in warm air to west of NW wind in cold air

   (*c*) E wind in cold air to north of SW wind in warm air

   (*d*) E wind in cold air north of weaker E wind in warm air

3   At a certain height above the ground a front separates air with a potential temperature of 295 from air having a potential temperature of 292.   The wind in the warmer air is from the southwest at 5 m sec^{-1} and in the cold air it is from the northwest at 10 m sec^{-1}.   The latitude is 45°N.   Find the slope of the front.

4   Examine the winds depicted in the life cycle of a cyclone in Fig. 10-7, and show qualitatively that the vorticity is positive in all the stages and increases through the stages.

5   It is found that air ascending over a warm front undergoes divergence of about 20 percent reduction in thickness with 100 mb of lifting.   What type of circulation should this induce above the warm front?

11

AIR MASSES, THEIR STRUCTURE AND MODIFICATION

We now find ourselves in the midst of considerations dealing with the surface and low-level weather maps. We see in them the picture of the characteristic weather and the changes experienced from day to day, especially at a location in middle latitudes. Cyclones and anticyclones, fronts, and major circulations are not the only aspects that impress us. We find that the weather of a particular day depends on the nature or thermodynamic structure of the air that envelops our area. Between the fronts, less spectacular yet significant processes are taking place.

The existence of a zone of sharp temperature changes at fronts requires that other parts of the atmosphere have lesser horizontal gradients. In their studies in the 1920s the Scandinavian meteorologists found that for purposes of synoptic analysis, it was advantageous to make the simplification that fronts are discontinuities in the field of atmospheric properties and that the areas between the fronts are occupied by horizontally homogeneous *air masses*. In this view the baroclinicity of the atmosphere is concentrated in the frontal zones and the rest of the atmosphere is barotropic. This simplifying model of the atmosphere represented one of the great forward steps in meteorology. The concept of fronts and air masses and the charting of their distributions forms an important part of practical meteorology today.

SOURCE REGIONS OF AIR MASSES

A rudimentary picture of the general circulation of the atmosphere shows a single front, the polar front, with two air masses, polar and tropical, on the two sides. This distribution is found in the middle and upper troposphere with only slight modifications, as is shown in Fig. 10-1. In the lower troposphere, the picture is more complicated for two reasons: (1) the perturbations (cyclones, anticyclones) are more complex and hence are capable of creating transitional air-mass types through circuitous passages of air and of forming intermediate, detached fronts; (2) the continents and oceans impart different properties to the overlying atmosphere and thus create contrasting air masses.

Different air masses are created because certain sections of the atmosphere are acted upon for long periods of time (days to weeks) by the radiation, convection, turbulent-exchange, and evaporation-condensation processes characteristic of a certain region of the earth. In the simplest picture described in the preceding paragraph, the polar air north of the polar front acquires properties characteristic of high latitudes and the air to the south has tropical or subtropical properties. The concentration of momentum in the jet stream is accompanied by a concentration of gradient at the front. Perturbations cause breaks in the front through which the air masses are mixed.

Regions in which air masses attain characteristic properties are called *air-mass source regions*. Air masses are given names according to their sources. In dealing with large sections of the earth, however, it is not possible to cling to the source name very long. Long trajectories of air masses over different parts of the earth, which are to be expected as a consequence of the air exchange of the normal circulation, subject the air masses to new source regions. Thus the source name refers only to the recent history of the air mass.

In the lower troposphere the continental and maritime source regions stand in sharp contrast, and during the extreme seasons of the year, summer and winter, they produce air-mass contrasts equal to or greater than the latitudinal effects. Another feature of the lower troposphere is the appearance of a third air mass, arctic air, behind a second front.

With the combined effects of (1) latitude and (2) continents versus oceans, the general classification of air masses is constructed as follows:

Type of surface	Latitude zone	Abbreviation
Maritime	Tropical	mT
	Polar	mP
	Arctic	mA
Continental	Tropical	cT
	Polar	cP
	Arctic	cA

The air mass designated "arctic" is from a more northerly and colder source than the "polar."

CHARACTERISTICS OF AIR MASSES

The temperature and moisture distribution in the vertical serves to characterize the air masses. As a starting point we look at the vertical soundings plotted in Fig. 4-4 on page 60. The soundings for Swan Island represent the source-region temperature characteristics of maritime tropical air. The winter soundings for Point Barrow show how the temperature is distributed with height in continental arctic air. At Omaha the mean winter sounding, although it might include several different air masses, is predominantly continental polar.

From any sounding a "characteristic curve" of the air mass present at the point of observation can be obtained by plotting the isentropic condensation points for each of the significant points of the sounding.[1] Potential temperature and mixing ratio determine the points in the characteristic curve, and these are properties that do not change in dry-adiabatic processes or, in the case of mixing ratio, with any temperature change without saturation. They are called "conservative" air-mass properties because they tend to be conserved as the air masses move away from their source regions. Characteristic curves for three air masses in winter and summer are shown in Figs. 11-1 and 11-2 plotted on tephigrams. These curves are constructed from the averages of soundings taken in the particular air masses at the indicated stations.

The great stability of the winter cA air is shown by the increase of 20° in potential temperature in the lowest 2 km. The mixing ratio changes little with height. The mP curve at Seattle shows, of course, higher moisture and higher temperature, but it is also quite distinct from the cA curve in showing much less thermal stability. The mP curve follows very closely a pseudoadiabatic line (omitted from the figures for the sake of simplicity) indicating nearly constant equivalent-potential temperature or wet-bulb potential temperature. This type of distribution indicates that the air mass is essentially in convective equilibrium with the ocean surface. In the mT air in winter, as represented by the San Antonio characteristic curve, still higher potential temperatures and mixing ratios are observed. It is also found that the equivalent-potential temperature decreases with height because of the existence of relatively dry air aloft. Note that at 4 km the mixing ratio is less than at Seattle. This upper dryness in North American tropical air is typical in winter except near fronts or in local

---

[1] See the discussion in Chap. 6, p. 121, where it is pointed out that the condensation point is a characteristic point, determining properties both before and after condensation, and establishing the equivalent-potential temperature as well as the wet-bulb potential temperature.

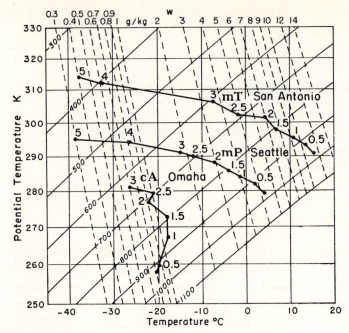

FIGURE 11-1
Characteristic curves on tephigram for winter air masses in North America.

zones of horizontal convergence.   It is not as evident in summer (Fig. 11-2) as in winter.

The differences between air masses in summer are not as striking as in winter. Arctic air is entirely missing in the coterminous United States but can be recognized as occurring with some frequency in Alaska and northern Canada.   The *mP* air from the Pacific is frequently found to be mixing with the *cP* air over the North American continent, but the contrast between the two air masses, especially in terms of water-vapor mixing ratio above 3 km, is sufficient for identification near the source regions.   The summer air masses over the United States have distributions of water vapor and potential temperature that correspond to a slight decrease of the equivalent-potential temperature with height.   The curves in Fig. 11-2 show stability in the lowest kilometer because the soundings were made in the early morning; in the afternoon the low-level stability disappears.   On characteristic curves the closeness of the points is a measure of the lapse rates in those parts of the curves above the stable near-surface layer. The summer curves of the polar and tropical air masses show greater compactness than in winter as a result of the effect of continental heating on the lapse rate.

In synoptic practice the air masses are identified not only by the characteristic properties that they carry with them from the source regions, but also by the contrasts

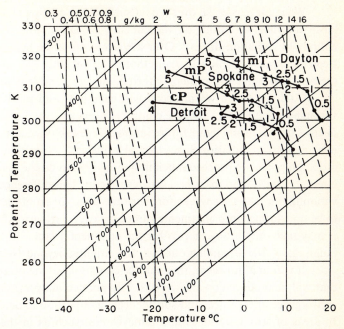

FIGURE 11-2
Characteristic curves on tephigram for summer air masses in North America.

that appear between them along the fronts on the synoptic charts.   Continuity in time between successive synoptic charts reveals the recent histories of the air masses.

PROCESSES DETERMINING AIR-MASS CHARACTERISTICS

As already implied, the air masses obtain their characteristics by radiation fluxes and by heat and water-vapor transport through the air-earth boundary layers.   The latitude and the nature of the underlying surface determine the relative importance of the various processes.

Thermal stability limits the vertical transport of heat and property, since the vertical exchange must be accomplished in air parcels carried by vertical eddies or convection currents.   If the temperature lapse rate is very stable, such as in the case of a temperature inversion, the radiation heat flux will exceed the eddy flux.   In the case of a moderate lapse rate, mechanical friction in the wind blowing across the surface of the earth can set up eddies which will transport heat and properties vertically. With superadiabatic or nearly adiabatic lapse rates the mixing is aided by thermally driven convection currents.

When air masses lie over a cold surface, the cooling from below creates a stable lapse rate virtually cutting off vertical eddy exchange. The air mass is cooled almost entirely by radiation fluxes. Over a warm surface the heating from below creates a steep lapse rate in the low levels so that turbulent eddy exchange and convection carry the heat and water vapor quickly through a considerable depth of the air mass.

Cold air over a warm ocean will have heat and moisture transported quickly through a great height. On the contrary, warm air over a cold surface may cool a few degrees until a strong temperature inversion is established at the ground, after which it must depend on radiation fluxes to cool it further. Numerical comparisons show that radiation processes proceed much more slowly than do the changes brought about by internal vertical motions over a warm surface. This difference accounts for the observation that it may take weeks for maritime-polar air from the North Pacific in winter to be transformed into continental-arctic air over northern Alaska and Canada, while it may take only a day or two for the continental arctic to be changed to maritime polar again after streaming out over the Atlantic beyond Newfoundland.

It is found, however, that surface conditions alone do not determine the effects that will be produced on an air mass. As pointed out previously (page 100), effects of convergence, divergence, and subsidence are important in determining the lapse rates. In areas of strong cyclones and anticyclones, such as in the eastern United States, these dynamic processes may act more prominently than the thermal ones.

OCEAN HEAT EXCHANGE AND EVAPORATION

By comparison with the land, ocean surfaces respond very little to seasonal change. Thus the oceans are warm in winter and cold in summer, relative to the land. The main reasons for the differences in heating and cooling of continents and oceans are, in descending order of importance, the following:

1 The oceans have a mixed or homogeneous layer extending for a number of meters below the surface. Heated or cooled water parcels at the surface are mixed to considerable depth and replaced by other water from below. Thus a mass of water a number of meters deep is involved in the heating or cooling. On the solid earth only the top few centimeters participate significantly in the heating or cooling; so only a relatively small mass is involved. From the relation

$$dQ = M_w s_w \, dT_w = M_e s_e \, dT_e \qquad (11\text{-}1)$$

where the subscript w refers to water and e to solid earth, we see that if the specific heats s_w and s_e are of the same order of magnitude and if M_w, the mass of water involved, is ten times the mass of earth involved M_e (both per unit area), then the increment of heat dQ added or removed from the land would raise or reduce its temperature $10°$ to every $1°$ for the same amount of heating or cooling over the ocean.

2   The specific heat of the sea water is about three times that of most types of land surface. This would make $s_w = 3s_e$ and, if $M_w = 10M_e$, then dT_e is 30 times dT_w. These are only approximate figures, useful in indicating magnitudes of the effects. It is obvious that not all the indicated temperature change would occur in the surface itself, since some of it would be transmitted to the atmosphere. The air, through eddies and convection, provides a "ventilation" effect which, combined with radiation fluxes, serves to limit the temperature extremes that any exposed surface can attain.

3   The ocean is a continuous source of evaporating water, while soil and vegetation are highly variable in this respect. The more the sun heats the surface layers, the greater the evaporation. Since each gram of water evaporated removes nearly 600 cal of heat from the evaporating surface, this process is very effective in preventing the oceans from getting very warm in summer.

4   The sun's rays penetrate to some few meters of depth in the ocean with appreciable intensity. Thus the solar energy is not entirely absorbed at the surface as it is in the case of the soil. The internal absorption helps preserve the homogeneity and therefore the mixed state of the upper layers of the ocean.

Differences in reflectivity might at first glance appear to be important, but they are not. Impressions of high reflectivity of water are gained from looking toward the sun at low angles of incidence. At normal angles the oceans have lower reflectivity than most land surfaces.

Evaporation at the earth-air boundary is more complicated than in a closed system such as is studied in the laboratory. The evaporation of a liquid in a partially filled, closed container proceeds at a rate proportional to the difference between the vapor tension of the liquid and the vapor pressure in the space above it. As vapor is added to the space by the evaporation, the rate must decrease until it reaches zero when the space is saturated with the vapor.

In the open atmosphere, it is possible for the water vapor to be carried away from the evaporating surface in the eddies or in the convection currents. The vapor pressure in the air above the water and therefore the evaporation rate for a given water temperature will be determined by the vertical flux of vapor, that is, the amount of water vapor flowing upward through a unit horizontal area in unit time. The eddy diffusivity, which varies with the temperature lapse rate, can have values several orders of magnitude greater than the ordinary (molecular) diffusivity of water vapor in air.

It is apparent that the same air-mass characteristics that favor the upward spread of water vapor through the atmosphere also enhance its evaporation from the surface.

Seasonal and annual charts of evaporation from the oceans prepared by Jacobs[1] show that the highest rates of evaporation occur in winter over the ocean near Japan where the water is warm and where the cold Siberian winter-monsoon air is pouring off the continent. In summer, this region has relatively little evaporation because the summer-monsoon air is warmer than the water and is laden with moisture. Condensation in the form of fog is prevalent in the early summer over the Japan Sea and the ocean areas to the northeast. On an annual basis, the greatest evaporation is from the tropical ocean regions. Evaporation involves a transfer of *latent* heat from the ocean to the atmosphere. Charts similar to those for evaporation show that the *sensible* heat taken up by the atmosphere is also, in general, at a maximum in regions of greatest evaporation.

THE NATURE OF VERTICAL FLUXES

It is obvious that in any medium heat flows from hot to cold, that is, down the gradient of temperature. In the physical sciences and engineering we learn that in heat conduction, a coefficient called the *thermal conductivity* serves as the proportionality factor between the heat flux and the temperature gradient. In any given solid it is constant through a wide range of temperatures. Similarly, a *coefficient of diffusivity* is determined for the diffusion of one gas in another, such as water vapor in air; and again the coefficient is a proportionality factor between the observed flux and the gradient.

In the atmosphere a proportionality factor between these fluxes and the gradient can be inserted, but it is of limited practical use. As a coefficient it is of interest only in a conceptual sense because it ranges in value through about three orders of magnitude, depending on the thermal stability, the vertical profile of the wind, wind speeds, and roughness of the terrain. The vertical transport in the atmosphere is accomplished by turbulent eddies and convection currents which derive their energies from thermal and mechanical processes related to the factors just mentioned. We already have hinted at this problem, especially in Chap. 7, where the property *viscosity* was applied to express the frictional effect on the wind, and it was pointed out that an eddy viscosity rather than the molecular viscosity would have to be used.

Advanced books and many contemporary scientific journal articles present analytical treatments of turbulent transport. These developments are beyond the intended scope of this book. For an understanding of processes determining or

[1] W. C. Jacobs, *Annals of the New York Academy of Science*, vol. 44, pp. 19–40, 1943

modifying the characteristics of air masses it is, first and foremost, necessary to appreciate the fact that the vertical exchange processes are dependent upon the thermal stability.   The more stable the air the less the vertical mixing, and in extremely stable air only the radiation fluxes remain.   The vertical transport of water vapor is much less dependent on the gradient of water vapor than on the gradient of temperature.

The properties of an air mass in the source region represent a balance with the condition of the underlying surface and with the net solar-terrestrial radiation of the location.   As the air mass moves away from its source to a different latitude or a different kind of surface, these balances change.   The balance as applied for mean conditions in the total atmosphere in Chap. 3 can also be applied to an individual air mass.   We may refer to Fig. 3-8 on page 51 for comparisons.   For example, in the source region of arctic air in winter there is no solar input, and the net outgoing flux of radiation from the surface makes the air very stable.   The fluxes E and C of Fig. 3-8 (evaporation-condensation and convection-turbulence) are approximately zero or negative (downward).   As the air mass moves southward, and especially if it passes over open water, E and C assume greater importance.   The latitudinal displacement also exposes the air to the effects of solar radiation.   Taking mT air as the opposite extreme, we find high solar input and a larger percentage of upward transport of heat and moisture than in Fig. 3-8.   The distributions result in a lapse rate near saturation equilibrium, although the air is rarely completely at saturation.

In addition to these balances or imbalances there are factors of large-scale divergence and convergence, subsidence and lifting, that affect air-mass structure. These modifying processes are common in the cyclones and anticyclones of middle latitudes.

Another factor that can be of importance under certain conditions and which needs to be taken into account in precise determinations of boundary-layer exchanges is the heat transferred to or from the atmosphere through the soil.   This process involves not only sensible heat but also the latent heat exchanged in freezing, thawing, evaporation, and condensation.   Consideration also must be given to the presence or absence of a snow cover.

SUMMARY OF AIR-MASS MODIFICATIONS

It is helpful to list the predominating types of modifying influences.   These may be classified as follows:

    A   Thermodynamic

        1 Heating from below
           a By passing from a cold to a warm surface
           b By solar heating of the ground

    2 Cooling from below
      a By passing from a warm to a cold surface
      b By radiation cooling of the earth's surface
    3 Addition of moisture by evaporation
      a From a water or ice and snow surface or from moist ground
      b From raindrops or other precipitation forms which fall through the air mass out of an overrunning saturated air current
    4 Removal of moisture by condensation and precipitation

  B Mechanical
    1 Turbulent mixing (eddies and convection)
    2 Large-scale dynamic effects on lapse rate
      a Divergence or outflow
      b Convergence (cyclones, etc.)
    3 Sinking
      a In subsidence and lateral spreading
      b Movement down from above colder air masses
      c Descent from high elevations to lowlands
    4 Lifting
      a Over colder air masses
      b To compensate for horizontal convergence
      c Over elevations of the land
    5 Advection of new properties aloft due to shearing action of the wind

These modifying influences seldom occur singly; usually two or more of the processes are combined, resulting in a change of the air-mass characteristics that sometimes can become fairly complicated. Consider, for example, an air mass moving from the Bering Sea out over the Pacific Ocean. It will be heated from below by passing from a cold to a warm surface; the resulting steep temperature lapse rate will cause considerable turbulent-convective mixing, and this combination will favor a rapid addition of moisture to the air by evaporation from the ocean. If it is winter, the air mass will be cooled from below by ground radiation as it moves inland over the North American continent, and furthermore it will undergo mixing and lifting over the western mountain ranges, which will cause condensation and precipitation of a large part of its moisture, so that it reaches the Middle West as a warm, dry air mass. The dryness may be accentuated by descent from the intermontane plateau, sometimes with the added effect of subsidence. During its travel from the source, the air has undergone the following types of modification: heating from below, lifting over mountains, more turbulent mixing, removal of moisture by condensation-precipitation, and then sinking. This is perhaps an extreme case of air-mass modification, but it happens regularly in maritime polar air masses from the Pacific Ocean entering the United States.

TEMPERATURE INVERSIONS

One of the most important characteristics of air masses is the development within them of temperature inversions. Their occurrence is so widespread that it is the rule rather than the exception to find them somewhere in the atmosphere. Whether the temperature actually increases with height or not, the general effect is the same, so long as there is pronounced thermal stability. In many cases the term is applied to describe a layer that does not always show a temperature increase, as in the trade-wind inversion, frontal inversions, subsidence inversions, etc. The potential temperature increases rapidly with height, however.

For convenience, inversions may be classified as either thermally or mechanically produced, with the frontal inversions, observed in the transition layer between a cold air mass and a warm one lying above it, forming a third group. Listed according to the processes that cause them, the inversion types are as follows:

1 Thermally produced
    a Radiation or contact cooling at the surface
    b Radiation cooling aloft
2 Mechanically produced
    a Turbulence or convection
    b Subsidence
3 Frontal inversions

The ground-radiation type is best exemplified by the well-known nocturnal inversions observed in the night and early morning, especially at land stations in light winds. It is caused by the rapid cooling of the earths' surface during the hours of darkness. It is emphasized in the polar darkness of winter, as already explained. In a middle-latitude night this cooling does not affect the air above about the first hundred meters. The result is an increase of temperature with altitude in the layer next to the ground—the nocturnal inversion.

Radiation cooling aloft is a relatively unimportant process. Water vapor, clouds, and atmospheric impurities (dust, smoke, haze) sometimes form a fairly effective radiating surface when concentrated in a well-defined layer. However, such stratification into definite haze and smoke lines depends on a preexisting inversion, so that radiation cooling in the free air is negligible in forming a temperature inversion but may intensify one that has already developed from another cause.

Mechanical processes are contributing causes of temperature inversions at altitudes above the surface. Turbulence and convection, if continued long enough, result in a thorough mixing of the atmosphere through the layers where the turbulent exchange exists. There is always a limiting height above which the turbulent or

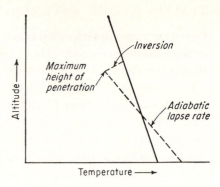

FIGURE 11-3
Development of turbulence inversion.

convective mixing does not penetrate, and it is at this altitude that temperature inversions are produced. In the turbulence layer, air parcels are brought downward from this maximum height of penetration to lower levels and air from below is carried upward in the general vertical mixing process. After this mixing has continued for some time, all the air in the turbulence layer will be air that has undergone adiabatic expansions and compressions because of change of level, so that an adiabatic or nearly adiabatic lapse rate will develop in which the air at the bottom of the turbulence layer will be warmer than formerly, and that at the maximum height of penetration colder than before. The transition from this cold upper part of the turbulence zone to the air above, with its temperature undisturbed, will comprise a temperature inversion. The process is shown in Fig. 11-3, where the solid line represents the temperature distribution existing before the development of the turbulence, and the broken line shows it after it has existed for some time. The approximate midpoint of the turbulence layer undergoes no temperature change, as is also the case in the warm air above the layer.

If clouds form in this type of condition, they will usually be of the stratus or stratocumulus variety. If the turbulence is accompanied by an addition of heat from the surface, cumulus or cumulonimbus may form. If it is an ordinary turbulence inversion near the ground, smoke, haze, and the lighter dust particles will be carried by turbulence up to the inversion, where they will spread laterally under the inversion to form a well-marked haze and smoke line, in the absence of clouds at that level.

The anvil-shaped tops of thunderstorm clouds are due to a temperature inversion or stable layer at altitudes through which the thunderstorm convection does not penetrate. Sometimes cumulonimbus clouds have sufficient convective energy to break through inversions or stable layers, and these are sometimes marked by a layer of altostratus forming a narrow girdle or appearing in bands around the cloud. Experience has shown that whenever stratiform clouds appear in the atmosphere one can be certain that they have some sort of stable layer above them. Sometimes cumulus clouds take on a stunted appearance because of this type of limitation of their vertical growth.

Subsidence, or slow sinking of the air in an area of high pressure, is an important air-mass modification and accounts for the development of a great many of the temperature inversions observed in the atmosphere, especially in regions where anticyclones remain stationary for long periods of time. The process of subsidence was outlined in its physical aspects in Chap. 5, where it was shown that sinking makes the air layers more stable than they were at their original higher levels. Furthermore, it was stated that in its slow movement downward the air heats at the rapid adiabatic rate with compression. The air near the ground does not participate in the subsidence because only the slightest degree of turbulence, which is nearly always present at low levels, can completely counteract the slow sinking. The lower atmosphere, then, acts as a sort of shield against the subsidence and shows no appreciable temperature increase, while above, the temperature at any level will slowly rise with the bringing down of potentially warmer air from greater heights. Between the shielding layer and the adiabatically heated air above there will then be observed a temperature inversion. The greater the amount of turbulence or vertical convection in the shielding layer the higher in the atmosphere the inversion will be found.

Subsidence is well developed in the Pacific anticyclone, the Azores-Bermuda anticyclone, and the stationary highs over the continents in winter. Often the combined process of a marked turbulence condition in the shielding layer with subsidence above produces an especially sharp temperature inversion, with radiation cooling of the cloud and dust- and smoke-laden stratum just below the inversion to intensify it further.

Frontal inversions display a variety of conditions depending on whether the front is an active upglide surface, is passive, or actually has descent along it. The chief differences are in the vertical distribution of water vapor through the front. Whereas the ordinary inversions in an air mass exhibit a sharp decrease of moisture content accompanying the temperature rise, active frontal types usually show an increase of mixing ratio in the inversion.

FOG

Because of their close association with temperature inversions and because of their tremendous importance in transportation, navigation, warfare, public safety, and human activities generally, fogs require serious consideration as an air-mass property.

For purposes of the present discussion, fog may be defined as a stratus cloud layer occurring at or very near the ground. Stratus clouds are characteristic of the lower part of the atmosphere when a well-developed temperature inversion or nearly isothermal layer exists there.   If the air below is sufficiently moist, a stratus layer will form, its top at the base of the inversion.   For the formation to take place as fog the base of the inversion must, then, be at the surface or very close to it.   A temperature inversion at or near the ground is merely an expression of cooling from below and is therefore characteristic of air masses that were originally warmer than the surface over which they are passing or resting.   The problem of investigating fog formation then reduces itself to the determination of the circumstances under which cooling of air masses at the surface, in the presence of high moisture content, can take place.   Also some fogs owe their immediate origin to the increase of the water-vapor content without appreciable cooling, such as in pre-warm-front rain.   However, the air must first be rendered stable by cooling.   Over continents in winter this stability due to cooling is nearly always present, and the formation of stratus or fog simply awaits the addition of sufficient moisture by rain from an overrunning air mass.

In the discussion which follows, fogs are classified according to the easily recognizable meteorological processes which cause them.

Sea fog is caused by advection of air over a cold ocean area from one that is warmer.   The summer sea fogs of the coasts of California, Peru-Chile, and northwest and southwest Africa, where the air moves over a cold, upwelled ocean current are examples.   Subsidence in the oceanic anticyclones accentuates the inversion. Another region of frequent occurrence is near Newfoundland, where warm Gulf Stream air encounters the Labrador Current.   With strong turbulent exchange the condensation may be in the form of stratus—the famous "high fog" of the San Francisco Bay area.

Land-and-sea-breeze fog is similar to sea fog, but the name usually is applied to regions such as middle-latitude east coasts where wind is normally off the land. If a fog forms over the water, the reversal with a sea breeze, usually coming up in the afternoon against the heated land, can bring the fog inland.   Fogs of the New England coast in spring and early summer and around the Great Lakes at this season are examples.

Tropical-air fog occurs mainly in winter when tropical air makes a pronounced northward excursion over colder water or land surfaces, especially if the circulation is anticyclonic under a strong temperature inversion.

Steam fogs are observed over lakes, streams, and open arctic waters when quite cold air settles over the water.   If the winds are light, the temperature contrast does not have to be very great, but if the temperature difference is very large, the steam will occur even in strong winds.

Ground fog is a radiation type occurring over the land, usually because of a single night's cooling.   It is most frequent in early fall when the air is still moist from summer vegetation and the nights are relatively long.   It is usually shallow and disappears shortly after sunrise.

High-inversion fog is a winter fog representing stronger radiation effects accompanied by a subsiding winter anticyclone. The cool air is so deep that the fog may last all day and, indeed, for several days at a time. This fog is characteristic in winter in the central valley of California, the intermontane valleys, and Western Europe. The air is originally from the ocean and, settling over the land, is modified by prolonged winter radiation effects, anticyclonic stagnation, and subsidence.

Advection-radiation fog is a radiation fog that occurs with nighttime cooling of air brought in by advection from warm, moist seas. It occurs a short distance inland from coasts such as the Gulf of Mexico and Atlantic coasts of the Southern United States. It also is common around the Great Lakes. It is likely to form in the fall when the waters are warm and nighttime radiation over the land has begun to strengthen.

Upslope fog is the fog of the higher Great Plains of the United States. Moist air from the east ascends from near sea level to 1000 m over the sloping plains to western Texas and to 1500 to 2000 m to places such as Denver or Cheyenne. The adiabatic cooling causes the condensation. The winds are predominantly from the west and downslope in these areas; so fogs are not common.

Prefrontal (*warm-front*) *fog* forms especially densely when the pressure gradient and winds are weak ahead of the front. This is often the case when a warm front is part of a secondary development south of the main center. The falling precipitation supplies moisture so that fog can form even if the temperature is increasing.

Postfrontal (*cold-front*) *fog* is seen when polar air spreads laterally and shrinks vertically behind a cold front moving over a moist land surface. The spreading and subsidence accentuate the temperature inversion in the polar air. Rain behind the front may augment the fog tendencies.

Front-passage fog is caused from a lowering of the clouds during a front passage in very moist conditions.

SMOG

The contraction "smog," designating a combination of smoke and fog, was first applied many years ago to the London fogs in which a pall of black smoke mixes with a dense fog. Near the turn of the century, measurements by Aitken and others showed that the condensation nuclei were of such a nature as to cause highly restricted visibilities at relative humidities of 90 percent or lower. In the 1940s, as the typical Southern California haze combined with city air, and pollution became a serious annoyance, Los Angeles newspapers began to apply the term "smog" to this condition. Popular usage has resulted in the application of the word to any situation of poor visibility in polluted air.

The Los Angeles haze is associated with the inversion typical of the California ocean areas in summer.   In the southern part of the region the condition continues to some extent into other seasons because of the persistence of the Pacific anticyclone. As the coast line curves eastward south of Points Arguello and Concepcion, a lee eddy of light westerly or southwesterly winds is found near shore, with the stronger winds from the northwest farther at sea.   The orientation of the Los Angeles Basin and the San Fernando Valley and the absence of low-level gaps to the east and north contribute further to the stagnation of air.   When the winds are light and of sea origin, the haze reduces the visibility to about 3 miles even on clear days over the ocean.   In city smoke, the visibility may be a mile or less even in the middle of a warm summer or fall afternoon.

Except for heavy traffic of motor vehicles, the pollution of the Los Angeles air is less than that of other cities its size.   The problem is severe because of the peculiar meteorological conditions, in part determined by the relief features of the area.   The ozone content of the lower tropospheric air is greater in the Pacific anticyclone than elsewhere, producing a strong oxidizing effect on atmospheric pollutants.   Some of the oxidized hydrocarbons in motor fuels have caused unpleasant eye irritation in the Los Angeles atmosphere.   Photochemical oxidations produce substances that are toxic to some plants.

Lethal fogs resulting in numerous deaths of aged persons and persons with respiratory troubles have occurred in the modern industrial age.   Notable examples are those of Liège, Belgium, in 1930; Donora, Pennsylvania, in 1948; and London, England, in 1952.   In these cases a dense fog under a stagnant anticyclone with subsidence inversion continued for several days, with industrial and domestic smoke and other pollutants entering the foggy air.   In the absence of normal atmospheric diffusion, the concentration of gases and aerosols increased to toxic levels.

EXERCISES

Because the material in this chapter is descriptive, exercises of the type given in preceding chapters cannot be presented.

12

OUTLINE OF SYNOPTIC ANALYSIS

The drama of the weather constantly unfolds in the synoptic laboratory or the fore-casting office.   Whether we study the weather maps and auxiliary charts transmitted by facsimile from national meteorological centers or prepare our own synoptic charts, we see in striking fashion the processes at work that we have studied in our course in meteorology.   From the global network of stations come the observations made simultaneously in every land.   Ships at sea, aircraft, and weather satellites provide a record in real time to fill in the details.   The developments in the weather, which are of such great interest that they receive prime-time coverage on television and radio, are all the more interesting and exciting to the meteorologist equipped to delve into the intricacies of the situations.   By means of laboratory practice or internship at a forecasting center the student meteorologist develops a familiarity with weather behavior and a proficiency in synoptic analysis that will provide him with forecasting and analysis skills.

The purpose of this chapter is to point out the general character of the problem of representing the atmosphere and its processes on charts, diagrams, and cross sec-tions.   The intricate craftsmanship required to construct a three-dimensional model of the fleeting structure of the atmosphere at a given instant can be done only for exhibit purposes; therefore, the representation is accomplished on various surfaces cutting through the atmosphere. These may be horizontal surfaces at various heights; vertical surfaces or cross sections; isobaric, isentropic, or frontal surfaces; and of course, the surface of the earth. The earth is represented on a map projection, preferably one that comes close to preserving angles and distances.

The task of analysis is essentially one of representing scalar fields by means of lines on the surfaces, and the recognition of sharp transition zones, maxima, and minima in these scalar fields. The lines are the intersections of one or more sets of surfaces with the surface represented on the chart. Thus a chart of an isobaric surface has lines corresponding to the intersections of various level surfaces, of temperature surfaces, of surfaces of equal humidity, and of other properties. Some vector fields, such as the wind field and the vorticity field, may be represented.

The various lines or surfaces each represent equal values of the scalar quantities. The general term for such a line or surface is *isogram* or *isopleth*, from the Greek *isos* meaning equal, *gramma* meaning weight, and *plethes* meaning quantity. The word *isoline*, a linguistic hybrid, is sometimes used. An incomplete list of isograms or isopleths, together with the quantities they represent, includes the following:

isobars—pressure
isotherms—temperature
isentropes—entropy or potential temperature
isotachs—speed
isokinetics—speed
isopachs—thickness

isohypses—height
isohyets—precipitation amount
isochrones—time of arrival
isosteres—specific volume
isopycnics—density
isogons—angle

In the government meteorological services machine analysis by appropriate programming of an electronic computer is practiced. A certain amount of manual correction and interpolation is necessary in order to ensure a product that is reasonably artistic and suggestive of the flow patterns. The analyses are transmitted by facsimile methods. Satellite and radar picturizations are also transmitted or read out directly by receivers at individual stations or centers.

In the following sections some of the principal ways of analyzing atmospheric structure and motions will be summarized. Not all of them are used in daily charting and forecasting, but all have an important place in study and research concerning atmospheric characteristics and behavior.

CROSS-SECTION ANALYSIS

Cross sections present a logical introduction to the vertical structure of the atmosphere.

The usual cross-section chart is quite simple, being based upon coordinates of

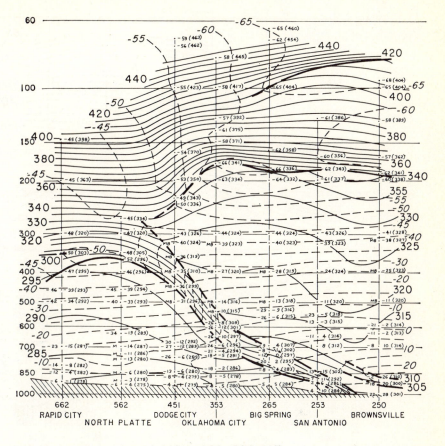

4 APRIL 1950, 1500 GCT

FIGURE 12-1

Example of a vertical cross section in a north-south direction. At each characteristic level at each station the temperature (°C) and potential temperature (K) are entered on the right and the dew-point temperature on the left. Isotherms of actual temperature are dashed lines, while the light lines are for potential temperature. Heavy lines represent frontal and tropopause surfaces.

horizontal distance and height or, preferably, against the more or less equivalent ln *p*. The vertical scale is exaggerated to about 250 times the horizontal scale, but the user is free to choose any scale he wishes. The cross section lies along an approximately straight line of aerological stations (they almost never lie in a straight line) and the significant points of each sounding are entered in their proper places over the locations of each station (see Figs. 12-1 and 12-2).

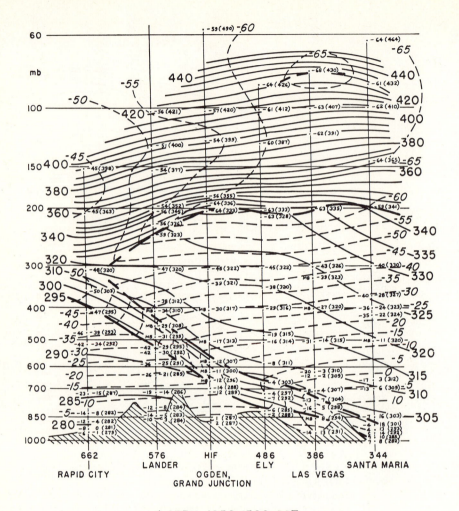

4 APRIL 1950, 1500 GCT

FIGURE 12-2

Same situation as Fig. 12-1 represented in east-west cross section.

On the cross sections, lines are drawn connecting equal values of temperature—isotherms—at intervals usually of 5°C, lines of constant mixing ratio or isentropic condensation temperature, potential temperature, and in addition, the locations of fronts, inversions, and tropopauses are entered.   In north-south cross sections the speeds of the zonal components of the winds are often represented by isotachs.   As in the cross sections in Figs. 12-1 and 12-2 and in others shown on previous pages, not more than two or three of the above-mentioned sets of isolines are drawn on a cross section; otherwise it would be too hard to read and interpret.

THE D SYSTEM

A system useful for analyzing the vertical distribution of highs, lows, ridges, and troughs was devised by J. C. Bellamy.[1] It provides for the charting of the departure of the pressure at a given height from that of a standard atmosphere at that height, or, conversely, the height departure for a given pressure. Called the D system, it is based on the height departure $D = Z - Z_s$, where Z is the true height and Z_s is the height for the corresponding pressure in the standard atmosphere. Bellamy chose the aeronautical standard atmosphere upon which altimeter readings are based; therefore, D is the altimeter correction at the given height and pressure.

Airplanes flying over the sea and equipped with both radio altimeters and pressure altimeters provide D directly as the difference between the readings of the two instruments. From ordinary upper-air soundings the meteorologist obtains Z and finds Z_s from a table or graph usually printed on his thermodynamic chart.

If an attempt were made to represent height profiles in a cross section in the x, p or y, p plane or pressure profiles in the x, z or y, z plane, it would be necessary to have the vertical scale expanded to many thousand times the horizontal scale. In the vertical exaggeration of some hundreds used in this chapter, the surfaces or lines would depart only slightly from the horizontal. With the D system, the high and low centers, ridges, and troughs are sharply revealed.

By definition, Z_s is constant on an isobaric surface; so

$$\left(\frac{\partial D}{\partial n}\right)_p = \left(\frac{\partial Z}{\partial n}\right)_p$$

While the Z surfaces cut the isobaric surface with the same spacing as the D surfaces, the former always intersect at a very low angle while the latter may intersect vertically. An appreciation of this fact can be gained from examining Fig. 12-3. The equality of the gradients on an isobaric surface permits the computation of the geostrophic wind, geostrophic vorticity, and related features by either system.

The D lines are useful for showing the vertical tilt of troughs and ridges and for indicating the extent to which the disturbances may be derived from conditions at the tropopause or higher. They are also useful in showing the presence of weak disturb-

---

[1] J. C. Bellamy, The Use of Pressure Altitude and Altimeter Corrections in Meteorology, *Journal of Meteorology*, vol. 2, pp. 1–79, 1945.

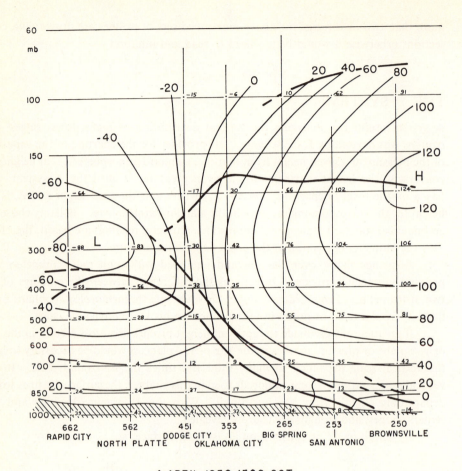

4 APRIL 1950, 1500 GCT

FIGURE 12-3
Cross section represented in the *D* system corresponding to Fig. 12-1. Values are in feet. The speed of the wind component normal to the cross section at any point is given by the *D* values at that point. At all points except in the low levels south of Big Spring this component would be from the reader into the page, i.e., westerly.

ances, such as are frequently found in the tropics. In cross sections, the horizontal gradient of *D*, or gradient along an isobaric surface, gives the geostrophic wind component normal to the section. Thus, on a north-south cross section the zonal wind components can be studied in detail for a chosen meridian.

ISOBARIC SURFACES

The use of isobaric surfaces has some advantages over the use of level surfaces in weather mapping, as was pointed out in Chap. 5. The 1000-mb surface approximates the surface of the earth except in elevated land areas. It is common practice at all stations to reduce the barometer readings to sea level, and in the process both the sea-level isobars and the heights of the 1000-mb surface can be obtained. Such reductions, of course, are poorly representative of the gradient flow at elevated stations. By international agreement the reports from aerological stations contain data for the standard isobaric surfaces of 850, 700, 500, 300 mb as well as, in some cases, surfaces in the stratosphere and mesosphere.

In forecasting practice the area analyzed to prepare the usual 1-day to 2-day forecast should be about the size of North America and adjacent ocean areas or larger, if practical. For predictions more than 2 days ahead, the entire hemisphere should be considered. This statement should not imply that the patterns of highs and lows, troughs and ridges maintain their identities around the hemisphere. Today's flow patterns may be hard to recognize 3 days hence. Changes only in the very broad features, such as total hemispheric wave number, major blocking actions, and total jet-stream energy, must be carefully followed in the analyses directed toward extended forecasting.

In Fig. 12-4 the surface (1000-mb) chart, with fronts and contours, is shown for the Northern Hemisphere for the same situation that is represented at 500 mb in Fig. 10-1 of Chap. 10. Note that the 1000-mb contour configurations are much more complex than at 500 mb and that the fronts are detached in a very complicated system difficult to relate to the simple jet-stream polar front at the 500-mb surface.

Analysis over the oceans and in other areas of sparse data can be accomplished by taking into consideration certain details of properties other than the scalar field being analyzed. For example, historical continuity in the systems must be observed to give a consistent change from one region to another and from one time to the next. In many cases wind data are available but pressure measurements have not been made. The geostrophic wind scale can be applied to the chart to estimate the direction and spacing of the contours. In regions of sparse data the satellite photographs are indispensable in locating storms and storm centers. In all cases it should be remembered that analysis is four-dimensional, requiring consistency in time as well as in the three space dimensions.

Perhaps the most useful check for consistency to be made in the preparation of charts for the various isobaric levels is the relation of the thickness patterns to the contour patterns. The thickness can be obtained from aerological stations by simple subtraction of the heights of the two isobaric surfaces being considered. One can then draw lines for this quantity on a map by the usual methods of scalar analysis.

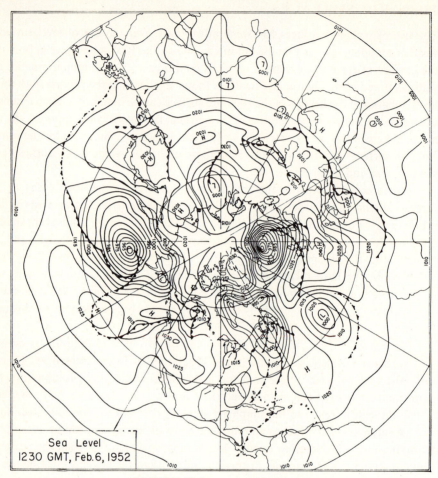

FIGURE 12-4
Surface chart corresponding with the 500-mb chart of Fig. 10-1.

From charts for two isobaric surfaces superimposed on a light table the thicknesses can be obtained immediately for any desired points.   One or both of the charts must be changed if they are not consistent with a reasonable thickness pattern.   The thickness pattern must also be consistent with the pattern of virtual temperature, since thickness between two isobaric levels depends only on the mean virtual temperature of the layer through which the thickness is measured.   Furthermore, the thickness pattern must be consistent with the thermal-wind pattern, since the gradient of virtual temperature and therefore of thickness determines the thermal wind.   In general, all surfaces in the atmosphere have fairly smooth contours and the gradients do not change sharply or vacillate widely.   If any of the sets of lines have to be strained into peculiar contortions or odd gradients in order to satisfy internal consistency, something is wrong.

Isotherms are usually drawn on the isobaric surfaces.   Since, as was pointed out in Chap. 8, these have the same distribution on an isobaric surface as isosteres, isopycnics, or isentropes, the baroclinicity or solenoidal field at that surface is shown.   On the lower isobaric surfaces, such as 850 mb, it is useful to draw lines representing the water-vapor distribution.   The tongues of moisture commonly found in summer weather show up well on this surface.

Beginners and also, to some extent, experienced meteorologists have difficulty locating fronts on the surface map.   A weather map with completely ideal frontal situations is virtually nonexistent, and the methods of dealing with frontal analysis can be learned only by hard experience.   As in all features of analysis, the fitting together of numerous factors in such a way that they will be consistent with each other establishes the work on a firm foundation.

One starts with a logical and consistent historical sequence of movement and frontogenesis (frontolysis).   The isobaric analysis locates the troughs in which major fronts often lie, but no rule requires that fronts be located in troughs.   All that is required is that there be a cyclonic wind shear *or* curvature in the vicinity of the front and that isobars crossing it have a cyclonic kink (often imperceptible in the coarse synoptic network).   Next, the various discontinuities are looked for—discontinuities in the wind, the temperature, the moisture, and the 3-hr pressure tendency as plotted at each station.   Clouds and precipitation areas and conditions at the 850-mb surface and lower help to confirm the locations of surface fronts.

ISENTROPIC ANALYSIS

In preparing an isentropic chart for an area, it is well to choose an isentropic surface which extends quite low in the warmer part of the area but which does not intersect the ground.   For North American isentropic charts the elevated lands around the northern border of Mexico are in the critical area; they have high surface potential temperatures and cannot be included on isentropic surfaces with lower θ values.   One should not go to the other extreme of selecting an isentropic surface that is too high, for then the influx of low-level moisture from the south cannot be studied and the northern portion is likely to be in the stratosphere.   The lowest usable potential temperatures for the entire United States are about 290 K in winter and 310 K in summer.

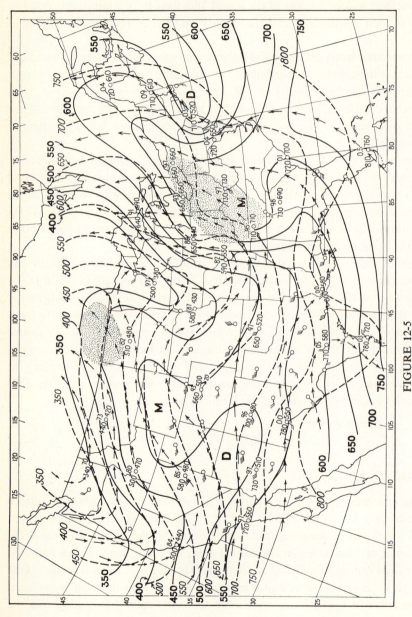

FIGURE 12-5

Isentropic chart, May 15, 1942, for potential-temperature surface of 307 K. Solid lines are for isentropic-condensation pressure; dashed lines for actual pressure; lines of arrows are streamlines. Shaded areas show saturation at this surface.

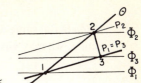

FIGURE 12-6

The two principal sets of lines drawn on the isentropic chart are those for pressure and those for isentropic-condensation pressure. The first show essentially the heights and slopes of the surface. The isentropic-condensation pressures give the moisture distribution. For a given isentropic surface the mixing ratio is uniquely determined by the isentropic-condensation pressure. Furthermore, the separation of these lines from the pressure lines measures the pressure interval through which the air must ascend to reach condensation. Corresponding values of the two sets of lines intersect at points which outline the area of saturation on the isentropic surface. Since appreciable supersaturation does not occur in the atmosphere, the lines do not cross. The saturation areas contain only the lines of actual pressure and are usually shaded on the chart.

The tongues of moisture maxima, labeled M in Fig. 12-5, are typically seen on isentropic charts, with dry areas, marked D in the figure, around them. If the isentropic surface does not intersect the ground outside of the tropics, these circulations may usually be considered as occurring above the fronts. Since isentropic-condensation pressure is a conservative property before saturation, the moist and dry tongues might be expected to show the trajectories of the air, and this they do. The moisture lines advance at a speed somewhat less than the wind speed at the isentropic surface, and this is believed to be due to loss of moisture by lateral mixing with the surrounding drier air. This process, on a broad horizontal scale, is similar to entrainment on the convective scale.

It is possible to represent geostrophic or gradient winds on an isentropic surface by means of a stream function ψ that can be developed from the geostrophic wind relations for an isobaric surface or a level surface. Consider the isentropic surface shown in cross section in Fig. 12-6 running through points 1 and 2. Isobaric surfaces and level surfaces, the latter in terms of geopotential Φ, are shown. We can write the difference in geopotential along the isentropic surface between 1 and 2 as

$$\Phi_2 - \Phi_1 = (\Phi_3 - \Phi_1) + (\Phi_2 - \Phi_3) \qquad (12\text{-}1)$$

In differential calculus we can write

$$\left(\frac{\partial \Phi}{\partial x}\right)_\theta \delta x = \left(\frac{\partial \Phi}{\partial x}\right)_p \delta x + \frac{\partial \Phi}{\partial p}(p_2 - p_1) \qquad (12\text{-}2)$$

Since $\partial\Phi/\partial p$ is the same along an isentropic surface as along any other surface in this cross section, we may write, after substituting from the hydrostatic equation $\partial\Phi/\partial p = -\alpha$,

$$\left(\frac{\partial\Phi}{\partial x}\right)_p \delta x = \left(\frac{\partial\Phi}{\partial x}\right)_\theta \delta x + \alpha \left(\frac{\partial p}{\partial x}\right)_\theta \delta x \qquad (12\text{-}3)$$

In an adiabatic process, that is, along an isentropic surface,

$$dq = c_p\, dT - \alpha\, dp = 0$$

and $\alpha\, dp = c_p\, dT$. This can be substituted in the last term, so that, after dividing through by δx, we have

$$\left(\frac{\partial\Phi}{\partial x}\right)_p = \left(\frac{\partial\Phi}{\partial x}\right)_\theta + c_p\left(\frac{\partial T}{\partial x}\right)_\theta \qquad (12\text{-}4)$$

$$\left(\frac{\partial\Phi}{\partial x}\right)_p = \left[\frac{\partial(c_p T + \Phi)}{\partial x}\right]_\theta = \frac{\partial\psi}{\partial x} = fv \qquad (12\text{-}5)$$

The last equality shows applicability to the geostrophic wind. The stream function is $c_p T + \Phi = \psi$. We have, then, on an isentropic surface,

$$\frac{\partial\psi}{\partial x} = fv \qquad (12\text{-}6)$$

$$\frac{\partial\psi}{\partial y} = -fu \qquad (12\text{-}7)$$

Both terms in ψ have the dimensions of ergs per gram or square centimeters per second per second. The value of c_p is $0.24 \times 4.186 \times 10^7$ ergs g^{-1} K^{-1}, and since T ranges roughly between 200 and 300 K in the atmosphere, $c_p T$ ordinarily lies between 2 and 3×10^9 ergs per g. Φ is 9.8×10^7 ergs per g at 1 km and 9.8×10^8 at 10 km. The values of ψ usually found in the usable isentropic surfaces range from about 2.9 to 3.2×10^9 ergs per g. The first term dominates, of course, in the low levels, but the second approaches it in magnitude near the tropopause. Normally the two terms vary inversely.

The streamlines in Fig. 12-5 are drawn for intervals of 10^7 ergs per g. As might be expected, they follow the isentropic isobars to a considerable extent. This relationship is in agreement with the fact that p on an isentropic surface uniquely determines T on that surface; also, Φ and p are related through the integral involving virtual temperature and pressure from sea level to the point in question, and T on the isentropic surface bears a relation, although a varying one, to the virtual temperature of the underlying column. The variations from place to place of sea-level pressure, temperature, and lapse rate in the underlying atmosphere cause the lack of coincidence of pressure and stream-function lines on isentropic surfaces.

The concept of air-mass movements and modifications implies that patterns of distribution of atmospheric properties are continually changing and being transported (advected) with the air currents.   Although isentropic motion of air particles means that their potential temperature is conserved and that they must remain on the isentropic surface, their paths may not follow the contours of the isentropic surface because the surfaces themselves are usually displaced.   Thus if a streamline intersects an isobar in the upslope sense, it does not necessarily follow that the air particles are ascending; the isobars may retreat at the same speed as the motion.   *Relative* motion must be considered.

Another type of relative motion—the geostrophic motion of air particles on an isentropic surface relative to that of air particles on a lower isentropic surface—was introduced by Starr.[1]   If the upper particles are moving relative to the lower ones and in an upslope direction, it is reasonable to assume that ascent is occurring.   The relative geostrophic velocities between the two have the components

$$u_2 - u_1 = -\frac{1}{f}\frac{\partial}{\partial y}(\psi_2 - \psi_1) \qquad (12\text{-}8)$$

$$v_2 - v_1 = \frac{1}{f}\frac{\partial}{\partial x}(\psi_2 - \psi_1) \qquad (12\text{-}9)$$

The quantity $\Delta\psi = \psi_2 - \psi_1$ is plotted on the lower of the two charts.   The quantity is the stream function of the geostrophic motion on the upper chart relative to that on the lower one.   The isopleths are streamlines of this relative motion.

The geostrophic wind scale as used on isobaric charts can be used in the same way on isentropic charts to obtain geostrophic winds, or the relative winds just described, from the spacing of the ψ lines.

KINEMATIC ANALYSIS

Up to this point the wind field has been considered as derived from the pattern of a scalar field, such as from heights on an isobaric surface or from stream function on an isentropic surface.   For a picture of the broad synoptic features this type of analysis, supplemented with plotted wind observations, is sufficient in many cases.   In studying smaller-scale details, in analyzing weather in relatively inactive areas such as the tropics, and in making the transition from the analysis to the forecast, certain analytical aids obtained from the field of motion are necessary.   Analyses of streamlines, trajectories, divergence, and vorticity are derived from a study of the winds themselves, or

[1] V. P. Starr, Construction of Isentropic Relative Motion Charts, *Bulletin of the American Meteorological Society*, vol. 21, pp. 236–239, 1940.

in some instances, these properties can be investigated from the geostrophic values derived from the conventional scalar analysis.

Wind speed is itself a scalar and can be represented on charts by means of *isotachs*. These can be based on the observed winds or the geostrophic values or a combination of the two. Streamlines, as defined in Chap. 8, can be used to represent the wind field as to direction. They are especially useful in the tropics where, owing to the relatively small value of the coriolis parameter, they are more revealing of the flow than the pressure analysis. The wind arrows or vectors plotted on the map are the tangent vectors which define the streamlines.

Two different types of streamlines are in use. One type, which we may call *directional* streamlines, consists of a set of lines drawn in such a way that the only requirement is that the wind be tangent to them throughout. The wind speed does not necessarily correspond with the spacing of these streamlines. At the windward edge of the map the spacing might be chosen to correspond with the speeds, but unless all changes in speed are accounted for in a nondivergent manner, a situation which is seldom the case over a large area, the relation of streamline spacing to speed is not preserved. In the second type, which we may call *flux* streamlines, the spacing is made to correspond with the wind speed. Each streamline can be labeled in terms of a stream function ψ, and the wind will then be proportional to the gradient of this function. The relation of wind to this streamline pattern would be analogous to the geostrophic relationship to isobars.

The directional streamlines have a characteristic not found in the scalar lines representing flow. They can branch, one from the other, join together, originate, or terminate at various places on the map. As long as they are not assigned stream-function values, they can begin or terminate at lines along which the wind is discontinuous, such as at fronts. They may spread outward or converge into a line of divergence or convergence. They diverge outward from the calm centers of anticyclones and converge spirally inward to centers of low pressure. The definition of such streamlines requires that where they join or separate they must do so at an infinitesimal angle. Some of these features are shown on the streamline chart of Fig. 12-7.

The flux streamlines can be made to represent divergence and convergence by the beginning and termination of the lines, although this requires that some liberties be taken with the definition of flux streamlines. The technique is illustrated in Fig. 12-8. Where new streamlines originate, the flow is characterized by horizontal divergence and where they terminate, the flow is convergent. The method is more pictorial than exact, for there are regions, designated by the dashed portions of the streamlines in the figure, where the spacing does not represent the speed. It will be shown presently that gradients of stream function, as well as the geostrophic relationship, are valid only for representing nondivergent flow.

FIGURE 12-7
Directional streamlines.

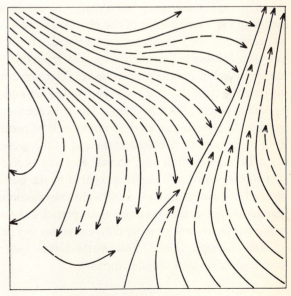

FIGURE 12-8
Flux streamlines for same flow as in Fig. 12-7.

It is sometimes useful to construct a map of *isogons*, that is, of lines of equal wind direction, expressed in scalar values of compass degrees. These are somewhat easier to analyze after the data have been plotted than are streamlines directly, and they aid in the construction of the streamlines. Isotachs may be drawn on the streamline chart to represent speeds. They are also useful in making certain computations of divergence, to be described presently.

In some forecasting and analytical applications it is desirable to determine the *trajectories* of the air. These are the actual paths followed by the air particles with respect to the surface of the earth. Streamlines give the instantaneous motion, but if the streamline pattern is changing or being displaced, the particles are subjected to a continually changing wind field and follow trajectories which may be quite different from the streamlines. The air acquires properties which depend on the type of surface over which it has been passing, that is, on the nature of the terrain covered by its past trajectory. For example, the study of trajectory will show whether the air has been over a body of water or not. Estimates of future trajectory are used in temperature and fog forecasting.

ANALYSIS OF DIVERGENCE

Since the divergence in the x, y plane is given by $\partial u/\partial x + \partial v/\partial y$, the most obvious way of analyzing for divergence on a map is to obtain these derivatives graphically. On one map the x component of the observed wind u is plotted and on another map for the same time the y component v is plotted. Isotachs of these respective components are then constructed on each map. By finite differences, the variation of u in the x direction and of v in the y direction are determined for as many points as are necessary to describe the field, and these two values are added together and plotted on corresponding points on a third map. Isopleths are then drawn to show the field of divergence. In practice this method is seldom used because of problems of the scale of the field of motion, as discussed at the end of this section.

In Eq. (8-79) it was shown that the horizontal divergence is also given by $1/A\ dA/dt$, where A is the area covered by the portion of air under study. A can be defined as the area outlined by straight lines connecting any three or more identifiable particles. Since balloons move as air particles, three or more balloons released simultaneously from three or more different points can be followed to obtain the variations in area between them. This method was used in determining divergence and convergence around thunderstorms.[1]

---

[1] H. R. Byers and R. R. Braham, Jr., "The Thunderstorm," Government Printing Office, Washington, D.C., 1950.

Divergence measurements made from synoptic charts can best be understood by considering the flow in streamline tubes, as defined in Chap. 8, which are capable of changing their vertical and horizontal dimensions by expanding in one dimension while contracting in the other, that is to say, by horizontal divergence compensated by vertical convergence and vice versa. In the derivation in Chap. 8 of the equation of continuity and the expression for divergence resulting therefrom, a cube of constant unit volume was considered. Let us now consider a section of a streamline tube which is not a cube and whose cross-sectional dimensions vary in the distance δs along the streamlines. Only the horizontal velocity c along the streamlines is considered. Then, after the manner of Eq. (8-60), we write

$$\text{Mass accumulation} = \text{inflow at } s_1 - \text{outflow at } s_2$$

$$\frac{dM}{dt} = \rho c \,\delta n \,\delta z - \left[\rho c \,\delta n \,\delta z + \frac{\partial(\rho c \,\delta n \,\delta z)}{\partial s}\,\delta s \right]$$

$$= -\frac{\partial}{\partial s}(\rho c \,\delta n \,\delta z)\,\delta s \qquad (12\text{-}10)$$

where δn and δz are the horizontal and vertical cross-stream dimensions, respectively. In expanded form this expression becomes

$$\frac{dM}{dt} = -\rho c \,\delta n \,\frac{\partial(\delta z)}{\partial s}\,\delta s - \rho c \,\delta z \,\frac{\partial(\delta n)}{\partial s}\,\delta s - \rho \,\delta n \,\delta z \,\frac{\partial c}{\partial s}\,\delta s - c \,\delta n \,\delta z \,\frac{\partial \rho}{\partial s}\,\delta s \qquad (12\text{-}11)$$

If the density is not changed and if vertical and horizontal shrinking and stretching compensate each other, the mass does not change. We separate the horizontal and vertical changes after setting $dM/dt = 0$ and $\partial\rho/\partial s = 0$, and write

$$\rho c \,\delta z \,\frac{\partial(\delta n)}{\partial s}\,\delta s + \rho \,\delta n \,\delta z \,\frac{\partial c}{\partial s}\,\delta s = -\rho c \,\delta n \,\frac{\partial(\delta z)}{\partial s}\,\delta s \qquad (12\text{-}12)$$

This expression indicates that horizontal stretching would be compensated by vertical shrinking.

Note that all the terms are in units of change in mass per unit time. If we divide by ρ, which is already assumed to be constant, we have the rate of change in volume and if, in addition, we divide by δz, we have on the left-hand side of the equation the rate of change in area of the floor (or ceiling) of the section of the tube, or

$$c \,\frac{\partial(\delta n)}{\partial s}\,\delta s + \delta n \,\frac{\partial c}{\partial s}\,\delta s = \frac{dA}{dt} = \frac{\partial(c \,\delta n)}{\partial s}\,\delta s$$

$$= c_2 \,\delta n_2 - c_1 \,\delta n_1 \qquad (12\text{-}13)$$

The horizontal divergence is expressed by $1/A \; dA/dt$, so the above expression divided by the area gives the horizontal divergence. Note that Eq. (12-13) divided by the area is precisely the same as Eq. (8-80).

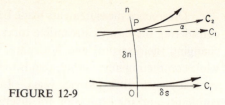

FIGURE 12-9

The method of computation on a map was shown in Fig. 8-14. The δn's are measured along a set of lines orthogonal to the streamlines. In obtaining A to divide into Eq. (12-13) one can set off equal downstream increments δs and multiply this constant number by the δn of the midpoint of δs.

It should be noted that if the velocity increases as δn becomes smaller, $c\,\delta n$ can remain constant ($c_2\,\delta n_2 = c_1\,\delta n_1$) and the wind is nondivergent. In the case of potential flow, $c\,\delta n = -\delta\psi = $ constant (ψ lines drawn for constant ψ interval), and the divergence is zero. Similarly, in geostrophic flow, $fc\,\delta n = -\alpha\,\delta p$, and if f and α do not change, the velocity is determined only by the closeness of the isobars. Therefore, divergence may be expected only if the wind is *ageostrophic*. Strong divergence or convergence is found only where winds depart markedly from the geostrophic, such as in thunderstorms, in sea breezes, at some fronts, and in developing cyclones.

A practical graphical method based on streamline analysis is possible through the use of natural coordinates. The wind vector c along s is tangent to a streamline at any point. Even if the streamline is curved, we note that as s approaches zero, we may write $\partial c/\partial s$ in place of $\partial u/\partial x$ in cartesian coordinates. In place of $\partial v/\partial y$ we can similarly write $\partial v/\partial n$, even though n may also be curved. The orientation of axes is seen in Fig. 12-9, where two diverging streamlines are shown with their tangent vectors at O and P, normal to n at their respective tangent points only. Over a finite distance δn the wind vector changes direction through the angle α, producing a component of velocity v in the n axis with respect to the base direction c_1. This component is given by

$$v = c \sin \alpha$$

We find that

$$\frac{\partial v}{\partial n} = c \cos \alpha\, \frac{\partial \alpha}{\partial n} + \sin \alpha\, \frac{\partial c}{\partial n} \qquad (12\text{-}14)$$

As δn approaches zero, $\cos \alpha$ approaches unity and $\sin \alpha$ approaches zero; so we may write

$$\frac{\partial v}{\partial n} = c\, \frac{\partial \alpha}{\partial n} \qquad (12\text{-}15)$$

We then have the horizontal divergence given by

$$\operatorname{div}_2 c = \frac{\partial c}{\partial s} + c\,\frac{\partial \alpha}{\partial n} \qquad (12\text{-}16)$$

Taking a pair of streamlines, one sketches an orthogonal n curve and uses the vector at the intersection on the right side of the streamline channel as the base for c, then determines the angle the vector at the streamline on the left makes with the right-hand one, as shown in the figure. The assumption is made that $\partial \alpha/\partial n$ is constant between the two streamlines. The velocity c is taken as the average of the two tangent vectors at O and P, and $\partial c/\partial s$ is determined by comparison with the average at a distance δs downstream, assuming that $\partial c/\partial s$ is constant through δs. An isogon analysis superimposed on the streamline analysis assists in the process. One obtains the difference in isogon values at both ends of the finite distance δn, multiplies by δn, and if isogons are in degrees, multiplies also by $\pi/180$ rad per deg.

As might be expected, the two terms tend to cancel each other in normal conditions, for a spreading of the streamlines downstream is normally accompanied by a downstream decrease in velocity. In potential or geostrophic flow, the two terms are of equal value and opposite sign.

As first pointed out by Panofsky,[1] the values of divergence obtained in the atmosphere are highly sensitive to scale of distance or area over which the measurements are made. Panofsky found that regardless of the scale used the change in wind was of the order of 1 to 10 m per sec, tending to give very little variation in the value of the numerator in the divergence expressions. The denominator depends on the scale chosen and the choice depends on the size of the phenomena considered. Thus, for an entire cyclone the divergence is of the order of -10^{-5} per sec (convergence) in the low levels while for a thunderstorm cell which may have a diameter $\frac{1}{100}$ of that of a cyclone, the value is of the order of $\pm 10^{-3}$ per sec. It is wise when quoting values of divergence to specify the size of the area considered.

ANALYSIS OF VORTICITY

The analysis of the wind field for vorticity is done in essentially the same way as for divergence. On a streamline chart with orthogonal lines drawn, and perhaps with isotachs and isogons included, the necessary information is available for vorticity analysis. For example, the same method that is employed for divergence measurement in Fig. 12-9 can be applied to vorticity computation. It can be shown, or by

[1] H. A. Panofsky, Large-Scale Vertical Velocity and Divergence, *Compendium of Meteorology*, 1951, p. 639.

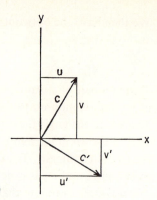

FIGURE 12-10

analogy with Eq. (12-16), it can be stated, that

$$\zeta = c\,\frac{\partial \alpha}{\partial s} - \frac{\partial c}{\partial n} \qquad (12\text{-}17)$$

It is seen that the vorticity is thus expressed as the sum of a shear term $\partial c/\partial n$ and a curvature term $c(\partial \alpha/\partial s)$, as in Eq. (8-39).

Another method is illustrated in Fig. 12-10. A vector c' is taken $90°$ to the right of the wind vector c, the two vectors having the components u', v' and u, v, as shown. It is noted that $u = -v'$ and $v = u'$ and it follows that

$$\frac{\partial u}{\partial y} = -\frac{\partial v'}{\partial y} \qquad \frac{\partial v}{\partial x} = \frac{\partial u'}{\partial x} \qquad (12\text{-}18)$$

The vorticity is given by

$$\frac{\partial v}{\partial x} - \frac{\partial u}{\partial y} = \frac{\partial u'}{\partial x} + \frac{\partial v'}{\partial y} \qquad (12\text{-}19)$$

Hence the vertical component of the vorticity is given by the horizontal divergence of the wind vectors rotated $90°$ to the right.

In Chap. 8 it was shown that the vorticity of the geostrophic wind is given by the curvature of the isobaric surface, such that it is positive (cyclonic) in curving portions of depressions and anticyclonic or negative around domes. The curvature of the surface can be computed from a large number of grid points. This may be done by an electronic computer as part of the forecasting problem.

MESOMETEOROLOGY

Some weather systems are too small to be followed in the standard synoptic network yet produce profound effects. Such organized systems as squall lines, masses of thunderstorms, sea-breeze systems, and large precipitation cells are in this category.

They are said to be on a *mesosynoptic* scale. As a rough classification based on scale we may list *micrometeorology*, which covers turbulence, diffusion, evaporation, and heat fluxes, especially in the layer next to the ground; *mesometeorology*, which is concerned with small-scale weather patterns over distances of perhaps 10 to 1000 miles; standard *synoptic meteorology*, which covers those phenomena observed in networks of stations 50 to 500 miles apart; and *macrometeorology*, in which the atmosphere is viewed on a hemispheric scale, with only the major waves, blocking actions, and broad circulation features considered. A fifth scale, called *cosmic meteorology*, might be added, concerned with the relation of the atmosphere to its interplanetary environment. All these scales require somewhat different analysis techniques.

Mesosynoptic analyses can best be accomplished where special, closely spaced station networks have been established for that or similar purposes. There are also a number of special airway and cooperative stations which are not included in the regular synoptic reports. Many of the special network stations are unmanned, since the cost of automatically transmitting the records to the analysis center is less than the cost of manpower. Either way the job is done, the communication load is such that most of the data have to be collected hours or days after the event from the autographic records. Thus the starting point of mesoanalysis is nearly always the autographic records—the barograph, thermograph, hygrograph, rain gage, wind, and other traces. These, of course, are supplemented by any visual, direct-reading, or upper-air observations made at manned stations.

An event such as a squall line or mass of thunderstorms leaves an unmistakable record at each station as it blusters in and works its way past the station. An example of the pronounced changes produced by a squall line is given in Fig. 12-11. Note that time increases from right to left in the figure. This reversal of the time scale when reproducing the autographic records is commonly practiced outside the tropics so that a time graph may be fitted with a space graph. The systems move from west to east, and west is always to the left. The conditions preceding the disturbance are those conditions found to the east of it, therefore to the right in map representation; so we find it convenient to have the prior events to the right and the later time to the left.

The time of arrival of the disturbance at each station can be recognized on the traces, and isochrones can be drawn on the map. Over short distances and short periods of time it can be assumed that the individual changes in properties of the disturbance are negligible, and that, taking p as an example,

$$\frac{Dp}{Dt} = \frac{\partial p}{\partial t} + c \cdot \nabla p = 0 \qquad (12\text{-}20)$$

$$\frac{\partial p}{\partial t} = -c \cdot \nabla p \qquad (12\text{-}21)$$

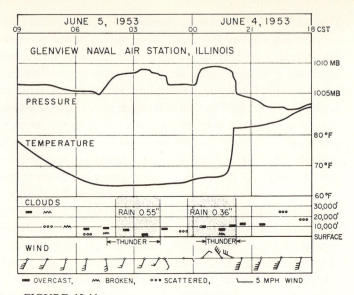

FIGURE 12-11

Autographic records during passage of a squall line. (*After Fujita.*)

In other words, the local change, as indicated on the autographic records, may be regarded as being entirely due to advection. Through this relationship it is possible to extrapolate and interpolate in short intervals of time and space. Time changes at a point can be substituted for space gradients through a knowledge of the rate-of-displacement vector c.

A variety of graphical methods, which will not be dwelt upon here, are employed to reveal features of the disturbances. One such method consists of obtaining from the surrounding, undisturbed stations the undisturbed pressure trace. In the disturbed areas only the pressure excess or pressure deficit as compared with the undisturbed field is treated in order to give a more striking picture of the important features. Other properties can be similarly represented.

Mesometeorological techniques were first directed toward an improved understanding of the mechanism of squall lines, that is, the lines of thunderstorms that do not occur along fronts. These disturbances are large enough to affect the analysis and forecasting on the standard synoptic scale, but not much more can be done on that scale than to indicate their presence. Fujita[1] showed that these lines, which may be several hundred miles long and may be accompanied by disturbed areas of as much as

---

[1] T. Fujita, Results of Detailed Synoptic Studies of Squall Lines, *Tellus*, vol. 7, pp. 405–436, 1955.

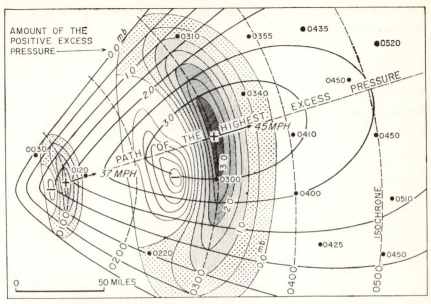

FIGURE 12-12
Spread of thunderstorm area with advancing squall line. Isobars of excess pressure are drawn for two times, 0100 and 0300, with hourly isochrones and envelopes of the various pressure values. (*After Fujita.*)

100,000 square miles, have a humble beginning in a single shower or thunderstorm. This growth from a point on the map is illustrated in Fig. 12-12. The squall line and accompanying thunderstorm area are shown in their 0300 CST position. The isochrones, represented somewhat like arcs centered on the point of origin, show the position of the squall front at other times. The radial lines at the ends of these isochrones envelop the maximum lengths attained by the squall system. Thus from the time of its beginning at 0030 CST to the time of the last isochrone represented here, the system has swept out an area bounded by the last isochrone and the outer radial lines.

Radar has its greatest usefulness in mesometeorology. Thunderstorms and squall lines show up markedly on the radarscopes, and in this type of analysis much use is made of photographs or digital displays of the radarscope presentations. In daily forecasting practice, severe local storms, usually in the nature of squall lines, are detected and followed by radar. Tornado alerts are issued on the basis of large squall systems, such as that represented in Fig. 12-12, as seen on the radarscopes over a range of perhaps 20 to 200 miles from the radar. The establishment of a network of weather radar stations in the United States has resulted in great reduction of losses of life and property from severe local storms.   Techniques have also been developed for using satellite surveillance.

EXERCISES

Exercises in meteorological analysis can be performed only with maps in the synoptic laboratory.   This chapter gives general guidance for such exercises.

13

TROPICAL METEOROLOGY

As we view the daily weather maps, we note that the perturbation activity diminishes rapidly toward the tropical borders of the charts. The impression is obtained that, except for the rare appearance of tropical cyclones (hurricanes or typhoons), the tropical atmosphere is characterized by monotonous inactivity. In terms of fronts, cyclones, and major wave disturbances this impression is correct. However, the tropical atmosphere is so heavily charged with water vapor that any slight low-level convergence will produce rain and squalls. Some of the smaller convergence areas cannot be detected on the standard synoptic scale, but many of them can. Tropical analysis and forecasting consist mainly in looking for minor disturbances and predicting their development and movement, especially with reference to the major rainstorms which they produce. In addition, of course, tropical cyclones are a major concern during the season and in the regions of their occurrence. Tropical meteorology also involves a variety of studies concerning heat and momentum transfers, meridional circulation components, monsoons, and other features of importance in the broad-scale flow patterns and in the general circulation.

ZONES OF CONVERGENCE

There are two principal convergence zones in the tropics—the *intertropic convergence zone* and the *monsoon convergence zone*. The former is the zone of convergence of the trade-wind systems of the two hemispheres. Unlike the polar-front zone, it has no significant, persistent temperature differences across it. The zone is one of weak air currents and, from nautical terminology, sometimes is called the *doldrums* or belt of calms. A net upward motion and therefore heavy clouds and rain are formed there. The intertropic convergence zone migrates seasonally as far as 15° or more north and south of the equator in some places, reaching its northernmost point in September and southernmost in March, corresponding to the seasons of highest temperatures in the oceans. These two extreme monthly mean positions are illustrated in Fig. 13-1. The reverse coriolis effect on trade-wind air from one hemisphere moving into the other accentuates the conflict in wind direction on the two sides on the zone. When it is at or near its seasonal maximum displacement, the intertropic convergence zone plays a part in the formation of tropical cyclones.

Monsoon-wind systems have convergence zones associated with them. Although monsoons are usually defined merely as winds that come in from the sea during summer and from the land during winter, the true monsoon amounts to more than that. In the Indian monsoon there is convergence between the very humid air from the Indian Ocean and the westerlies coming from the Mediterranean across the Arabian Desert and the Arabian Sea, and the heaviest monsoon rains occur in this convergence zone. Farther east, the monsoon is linked with the intertropic convergence zone, the monsoon air being Southern Hemisphere air from the South Indian Ocean. Winter monsoon rains occur in areas around the South China Sea from air that originally came out of Siberia and has become laden with moisture in an anticyclonic trajectory over warm water, a process accentuated by convergence toward stagnant air over the mainland and ascent of the mountains. In summer, all along the China coast the monsoon winds from the south converge with air from the zone of the westerlies. In Central America a weak summer monsoon on the Pacific side converges against the trade winds of the Atlantic and Caribbean, with mountain accentuation. A similar situation, where the monsoon and intertropic convergence zones are tied together, occurs in West Africa.

Rather than having directional convergence, the intertropic and monsoon convergence zones may be of the shear-line type at some points. Cyclonic shear in straight flow can produce positive vorticity, as shown on page 195, and if the flow is essentially zonal and the vorticity not decreasing, it must be accompanied by convergence. In addition to the intertropic convergence zone, other east-west shear lines of lesser magnitude are found in the tropics. Often they are the remnants of polar fronts swept into the tropics with the flow still exhibiting a shear although density

FEBRUARY RESULTANT WINDS

SEPTEMBER RESULTANT WINDS

FIGURE 13-1

Resultant winds over the oceans at the height of the northern and southern hurricane seasons, showing positions of convergence zones over the oceans.

differences have disappeared. In other cases they result from peculiar configurations of the flow imposed by crowding of the subtropical highs by extratropical disturbances.

The often-dramatized rainy seasons at tropical locations are associated with seasonal displacements of the convergence zones. Some tropical stations have two distinct rainy seasons, one when the intertropic convergence zone crosses the area going northward and the other when it passes returning south. The dry seasons between these periods can be as dramatic as the rains. Contrary to popular opinion, the greater part of the land areas of the tropics are grassy or only sparsely tree-covered because of the climatic stress of long dry periods each year.

Temporary moving convergence zones associated with moving disturbances will be taken up in later sections.

VERTICAL STRUCTURE OF THE WINDS

In the discussion of the general circulation in Chap. 9 the tropics were characterized as a region of easterlies at all heights. This picture is somewhat simplified. The subtropical (horse-latitude) ridges of high pressure are displaced equatorward with height, leading to a corresponding tilt in the boundary between the westerlies and the easterlies. The boundary may have a slope of about $\frac{1}{600}$ in the low levels so that a place in the easterlies 10° equatorward from the surface ridge line will have westerlies beginning at a height of about 2 km. The slope increases with height and distance from the surface position of the ridge to such an extent that 20° equatorward from the subtropical ridge easterlies are found throughout the troposphere. The seasonal shifting of the entire system toward the warmer hemisphere results in wide seasonal variations in depth of the easterlies, at least at stations 15° or more from the equator.

Across large areas of the deep tropics, *equatorial westerlies* are found seasonally or intermittently, often connected with monsoon circulations. These are most pronounced on the average across the Indian Ocean and into the western Pacific on both sides of the equator, but they have been observed on occasions in many tropical locations. Except where mountains affect the circulation, these westerlies are nearly always confined to the lower levels; they are thus quite distinct from the upper westerlies that are beyond the subtropical ridge, and from which they are separated by a broad and deep mass of easterlies.

The easterlies, or trade winds, have an equatorward component in the low levels usually explained by friction. But deep winter monsoon winds from the cold hemisphere, especially from the Asiatic continent, feed cold air into the trades to considerable heights and account for an appreciable meridional exchange of heat.

Because the tropical atmosphere is essentially barotropic, thermal-wind components are negligible in the deep easterlies. Near the outer limits of the trades, the winds often decrease with height above the friction layer, but this does not seem to be true in the deep easterlies. As the coriolis effect disappears near the equator, the thermal-wind component no longer has a real meaning, since it is based on the geostrophic assumption.

CHARACTERISTICS OF TRADE-WIND AND MONSOON AIR

In the brief discussion of the oceanic anticyclones in Chap. 9 it was pointed out that the air which they feed into the trade winds has undergone strong subsidence and is therefore dry aloft and has an inversion of temperature in the low levels. The temperature inversion is accentuated in the eastern parts of the oceans by cooling from below over cold ocean currents. Through the trade-wind belt the transition is made from a low and very strong inversion with stratus clouds or fog in the eastern part of the ocean to a disappearance of the inversion in the western part. Subsidence is likely to continue in the trade-wind region despite the addition of heat from below as the air moves over the warm tropical waters; so the effects of the inversion are seldom completely obliterated. Over the Caribbean, for example, a stable layer is usually found at about 2000 to 2500 m above the sea. Above it the air is remarkably dry. The depth of the moist layer varies with changes in the low-level divergence-convergence pattern. This stable layer is well within the easterlies and does not correspond with the boundary of the westerlies; in fact there is seldom much change in wind through it.

Where no large-scale convergence has taken place in the low levels, a characteristic type of convective cloud known as the *trade-wind cumulus* occurs. It does not have the "hard" rounded top found in a continental cumulus. It has its base at about 600 m and top at 1500 to 2500 m. Characteristically about one-third to one-half of the ocean area is covered by these clouds. The convection currents in them are weak, but nevertheless they are capable of producing showers of fine rain. In areas of convergence and over islands in the daytime they mass together to produce large towering clouds and copious downpours of rain.

The monsoon air and the air of the equatorial westerlies is generally more moist aloft than the typical trade-wind air, although there are some striking exceptions. In Fig. 13-2 two soundings are plotted to illustrate the difference between monsoon air and trade-wind air. The monsoon sounding for the Indian station shows higher temperatures and moistures than the mean Caribbean atmosphere, characteristic of the trade-wind region.

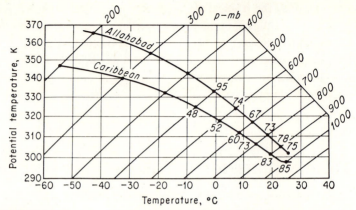

FIGURE 13-2
Distinction between monsoon air and trade-wind air as shown by mean monsoon soundings at Allahabad, India, and mean soundings in the Caribbean.

TRANSVERSE WAVES

Transverse waves, with axes across the trade-wind current, are nearly always found moving through the tropics. They are essentially of two types: (1) waves in the deep easterlies or (2) waves in the upper westerlies over shallow easterlies. While neither of these looks impressive on isobaric or streamline charts, the high moisture content of the air results in the production of copious rains. The troughs of these waves appear only as poleward bulges of the isobars or streamlines. In this sense the two types have the same appearance at the surface. Their differences are found aloft and also in their direction of movement. The waves in the deep easterlies exhibit only perturbed easterlies at all heights. The waves in the upper westerlies are found where westerlies extend aloft equatorward over the easterlies, and they progress from west to east, against the low-level winds.

The transverse waves cresting away from the equator establish patterns of convergence and divergence depending on the meridional component of the disturbed flow. This effect can be shown by considering the relationship of the vertical component of absolute vorticity and the horizontal convergence as given in Eq. (8-87), or

$$\frac{1}{f + \zeta} \frac{d}{dt} (f + \zeta) = - \text{div}_2 c \qquad (13\text{-}1)$$

We will consider Northern Hemisphere conditions, but with suitable interpretation of signs the equations will be applicable to the Southern Hemisphere. The variation of the coriolis parameter may be written as

$$\frac{df}{dt} = \frac{df}{dy}\frac{dy}{dt} = \beta v \qquad (13\text{-}2)$$

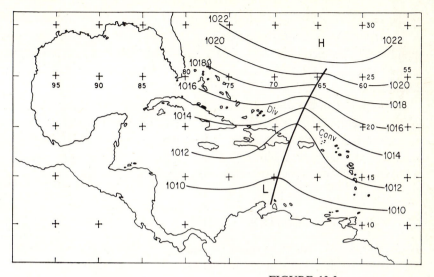

FIGURE 13-3
Transverse wave in the easterlies.

where β is the increase in f with linear distance northward. This substitution and a rearrangement of terms produce

$$\mathrm{div}_2\, c = -\frac{1}{f+\zeta}\left(\frac{d\zeta}{dt} + \beta v\right) \qquad (13\text{-}3)$$

The term β is always positive; f is positive and, near the equator, quite small. The denominator of the right-hand term can be negative only if the relative vorticity ζ is anticyclonic (negative) and greater in absolute value than f, which is usually not the case in these waves. It turns out that if the wave is not moving too rapidly, all the terms are positive to the east of the wave axis. This can be seen from the typical example of such a wave in Fig. 13-3. Where the winds have a southerly component, that is, where βv is positive, ζ is increasing. On the western side, ζ is decreasing as the air moves back southward again; so both terms are negative, indicating divergence.

The weather connected with the waves reflects the field of horizontal divergence, exhibiting large, massive clouds and showers or continuous rain to the east and in the center. To the west, divergence and subsidence suppress the convection, the moist air is shallow, and the stable layer characteristic of the trade-wind region is quite well marked.

Over the tropical oceans, at least in their western parts, there is a more or less continual succession of these waves, producing alternate periods of bright days and heavy rains. Over land masses and larger islands they are difficult to detect because of the overpowering orographical and diurnal effects.   The differences in weather during their passage are accentuated at places on the southern sides of mountainous islands.   The protection of the mountains makes the air unusually hot, dry, and clear in the current from the north, but in the southerly winds the convective activity is accentuated as the air piles up against the mountains.

The waves in the deep easterlies progress westward at a speed of less than that of the basic current.   The troughs usually tilt slightly toward the east with height and most often have their maximum intensity somewhere between 1500 and 5000 m above sea level.   As shown by Riehl,[1] these disturbances often show up in a marked manner even on the 200-mb surface, roughly at 12,000 m.

TROPICAL CYCLONES

Tropical cyclones, called *typhoons* in the western North Pacific and *hurricanes* in the North Atlantic and Caribbean, are quite different from extratropical cyclones. The following distinguishing features are most prominent:

1   They have no fronts associated with them; they form in a barotropic atmosphere.

2   They are found only at certain seasons.

3   They are found only in certain regions of the tropics.

4   They do not form with any regularity, even in their appropriate season and region, which suggests fortuitous circumstances in their origin.

5   They form only in those ocean areas having a high surface temperature, 26 or 27°C being the lowest such temperature observed at the time and place of formation.

6   Pressure and other properties are fairly symmetrically distributed around the center, with gradually rounded and nearly circular isobars.

7   They are not associated with moving anticyclones.

8   They derive their energy from the latent heat of condensation.

9   They are usually about one-third the diameter of extratropical cyclones.

10   They are many times more intense than extratropical cyclones, occasionally having a central sea-level pressure of 900 mb or lower and surface winds often exceeding 100 knots.

11   They can exist only over the oceans and die out rapidly on land.

12   They have a central core of calm or very light winds.   This central region, called the *eye*, has a diameter, on the average, of about 20 km and is largely free of heavy clouds.

[1] H. Riehl, On the Formation of Typhoons, *Journal of Meteorology*, vol. 5, pp. 247–267, 1948.

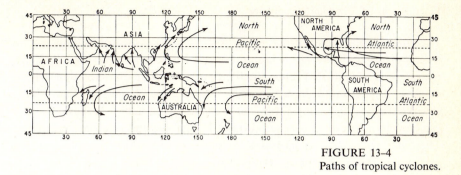

FIGURE 13–4
Paths of tropical cyclones.

13 They are seldom observed within five degrees of the equator, thus indicating the importance of the coriolis force in their development and maintenance.

14 The centrifugal effect in the gradient wind is of great importance; the cyclostrophic term is about thirty times the geostrophic term at a radius of 25 miles from the center at lat 15° with a 100-mph wind.

Geographical and Seasonal Distribution

The map in Fig. 13-4 shows the tropical-cyclone regions of the world and the principal paths of the storms. It will be noted that there are eight regions of tropical cyclones: one in the North Atlantic, two in the North Pacific, two in the region of India, one in the South Pacific, and two in the South Indian Ocean. There are great areas of the tropical seas, notably the entire South Atlantic and the eastern South Pacific, that are entirely free of tropical cyclones.

Of all the regions of the earth, the southwestern North Pacific has by far the greatest number of tropical cyclones. The western North Atlantic and Caribbean Sea is perhaps the best-known hurricane region of the world. The hurricanes of this region, although surpassed by the tropical storms of the Far East in intensity, have attracted a great deal of attention because of the damage they have caused in the West Indies and on the highly developed Atlantic Coast of North America. The South Indian Ocean and the South Pacific have frequent, long-lived tropical cyclones. The frequency in the South Indian Ocean is exceeded only by that of the western North Pacific.

Late summer and early autumn is the time of greatest tropical-cyclone activity. In the Far East the season is longer; also, off-season hurricanes are sometimes encountered in mid-ocean in the tropical North Atlantic. The peak frequency corresponds to the period of highest ocean-surface temperatures and also to the time of maximum poleward displacement of the intertropic convergence zone.

Formation of Tropical Cyclones

The processes leading to formation of tropical cyclones are not well known, although several plausible theories have been presented. Since they form regularly in the same regions, it is obvious that some peculiarity of circulation or thermodynamic processes in those regions must be associated with their development. These regions are characterized by high temperature and water-vapor content. It is found that tropical cyclones do not develop over sea surfaces having a temperature less than about 26°C. This kind of heat source in itself is not sufficient to start a hurricane going, although the heat of condensation supports the process once it is started.

Tropical cyclones are generated in disturbances along the intertropic convergence zone, in transverse waves, or under superimposed upper disturbances. In these situations, however, only a fraction of the disturbances lead to fully developed hurricanes or typhoons. Prediction of the formation would be easy if it were clearly apparent that similar synoptic situations produce similar dynamic processes in these tropical disturbances, but positive indications of future development do not come that easily in the tropics. One dynamic requirement, unquestionably the most important one, is that upper-air divergence must exceed low-level convergence in order for surface pressures to decrease. This "deepening," as it is called, is expressed by the tendency equation (8-93), page 214, integrated from 0 to ∞.

In order for development to take place, the proper combination of circulation, divergence, and convergence must be maintained over a considerable period of time on a proper scale. The marked selectivity of the tropical cyclone in the scale on which it will operate can be partially explained in theoretical models. In its initial stages the disturbance seems to need an upper divergence pattern producing a suitable vertical circulation over a distance of the order of 100 miles. It is probable that the so-called "minor" disturbances, those that have well-defined centers but never develop hurricane winds, do not have the right patterns of divergence, or if the patterns are suitable they are either too large or too small in scale.

The difference between a minor disturbance and a tropical cyclone also often appears in the vertical temperature distribution. The minor disturbances may have temperatures in their active centers lower than in the surrounding tropical environment, whereas the tropical cyclone is nearly everywhere warmer than the surrounding, undisturbed tropical air. In the former, the convection, with entrainment, may be so cellular in character that there may be no total net upward flow of heat. In the tropical cyclone there are large areas of well-organized ascent. Energy is derived from the massive liberation of the latent heat of condensation. When a minor disturbance originally cold aloft is transformed into a vigorous cyclone, it is invariably noted that it changes to a warm-core system.

Vertical Structure of Tropical Cyclones

Soundings in the rain areas which comprise the major part of a tropical cyclone show that the values of temperature and temperature lapse rate are close to those expected from the ascent of a parcel from the surface.   In the eye of the storm much higher temperatures are observed, which can be accounted for only by a certain amount of dry-adiabatic descent.   The simplified thermal picture of the cyclone can be depicted by three vertical temperature curves: one representing the tropical environment, one for the ascending air of the rain area, and another taken in the eye.   See, for example, the distributions in Fig. 13-5.

Actual measurements usually show that the rain area can be represented by various values of temperature, indicating that temperatures near the center are higher than those in the outer part of the rain area.   As the air spirals inward, it is subjected to sea-level pressure decreases that could lower the surface temperature by 3°C or more if the process were adiabatic.   The ocean is such a massive heat reservoir that it is able to supply enough heat to maintain the high temperature in the air immediately above it; so the pressure is reduced isothermally.   The heat added is equal to the work done by the air in expanding.   This added heat is reflected in the soundings.   Furthermore, the capacity of the air for water vapor is greater at the lower pressure.   Thus the wet-bulb-potential temperature is increased, prescribing higher-valued pseudo-adiabatic ascent curves and accordingly greater buoyancy.

The three cross sections through hurricane Hilda, 1964, shown in Figs. 13-5, 13-6, and 13-7 as presented by Hawkins and Rubsam,[1] are typical.   They show temperature anomalies, D-values, and winds, respectively.   The temperature anomaly with respect to the mean annual tropical atmosphere is positive up to about 125 mb, having its maximum at 250 to 300 mb.   The eye is warmest, but it is also warm outside the eye at all heights, especially between about 200 and 500 mb.   The high, cold tropopause is revealed in the negative anomalies from about 125 mb upward.

The D-values, based on a normal tropical atmosphere, show, obviously, large negative values in the eye, reaching 0 at about 220 mb there.   Even though positive D-values are found in the upper part of the hurricane, the slope of the lines shows that the pressure at all levels in the hurricane is lower than to the east or west in this particular cross section, because the pressures outside are higher than normal.

The winds in Fig. 13-7 show the nearly vertical eye "wall" with a calm eye all the way up surrounded by a sharp wind maximum, most pronounced in the low levels.

The three quantities represented in the three cross sections are, of course, interrelated.   Hydrostatically, the temperature anomalies account for the pressure-height

---

[1] H. F. Hawkins and D. T. Rubsam, Hurricane Hilda 1964, *Monthly Weather Review*, vol. 96, pp. 617–636, 1968.

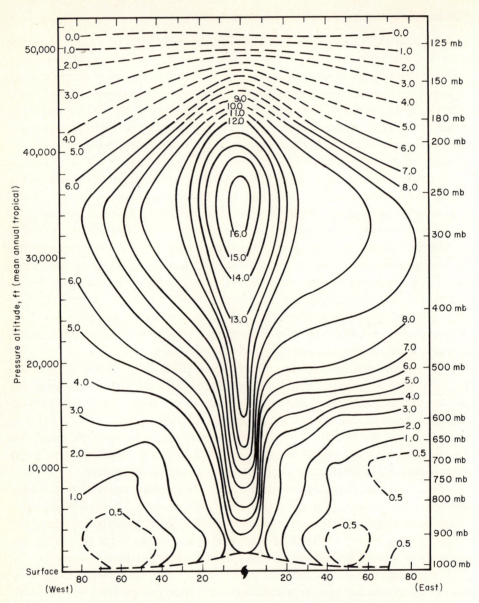

Radial distance in nautical miles from geometrical center of hurricane eye

FIGURE 13-5
Vertical cross section of temperature anomaly (°C) for hurricane Hilda, 1964.
The anomalies are based on the mean annual tropical atmosphere.

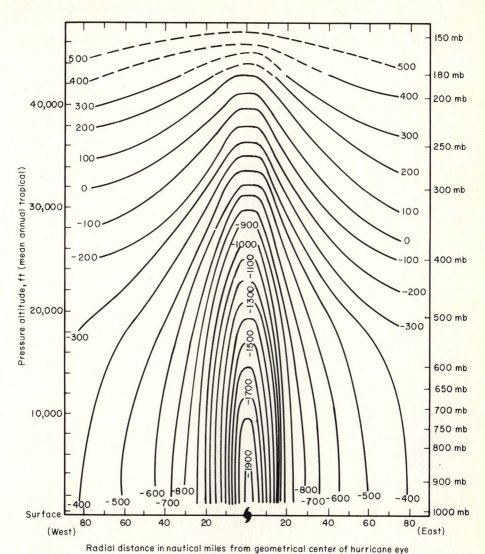

FIGURE 13-6

D-values (ft) represented as departure from mean tropical atmosphere, for hurricane Hilda, 1964.

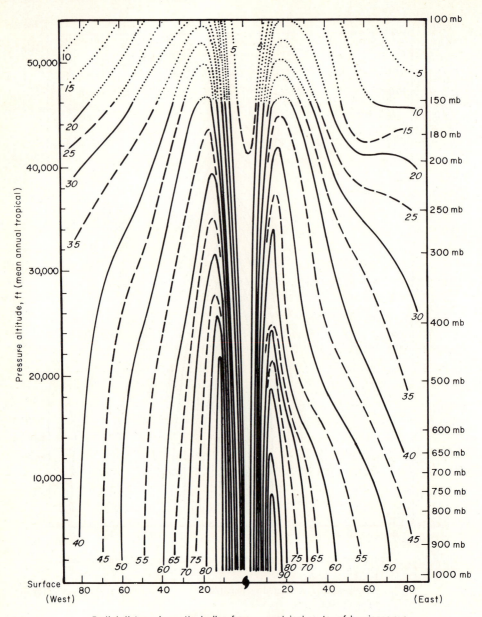

FIGURE 13-7
Wind speeds relative to the moving hurricane eye, Hilda 1964.

anomalies represented by the D-values.   The horizontal gradients of the latter are related to the winds, but in the areas near the eye the balance is cyclostrophic, as defined in Eq. (7-43) in Chap. 7.

The temperature cross section indicates that the middle and upper troposphere gains heat from the condensation-precipitation process.   Although the greatest positive anomalies are at 300 mb, the strongest horizontal temperature gradients are at 500 mb and lower.   A "wall" of cumulonimbus cloud extends upward around the eye, forming an essentially vertical solid boundary to the eye.   The eye wall is the zone of greatest temperature gradient, strongest winds, and strongest, most persistent upward motion.   Flights through hurricanes show that the wall cloud extends typically to about 15 km (ca. 150 mb).   A huge anvil extending outward from this giant cloud 100 km or more represents the outflow which completes the more or less steady-state circulation of inflow in low levels, ascent in the wall cloud, and outflow above 500 mb. Smaller clouds farther out may penetrate the anvil, but they have downdrafts in them that approximately balance the updrafts.   Weak descending motion beyond 50 km from the center completes the vertical circulation.   All the convective cloud forms occur in bands.

The large, dominant wall cloud may be highly developed in only a portion of the area surrounding the eye.   In a study of hurricane Gladys 1968, Gentry, Fujita, and Sheets[1] used photographs from manned spacecraft, radar, and aircraft data to obtain details of cloud structure.   They found the cap of the giant off-center cloud to be circular and to have outflowing winds.   They called it the "circular exhaust cloud" and surmised that it was a characteristic of a developing hurricane.   In 4 hours and 47 minutes, the diameter of the cloud increased from 55 to 120 km.   The spreading top was fed by a relatively narrow tower of ascending air, having a maximum updraft speed of 10 m sec^{-1} at 6 km.

Figure 13-8 is a photograph taken from the Apollo 7 manned spacecraft of hurricane Gladys 1968.   It shows the circular exhaust cloud and the spiral patterns of lesser clouds around the center.   At this stage Gladys had hardly reached hurricane intensity.   That is why the picture is so revealing.   Photographs taken of fully developed hurricanes often show such solid cloudiness that the dynamic features are not evident.

Precipitation

Rain is distributed around a tropical cyclone in bands spiraling inward toward the center.   The bands exhibit a cross-wind component of about 20° to the left in the

[1] R. C. Gentry, T. Fujita, and R. C. Sheets, Aircraft, Spacecraft and Radar Observations of Hurricane Gladys, 1968, *Journal of Applied Meteorology*, vol. 9, pp. 837–850, 1970.

FIGURE 13-8
Photograph from Apollo 7 manned spacecraft of hurricane Gladys at 1531 GMT October 17, 1968. The saucer-like top of the exhaust cloud and the spiral bands of convective clouds are plainly visible. The hurricane center is just beyond the exhaust cloud at the right-hand edge of the cloud. The picture was taken from a point above the northeastern Gulf of Mexico looking toward western Cuba, or toward the southeast.

Northern Hemisphere, to the right in the southern. Radar and satellite presentations show the rain bands as a prominent identifying feature of tropical storms, as shown in Figs. 13-9 and 13-10.

Lines or bands of convective clouds are found not only in tropical cyclones but throughout the world, especially over the oceans. Squall lines are pronounced examples of this type of distribution, and as in the case of hurricane bands, they form at a small angle across the winds in which they are embedded. In tropical cyclones the curvature of the streamlines requires that the rain bands also be curved. The bands terminate in a converging pattern around the center, but the eye remains rainless and almost cloudless. The eye appears as a dark core on radarscopes.

Some of the heaviest rains of record in low-latitude land areas have been experienced with slowly moving tropical cyclones; fast motion reduces the duration and therefore the accumulation at a fixed point of measurement. Near the center where

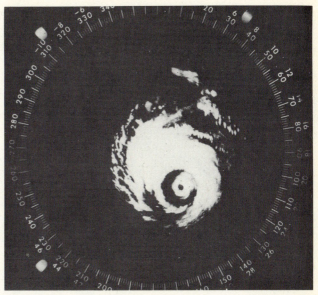

FIGURE 13-9
Photograph of radarscope at Brownsville, Texas, as hurricane Beulah 1967 was enveloping the station.   The central eye is to the southeast of Brownsville.   The radar is at the exact center of the radarscope.   Rain bands have merged into a uniform mass near the station.   (*NOAA photograph.*)

the bands are joined, the rain is heavy and essentially continuous, although the eye is free of rain.   Outward from the center the bands are farther apart, and the rain is likely to have a repetitive squall character as the successive bands come past, especially if the storm center is moving toward or away from the point of observation.

Late Stages and Dissipation

As long as a tropical cyclone exists over warm waters, it has the possibility of increasing in intensity, at least up to a certain theoretical maximum when the frictional forces and work against the surrounding atmosphere use up all the energy the storm can muster.   Many tropical cyclones follow the wind circulations of the upper atmosphere and break out of the tropics into the middle-latitude westerlies at the western ends of the oceanic high-pressure cells.   As they do this, they are said to *recurve* at the moment in their life histories when they cross the horse-latitude ridge of high pressure.   It so happens that the warm ocean currents are also driven into high latitudes on the western sides of the high-pressure cells; so the storms after recurvature may still follow a path over warm water.   Thus in the North Atlantic, for example, they can

FIGURE 13-10
Picture transmitted from a satellite showing hurricane Beulah 1967, lower right, in the Gulf of Mexico.   At this time it has just crossed the Yucatan Peninsula and is headed toward the Texas-Mexico coast.   The spiral rain bands are prominently shown.   Figure 13-9 shows the same storm as it moved into Brownsville, Texas.   (*NOAA photograph.*)

still maintain hurricane intensity as far north as New England and the principal northern steamer tracks. In passing over the cold waters near Newfoundland, they decrease rapidly in intensity.

Tropical cyclones also dissipate quickly when they move over large land masses, even in the tropics. As shown in the examples of mesometeorological analysis in the preceding chapter, convective rains cool the air in the low levels. The rain water, which arrives at the ground with about the same temperature as the cool air, covers the ground and holds the temperature down while the cloud cover prevents heating by the sun. Over the warm ocean, of course, the amount of rain water is infinitesimal when compared with the sea water and therefore does not cause a measurable decrease of the ocean temperature. (A further explanation of the relatively cool air under showers will be given in the chapter on thunderstorms.) Thus, over land, the hurricane loses the heat source necessary to keep the air buoyant, and the thermo-convective driving mechanism disappears. The added surface friction over the land also has a considerable effect.

While still quite intense over warm ocean currents of middle latitudes, tropical cyclones frequently draw fronts into their circulations. There is evidence that this event causes a brief increase in intensity as a new source of potential energy is added, but in the cool air outside the newly created warm sector, conditions are no longer favorable for convective ascent. The storm gradually takes on more of the character of an extratropical cyclone, becoming larger but less intense. The advent of fronts in the system is usually accompanied by the superposition of a polar trough to the west which also helps the intensification. Such a situation existed in the New England hurricane of Sept. 21, 1938, and in hurricane Hazel of 1955.[1]

EXERCISES

1 Plot the sounding in Fig. 13-2 for monsoon air at Allahabad on a thermodynamic diagram and construct an energy diagram as in Fig. 6-7 or 6-8 of Chap. 6. Assume that a dry-adiabatic lapse rate is achieved from 870 mb to the surface and use the mean humidity of that layer for an ascending parcel. Compare with a similar diagram for the Caribbean mean sounding, assuming dry-adiabatic conditions over an island below 850 mb.

---

[1] The custom of applying girls' names to tropical cyclones was started in World War II, apparently suggested by George Stewart's novel "Storm," in which the junior meteorologist at the San Francisco office of the U.S. Weather Bureau had the whimsey of applying girls' names to storms in the Pacific. He was dealing with extratropical cyclones, however. C. H. Pierce, The Meteorological History of the New England Hurricane of Sept. 21, 1938, *Monthly Weather Review*, vol. 67, pp. 237–285, 1939. L. A. Hughes, F. Baer, G. E. Birchfield, and R. E. Kaylor, Hurricane Hazel and a Long-wave Outlook, *Bulletin of the American Meteorological Society*, vol. 36, pp. 528–533, 1955.

2   In a wave in the easterlies a reasonable value for the v component on the eastern side is 5 m sec^{-1}. ζ can be of the order of 2×10^{-5} sec^{-1}, mainly from shear, while its time derivative may be of the order of 10^{-8}.   With these values apply Eq. (13-3) to obtain the horizontal convergence at latitude 18°N.   Show that the term βv can be neglected in cases of this kind so that $\text{div}_2 c = -(d\zeta/dt)/(f + \zeta)$.

3   Find the ratio of the centripetal term to the coriolis term in a circular hurricane at 15° latitude having a wind speed of 50 m sec^{-1} at a radius of 50 km.

14

QUANTITATIVE APPROACHES TO
WEATHER PREDICTION

How do you forecast the weather? The answer to that question is vastly different today from what it was twenty or thirty years ago. The change has come about by virtue of the introduction of advanced designs of electronic computers in weather prediction. This technological revolution in the ancient art of weather forecasting has not lessened the importance of the individual forecaster's functions in the local areas. The large-scale prognosis comes to him from the computer-equipped National Meteorological Center, and he refines and adapts the machine-made prognostic chart to make predictions for his region. He uses many of the concepts treated in previous chapters of this book as background knowledge of processes that he must predict. The alert student probably has recognized a number of quantitative and qualitative factors that have been discussed in the preceding chapters that seemed to be of value for weather prediction.

In this chapter only the quantitative approaches to general weather predictions will be treated, and these only in outline form. The qualitative or semiquantitative prediction methods can best be understood through direct contact with synoptic·aerologic charts and through experience.

Before describing the quantitative methods used in weather prediction, it is appropriate to consider some of the reasons why progress in this important area of meteorology is so difficult. A large part of the atmospheric circulation is made up of perturbations going through varied cycles of development and dissipation, giving a character of randomness and instability to the flow, which sometimes appears as hemispheric-scale turbulence. The movement and development of the perturbations (cyclones and anticyclones, troughs and ridges) can be extrapolated from a series of current synoptic charts with some success, but accelerations and intensifications are difficult to predict. For example, net divergence effects or deepening, computed by the tendency equation of Chap. 8 [Eq. (8-93)] come out as a very small difference between very large quantities which themselves may not be determined within an accuracy of 10 percent.

The generation of heavy cloud and precipitation systems depends on the concentration of upward motion in certain parts of the atmosphere where such motions are seldom measured. Or, rain may be associated with a mesoscale system inadequately detected on the regular synoptic scale available to the forecaster.

Despite these difficulties, meteorologists produce predictions which governments and businesses consider to be economically valuable and which the general public finds useful. An important ingredient in this partial success has been the subjective judgment of the forecaster based on knowledge and experience. A variety of computed prognoses emanating from the national center has reduced the number of value judgments that the forecaster must make.

Forecasting can be considered in two steps: (1) the prediction of the flow pattern (isobar-contour pattern) and (2) the determination of the quantities in which the users are interested, such as times and places of precipitation, temperature changes, and windstorms. The first step lends itself more easily to quantitative treatment than the second, although these latter weather elements are, indeed, related to the flow pattern. It is feasible to develop a system that combines the two steps, especially for those elements most directly derived from the predicted field of motion, such as upper-air winds, precipitation areas resulting from broad-scale vertical motions, and temperature changes caused by advection.

KINEMATICS OF THE SURFACE CHART

Surface synoptic observations include the pressure tendency as revealed by the barograph trace for the 3 hr immediately preceding the observations. This quantity, considered in connection with the pressure gradient, provides essentially current information concerning the rate of displacement of pressure systems. For local predictions this information can be used to extrapolate the movement of surface isobars, troughs, and ridges over short periods of time.

Station pressure changes always result from some form of movement or deformation of isobars. The pressure change in 3 hr at a station is the number of unit isobars crossing the station in that period.   The average speed of the isobars in a 3-hr period is given by

$$U = \frac{T}{N} \qquad (14\text{-}1)$$

where T is the 3-hr pressure tendency and N is the number of isobars per unit distance along the line of displacement.

If T is negative, the movement of the isobars is toward higher-pressure regions (lower pressure moving in).   If it is positive, the movement is toward lower pressure (higher pressure moving in).   If this rule is kept in mind, the sign of the tendency should be easily interpreted.   Another method is to consider that $N = -\Delta p/\Delta L$, where L is the unit of distance in the direction normal to the isobars, chosen positive in the indicated sense of the motion.   If this convention is followed, U will be positive when T and N have the same sign.

To obtain the speed of a trough or ridge, the relation could be applied to any set of isobars in the vicinity.   The displacement of a symmetrical trough or ridge is given simply by the average of the two speeds, front and rear.   For an asymmetrical system, that is, one in which the trough or ridge line does not bisect the angle the isobars make with it, a correction is necessary, but it is of consequence only in cases of extreme asymmetry.

This "old-fashioned" method of extrapolation is useful to the local forecaster in issuing short-period predictions of the time of passage of a cold front at places under his surveillance.   Many times these fronts do not have enough precipitation associated with them to be traceable on the radar; so the tendency indications may be the only information the forecaster has to work with.   Such activities as recreational boating, offshore drilling, and construction work often need short-term warnings of cold-front passages.

MOVEMENT OF THE UPPER WAVES

The waves in the upper westerlies, such as those appearing on the 500-mb chart, can be treated by simplified equations which are of forecasting value.   The formulas were developed by Rossby,[1] and they refer to a class of waves sometimes called *Rossby waves.*

---

[1] C.-G. Rossby and Collaborators, Relation between Variations in the Intensity of the Zonal Circulation of the Atmosphere and Displacement of the Semipermanent Centers of Action, *Journal of Marine Research*, vol. 2, pp. 38–55, 1939. C.-G. Rossby, Planetary Flow Patterns in the Atmosphere, *Quarterly Journal of the Royal Meteorological Society*, vol. 66 (Supplement), pp. 68–87, 1940. Kinematic and Hydrostatic Properties of Certain Long Waves in the Westerlies, *University of Chicago Department of Meteorology Miscellaneous Reports*, no. 5, 1942.

The following simplified assumptions are made: that the wave is sinusoidal in shape and does not change its speed, shape, or amplitude during the period under consideration; that the streamlines coincide approximately with the contours of the isobaric surface; that the motion is horizontal, adiabatic, at constant speed, and non-divergent; that the wave and air motion may be represented in a system of rectangular coordinates and are essentially one-dimensional in character; that the axes of the troughs and ridges lie in the north-south direction; and that $\beta = df/dy$ is a constant through the relatively small latitude range included in the amplitude of any given wave.

These simplified assumptions are desirable if the relations are to be useful in the rush of making a prediction, and surprisingly enough, it is found in practice that serious errors are usually not introduced as a result of the simplifications. Fortunately the 500-mb surface is commonly near the "level of nondivergence," between the upper and lower compensating divergent-convergent flows, but appreciable errors due to rigid adherence to the 500-mb surface are not infrequent.

The equation of a sine curve in x, y coordinates is

$$y = A \sin \frac{2\pi}{L} x \qquad (14\text{-}2)$$

where A is the amplitude and L is the wavelength. From elementary mechanics it is shown that at any time t, the equation of the curve for a sinusoidal wave (such as in a rope) moving in the x direction with speed C, is

$$y = A \sin \frac{2\pi}{L} (x - Ct) \qquad (14\text{-}3)$$

or, for sinusoidal streamlines,

$$Y_s = A_s \sin \frac{2\pi}{L} (x - Ct) \qquad (14\text{-}4)$$

Unlike the case of a wave in a rope, the streamlines represent air motion through the sinusoidal course.

A useful quantitative method was derived by Rossby from the concept of the conservation of vorticity in a nondivergent field of wave motion.

If the absolute vorticity is conserved and the divergence is zero, the individual change in the relative vorticity balances the local vorticity of the earth, that is,

$$f + \zeta = \text{const}$$

$$\frac{D\zeta}{Dt} + \frac{Df}{Dt} = 0$$

following along with the particles, and

$$\frac{D\zeta}{Dt} = -\frac{Df}{Dt}$$

$$\frac{Df}{Dt} = v\frac{df}{dy} = \beta v \qquad (14\text{-}5)$$

where $\beta = df/dy = df/R\,d\varphi = 2\Omega \cos \varphi/R$

$$\frac{D\zeta}{Dt} = -\beta v \qquad (14\text{-}6)$$

Consider the mean zonal current U constant in x, y, and t, and introduce perturbation velocities u', v' so that the actual wind components are $u = U + u'$ and $v = v'$. It is apparent that U does not contribute to the vorticity. The expression for the relative vorticity is

$$\zeta = \frac{\partial v}{\partial x} - \frac{\partial u}{\partial y} = \zeta' = \frac{\partial v'}{\partial x} - \frac{\partial u'}{\partial y} \qquad (14\text{-}7)$$

The individual change of relative vorticity, in expanded form, is

$$\frac{D\zeta}{Dt} = \frac{\partial \zeta}{\partial t} + u\frac{\partial \zeta}{\partial x} + v\frac{\partial \zeta}{\partial y} = -\beta v$$

$$= \frac{\partial \zeta}{\partial t} + U\frac{\partial \zeta}{\partial x} + \left(u'\frac{\partial \zeta}{\partial x} + v'\frac{\partial \zeta}{\partial y}\right) = -\beta v' \qquad (14\text{-}8)$$

According to Rossby the terms in the parentheses in this equation are small terms of the second order; so we may neglect them and write

$$\frac{\partial \zeta}{\partial t} + U\frac{\partial \zeta}{\partial x} = -\beta v' \qquad (14\text{-}9)$$

As shown previously, the local change vanishes in a stationary wave or in terms of coordinates moving with the wave; so the local change at a fixed point must be due to the advection of the wave motion only, or

$$\frac{\partial \zeta}{\partial t} = -C\frac{\partial \zeta}{\partial x} \qquad (14\text{-}10)$$

Eq. (14-9) then becomes

$$(U - C)\frac{\partial \zeta}{\partial x} = -\beta v' \qquad (14\text{-}11)$$

In the case of perturbations that are independent of y (no $\partial u'/\partial y$ term),

$$\zeta = \frac{\partial v'}{\partial x} \qquad (14\text{-}12)$$

so (14-11) becomes

$$(U - C)\frac{\partial^2 v'}{\partial x^2} = -\beta v' \qquad (14\text{-}13)$$

The perturbation velocity of a sinusoidal wave has the form, as in Eq. (14-3),

$$v' \approx \sin \frac{2\pi}{L}(x - Ct) \qquad (14\text{-}14)$$

The second derivative of this with respect to x is

$$\frac{\partial^2 v'}{\partial x^2} = -\frac{4\pi^2}{L^2}\sin \frac{2\pi}{L}(x - Ct)$$

$$= -\frac{4\pi^2}{L^2}v' \qquad (14\text{-}15)$$

which may be substituted in (14-13) to give

$$U - C = \frac{\beta L^2}{4\pi^2} \qquad (14\text{-}16)$$

$$C = U - \frac{\beta L^2}{4\pi^2} \qquad (14\text{-}17)$$

Thus the wave speed can be determined from the wind speed and wavelength. Since β and L are always positive, it is apparent that C cannot exceed U. It is interesting to note that $C = 0$ when $U = \beta L^2/4\pi^2$. The stationary wavelength ($C = 0$) is seen to be

$$L_s = \sqrt{\frac{4\pi^2 U}{\beta}} = 2\pi\sqrt{\frac{U}{\beta}} \qquad (14\text{-}18)$$

The graph in Fig. 14-1 is a convenient one for computing the various relationships. From the wavelength and the latitude, $U - C$ is obtained immediately, as at B. Since U is measured by upper-air soundings, C is then found as $U - (U - C)$. The stationary wavelength for any given U and latitude (or β) is that in which $U = U - C$; so by using the scale $U - C$ for U, one can read the value of L_s (72° for U of 13° per day at lat 47°). The speed C is the difference between points A and B, or 5° per day.

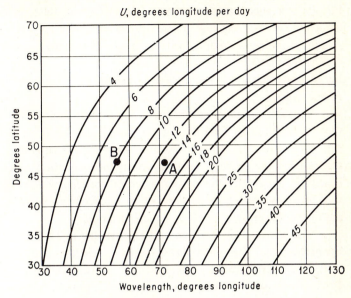

FIGURE 14-1
Diagram for computing displacement of Rossby waves.

NUMERICAL PREDICTION

A computation of the future state of the atmosphere from a given initial state by the application of the laws of motion was first attempted in 1922 by L. F. Richardson.[1] As might be expected in those days, he found that the computation work was so time-consuming that the job could not be completed until long after the predicted weather had occurred.   This difficulty has now been overcome by high-speed electronic computers.   Richardson also found that all changes in pressure and motion resulted from small differences between very large values, and since his equations were oversensitive to certain effects, some types of minor influences and errors were magnified to produce spurious results.   As a result of the work of Charney and others,[2] the equations have been made less sensitive to small effects.

The methods introduced in the early 1950s were based on the relation between the divergence and changes in the absolute vorticity—essentially the Rossby effects treated in the preceding pages.   The vorticity was advected by the geostrophic wind,

[1] L. F. Richardson, "Weather Prediction by Numerical Processes," Cambridge University Press, London, 1922.

[2] J. G. Charney, *Journal of Meteorology*, vol. 6, 1949.   J. G. Charney and N. A. Phillips, *Journal of Meteorology*, vol. 10, 1953.   J. G. Charney, R. Fjörtoft, and J. von Neumann, *Tellus*, vol. 2, 1950.

but the divergence, an ageostrophic condition, was related to the changing vorticity. To avoid small effects such as gravity waves, the equations have to be "filtered" by making certain approximations in them. As refinements are made through reducing the number of approximations, the equations become too complex, requiring integrations that are time-consuming beyond the real-time requirements of weather forecasting.

By 1960 a number of workers[1] were experimenting with the primitive equations, that is, the Eulerian equations of Newtonian accelerations of the form of Eq. (8-59). They are called "primitive" because they are the original equations of motion in contrast to the derived vorticity-divergence equations. The operational models based on an equation for conservation of vorticity which dominated numerical prediction in its first 15 or 20 years were replaced in 1966–1967 in the United States national services by a primitive-equation model.

As described by Shuman and Hovermale[2] in 1968, the United States model in use in the early 1970s is a six-layer model consisting of the planetary boundary layer, three layers in the troposphere, and two layers in the stratosphere, as shown in Fig. 14-2. The vertical coordinate σ is a generalization of a system introduced by Phillips[3] and has the form

$$\sigma = \frac{p - p_U}{p_L - p_U} \qquad (14\text{-}19)$$

where p_U is the pressure at a given quasi-horizontal surface above the point in question and p_L is the pressure at a surface below. Its chief advantage is that, being relative, its starting pressure can be at any ground elevation or pressure, thus eliminating a meaningless standard base pressure such as 1000 mb. Note that in a layer from p_L to p_U, σ varies from 1 to 0. In Fig. 14-2 the symbolic values of σ for layers numbered -1 at the top to 6 at the bottom are shown. The layer between 0 and -1 is included for computational properties rather than for its meteorological significance, and is not counted as one of the six layers. Vertical velocity within a layer can be represented as $d\sigma/dt$ or, for simplicity, $\dot{\sigma}$, positive downward. Horizontal coordinates can be

[1] K. Hinkelmann, Ein numerisches Experiment mit der primitiven Gleichungen (A Numerical Experiment with the Primitive Equations), in B. Bolin (ed.), "The Atmosphere and the Sea in Motion," Rockefeller University Press, New York, 1959. N. A. Phillips, Numerical Integration of the Primitive Equations on the Hemisphere, *Monthly Weather Review*, vol. 87, pp. 109–120, 1959. F. G. Shuman, Numerical Experiments with the Primitive Equations, *Proceedings of the International Symposium on Numerical Weather Prediction*, Tokyo, November 7–13, 1960 (published by the Meteorological Society of Japan, 1962). J. Smagorinsky, A Primitive Equation Model Including Condensation Processes, *ibid.*

[2] F. G. Shuman and J. B. Hovermale, An Operational Six-Layer Primitive Equation Model, *Journal of Applied Meteorology*, vol. 7, pp. 525–547, 1968.

[3] N. A. Phillips, A Coordinate System Having Some Special Advantages for Numerical Forecasting, *Journal of Meteorology*, vol. 14, pp. 184–185, 1957.

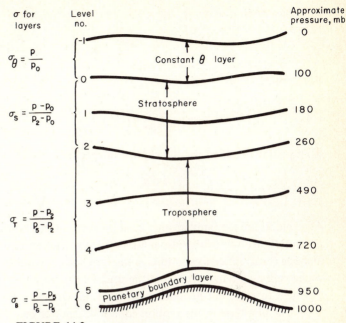

FIGURE 14-2
Schematic diagram of the vertical structure of the six-layer model.

cartesian on a conformal-map projection on which the ratio of a small distance on the projection to distance on the earth is designated by m. Since pressure is represented through potential temperature and temperature, a quantity is introduced given as $\pi = (p/p_0)^{R/c_p} = T/\theta$.

The complete equations probably appear formidable to those unfamiliar with the methods of dynamic meteorology, but they are displayed here in order that the various terms that enter into the numerical prediction can be explained. The equations are:

For motion

$$\frac{\partial u}{\partial t} + \dot{\sigma}\frac{\partial u}{\partial \sigma} - v\left(f - v\frac{\partial m}{\partial x} + u\frac{\partial m}{\partial y}\right) + m\left(u\frac{\partial u}{\partial x} + v\frac{\partial u}{\partial y} + \frac{\partial gz}{\partial x} + c_p\theta\frac{\partial \pi}{\partial x}\right) + F_x = 0 \quad (14\text{-}20)$$

$$\frac{\partial v}{\partial t} + \dot{\sigma}\frac{\partial v}{\partial \sigma} + u\left(f - v\frac{\partial m}{\partial x} + u\frac{\partial m}{\partial y}\right) + m\left(u\frac{\partial v}{\partial x} + v\frac{\partial v}{\partial y} + \frac{\partial gz}{\partial y} + c_p\theta\frac{\partial \pi}{\partial y}\right) + F_y = 0 \quad (14\text{-}21)$$

For hydrostatic equilibrium

$$\frac{\partial gz}{\partial \sigma} + c_p\theta\frac{\partial \pi}{\partial \sigma} = 0 \quad (14\text{-}22)$$

For conservation of energy (first law of thermodynamics)

$$\frac{\partial \theta}{\partial t} + \dot{\sigma} \frac{\partial \theta}{\partial \sigma} + m \left(u \frac{\partial \theta}{\partial x} + v \frac{\partial \theta}{\partial y} \right) + H = 0 \qquad (14\text{-}23)$$

For continuity

$$\frac{\partial}{\partial t} \left(\frac{\partial p}{\partial \sigma} \right) + \frac{\partial}{\partial \sigma} \left(\frac{\partial p}{\partial \sigma} \dot{\sigma} \right) + m \left[\frac{\partial}{\partial x} \left(\frac{\partial p}{\partial \sigma} u \right) + \frac{\partial}{\partial y} \left(\frac{\partial p}{\partial \sigma} v \right) \right] - \frac{\partial p}{\partial \sigma} \left(u \frac{\partial m}{\partial x} + v \frac{\partial m}{\partial y} \right) = 0 \quad (14\text{-}24)$$

An equation is added for the change in specific humidity as follows:

$$\frac{\partial q}{\partial t} + \dot{\sigma} \frac{\partial q}{\partial \sigma} + \left(u \frac{\partial q}{\partial x} + v \frac{\partial q}{\partial y} \right) + C = 0 \qquad (14\text{-}25)$$

All symbols are the familiar ones used throughout this book except those noted above. F_x and F_y are the accelerations arising from friction and turbulence; H represents diabatic (heat) influences; and C is the influence of condensation and evaporation on water-vapor content.   To avoid the use of too many Greek letters in the equations, z is used to represent geopotential height, designated elsewhere in this book by Φ.

   The first term in the motion equation, $\partial u/\partial t$ or $\partial v/\partial t$, is the unknown; that is, it gives the change in the air motion at a specified point which, when computed for the many grid points over the map and combined with the other solutions, will produce the predicted flow patterns.   The second term, $\dot{\sigma}(\partial u/\partial \sigma)$, is the local rate of change in the u component of velocity with upward movement through the σ surfaces.   The third term, $-vf$ or $+uf$, is the component of the coriolis force acting in the negative x direction on a $-v$ component and in a negative y direction on a $+u$ component.   The next two terms in the same parentheses as f give the effect of air moving away from the latitude where the projection represents the true distances.[1]   There are two processes represented in the second set of parentheses.   The term $u(\partial u/\partial x) + v(\partial u/\partial y)$ and the similar term in the second equation represent the advection of the velocity field.   Then there are two terms related to the pressure distribution.   The term $\partial gz/\partial x$ gives the rate of change of geopotential with distance along a σ surface.   The g is not constant in σ space because it is taken only in that component of g which is normal to the σ surfaces.   These are curved with respect to the constant g (level) surfaces.   The g component normal to a σ surface changes as the geometric slope of σ changes.   The last term in the parentheses is the pressure-gradient force in the σ surfaces.   Finally F_x and F_y represent the surface frictional stress applied to the equation from considerations of eddy viscosity.

---

[1] The projection used in the U.S. Numerical Weather Prediction (NWP) system is polar stereographic, true at 60°N, for which $m = (1 + \sin 60°)/(1 + \sin \varphi)$, where φ is north latitude.

All gradients or changes with distance are derived from maps of the standard isobaric levels.   The solution for the predicted velocity, for example, u_1, may be computed as $u_1 = u_0 + \Delta t(\partial u/\partial t)$, where higher-order derivatives with respect to t are justifiably neglected.   The same applies to the five variables u, v, θ, $(\partial p/\partial \sigma)$, and q.

The expression for hydrostatic equilibrium is recognizable from previous forms we have used, considering the second term as equivalent to $\alpha(\partial p/\partial z)$.   The expression for the conservation of energy establishes the local change as comprised of advection of potential temperature horizontally and vertical changes through ascent, plus heat that may be added from the surface.   This last factor can be handled easily only over the oceans where surface temperature varies but little and is well established by observation.   Heat may also enter into the relation from condensation expressed in Eq. (15-25).

The continuity equation states, in terms of σ surfaces, the divergence and convergence resulting from shrinking and stretching.   It is analogous to the expression we have used in Chap. 8, Eq. (8-75), relating changes in area to changes in depth.   The area change is sensitive to problems of measuring distances on a map projection; so the gradient of m figures prominently in the equation.

Finally, the change with time of the water-vapor content $\partial q/\partial t$ is related to the advective-convective changes plus phase transition in condensation C.   The computation must not allow q to exceed the saturation value.   After saturation, the heat of condensation contributes to the heat H in Eq. (14-23).

It is noted that the hydrostatic equilibrium expression is the only one that does not contain a time derivative.   It is used as a check on consistency.   It is also the one expression in which an important approximation is made.   The pressure-height relationship is taken from the soundings.   If there is a vertical acceleration other than that of gravity, such as in convection, hydrostatic equilibrium fails.   The values require smoothing, which is justifiable considering the fact that convection itself produces mixing which corresponds to a smoothing process.

In the U.S. Numerical Weather Prediction (NWP) system of 1970–1973 the horizontal grid used in the computation is a mesh of squares on the stereographic projection consisting of 53×57 points centered on the North Pole.   Grid points are 381 km apart at the true latitude of the projection, that is, at $60°$.   With the seven points in the vertical needed to separate the six layers, there are then a total of $53 \times 57 \times 7 = 21{,}147$ grid points.   In the daily computations the changes with time of the five variables are determined at these points through many time steps.   On the most advanced 1965–1972 computers the process can be handled only in sections.

In the early 1970s, with the advent of a new generation of computers, research meteorologists in the NWP center and other meteorological computing groups began experimenting with the use of spherical coordinates on a routine basis.   While this eliminates the map-projection problem, it involves other problems that have to be solved in a practical way. For the most part the transfer from cartesian to spherical coordinates is not very difficult, but it is usually considered in more advanced textbooks. In numerical computation, the grid in spherical coordinates is treated in essentially the same way as in cartesian coordinates.

FINITE-DIFFERENCE METHODS

The derivatives with respect to time are evaluated numerically, and the variables are then extrapolated through many short time steps. Thus the equations are no longer differential equations. Symbolically the extrapolation is stated in its simplest form as

$$\frac{\Delta \alpha}{\Delta t} = \frac{\alpha^{t+\Delta t} - \alpha^{t-\Delta t}}{2\Delta t}$$

where α represents one of the five variables changing in time Δt. In terms of distributions in atmospheric space, finite differences are taken around a center grid point among adjacent grid points by interpolation. The grid points form a pattern of squares, and each point has four squares adjacent to it. To obtain the field of the quantity α at each point, the difference method has to be applied to the four surrounding squares. The manner in which the finite differences are obtained affects the computational stability; so much attention has been paid to differencing routines.

In the process of numerical treatment there are a number of details having to do with such matters as smoothing, correcting, solving special computational problems, accommodation to surface relief, radiation effects, condensation phenomena, boundary-layer complications, and frictional effects. These are discussed in *Technical Memoranda* of the NOAA–National Weather Service and a specially distributed series of *Technical Procedures Bulletins*.

General Applications

Other models ranging from global to mesoscale are in use. The global models are useful in forecasts of reentry of space vehicles, in testing theories of the general circulation and its relation to the various heat sources and sinks of the earth and its atmosphere, theories of the conservation of absolute angular momentum and vorticity, effects of various factors in climatic change and long-term weather, and methods of long-range forecasting. Satellite data inputs are important on a global scale, especially over oceans and areas of sparse observation networks. On the mesoscale a so-called "subsynoptic advection model" (SAM) was put into practical use by NOAA–NWS in the late 1960s. The lowest three layers of the six-layer model are used. The end products, obtained partly from the full six-layer model, are predictions of probability of precipitation, including areas of excessive amounts, maximum and minimum temperatures at the ground, surface winds derived from the geostrophic, probability of freezing rain or snow, air trajectories, and forest-fire spreading rate. A special system for predicting sea swell and waves has been evolved.

It was ten to fifteen years after the introduction of routine NWP systems in the early 1950s before these predictions were definitely established as superior to those made in a partially subjective manner by the most experienced and talented fore-casters. In the previously cited Shuman and Hovermale 1968 presentation of the operation of the six-layer primitive-equation model, comparable skill scores were tabulated between pre-NWP prognostic surface and 500-mb values for 30, 36, and 48 hr in advance. NWP had superior scores, showing within its development a notable improvement with the inauguration of the six-layer primitive-equation model. For longer periods in the prognosis of sea-level daily pressures, a comparison was made with the "subjective" predictions of the Extended Forecast Division of NOAA–NWS. NWP was slightly superior through 96 hr, after which there seemed to be no significant difference, each being then only a little better than a total persistence forecast. When the "subjective" forecasters had the advantage of examining the NWP forecast before predicting, the result was notably better than either the machine or the men alone could produce. The computation might show the forecaster some developments of factors that he discounted too much, or it might place less emphasis than did the forecaster on other factors. Because of computational and other difficulties the computer may produce some details that the forecaster at once recog-nizes as preposterous, especially near the boundaries of the grid, and he can correct for them. On the average, a compromise between NWP and human efforts would probably produce a result showing an improvement over either one taken alone.

As far as the consumer is concerned, only the long-distance flight planner cares what the 500-, 300-, or 200-mb pattern will look like. The television viewer obtains a fuzzy understanding of a sketchy surface chart. But the ordinary user of the forecast needs to plan on the basis of quite local details, and it is here that the most difficult on-the-spot forecast has to be interpreted by the local forecaster from the NWP prod-uct. If a forecaster errs seriously, he is criticized by the entire community. An unpredicted snowstorm in a large city is a terrible experience for a forecaster. It is little comfort that he may have missed the edge of a large snow area by only 20 miles.

PREDICTION FROM VORTICITY ADVECTION

The relationship between vorticity changes and their effect on the contours of isobaric surfaces and on vertical velocities, which formed the basis of earlier official prediction schemes, continues to be used as a diagnostic and research method for understanding synoptic-scale processes. It has value in studies of forecasting practices because various steps can be examined easily, and in fact, much of the work can be done graphically from certain types of charts. For these reasons, the method is presented here.

The relation expressed in Eq. (8-87) may be expanded into the individual, local, and advective changes, written as

$$\frac{D}{Dt}(f+\zeta) = \frac{\partial}{\partial t}(f+\zeta) + c\frac{\partial}{\partial s}(f+\zeta) = -(f+\zeta)\mathrm{div}_2\,c \qquad (14\text{-}26)$$

The advection is along the streamlines; so only the component of the vorticity gradient along s is involved. We will designate the geostrophic absolute vorticity $f + \zeta_g$ as Q_g, but in the divergence term on the right we will regard ζ as negligible in comparison with f, and write

$$\frac{\partial Q_g}{\partial t} + c_g\frac{\partial Q_g}{\partial s} = -f\,\mathrm{div}_2\,c = f\frac{\partial w_p}{\partial p} \qquad (14\text{-}27)$$

the last expression on the right being in accord with Eq. (8-70).

The advection term can be divided into components of x and y as follows:

$$c_g\frac{\partial Q_g}{\partial s} = \frac{1}{f}\left(\frac{\partial\Phi}{\partial x}\frac{\partial Q_g}{\partial y} - \frac{\partial\Phi}{\partial y}\frac{\partial Q_g}{\partial x}\right) \qquad (14\text{-}28)$$

where, for an isobaric surface, $\partial\Phi/\partial x = fv_g$ and $\partial\Phi/\partial y = -fu_g$. Similarly, since

$$Q_g = f + \frac{\partial v_g}{\partial x} - \frac{\partial u_g}{\partial y} \qquad (14\text{-}29)$$

we substitute the geostrophic relation for the velocity components to obtain

$$Q_g = \frac{1}{f}\left(\frac{\partial^2\Phi}{\partial x^2} + \frac{\partial^2\Phi}{\partial y^2}\right) + f \qquad (14\text{-}30)$$

and the rate of change in the absolute geostrophic vorticity becomes

$$\frac{\partial Q_g}{\partial t} = \frac{1}{f}\left(\frac{\partial^2}{\partial x^2} + \frac{\partial^2}{\partial y^2}\right)\frac{\partial\Phi}{\partial t} + \frac{\partial f}{\partial t} \qquad (14\text{-}31)$$

but the local change of f represented by the last term is zero.

We now have the two terms on the left of (14-27), that is, the local and the advective changes, expressed in terms of the geopotential heights of the pressure surfaces and their changes with time. So we substitute them into the basic expression (14-27) and multiply through by f, with the result

$$\left(\frac{\partial^2}{\partial x^2} + \frac{\partial^2}{\partial y^2}\right)\frac{\partial\Phi}{\partial t} - f^2\frac{\partial w_p}{\partial p} = \frac{\partial\Phi}{\partial y}\frac{\partial Q_g}{\partial x} - \frac{\partial\Phi}{\partial x}\frac{\partial Q_g}{\partial y} \qquad (14\text{-}32)$$

It is to be noted that in the derivation of this equation the geostrophic condition is applied in the vorticity term but the horizontal divergence is nongeostrophic. For this reason the method is called *quasigeostrophic*.

The terms on the right can be obtained from the isobaric contour chart. Those on the left, namely, the height tendency $\partial\Phi/\partial t$ and the vertical-velocity gradient, are both considered as unknowns; so an additional equation is required. This equation is obtained from thermodynamics and hydrostatics. It is assumed that all changes are adiabatic and hence that potential temperature is conserved. Again, we express individual, local, and advective changes applied to potential temperature, seen as

$$\frac{1}{\theta}\frac{D\theta}{Dt} = \frac{1}{\theta}\frac{\partial\theta}{\partial t} + \frac{1}{\theta}c\cdot\nabla\theta = 0 \qquad (14\text{-}33)$$

The individual change is zero for this adiabatic process. We may divide the advective term into a vertical component and a geostrophic horizontal component to obtain

$$\frac{1}{\theta}\frac{\partial\theta}{\partial t} + \frac{1}{\theta}c_g\frac{\partial\theta}{\partial s} + \frac{1}{\theta}w_p\frac{\partial\theta}{\partial p} = 0 \qquad (14\text{-}34)$$

On an isobaric surface, $(1/\theta)(\partial\theta/\partial t) = (1/\alpha)(\partial\alpha/\partial t)$ and $(1/\theta)(\partial\theta/\partial s) = (1/\alpha)(\partial\alpha/\partial s)$. From the hydrostatic equation $\alpha = -\partial\Phi/\partial p$. Thus

$$\frac{1}{\theta}\frac{\partial\theta}{\partial t} = -\frac{1}{\alpha}\frac{\partial}{\partial t}\left(\frac{\partial\Phi}{\partial p}\right) = -\frac{1}{\alpha}\frac{\partial}{\partial p}\left(\frac{\partial\Phi}{\partial t}\right) \qquad (14\text{-}35)$$

$$\frac{1}{\theta}c_g\frac{\partial\theta}{\partial s} = -\frac{1}{\alpha}c_g\frac{\partial}{\partial s}\left(\frac{\partial\Phi}{\partial p}\right) \qquad (14\text{-}36)$$

In putting these values into Eq. (14-34), we can change signs and multiply through by α, to obtain

$$\frac{\partial}{\partial p}\left(\frac{\partial\Phi}{\partial t}\right) - \frac{\alpha}{\theta}w_p\frac{\partial\theta}{\partial p} + c_g\frac{\partial}{\partial s}\left(\frac{\partial\Phi}{\partial p}\right) = 0 \qquad (14\text{-}37)$$

Then, with the u_g, v_g, x, and y components substituted in the last (advective) term through the geostrophic equations for u_g and v_g, we have the expression

$$\frac{\partial}{\partial p}\left(\frac{\partial\Phi}{\partial t}\right) - \frac{\alpha}{\theta}w_p\frac{\partial\theta}{\partial p} = \frac{1}{f}\left[\frac{\partial\Phi}{\partial x}\frac{\partial}{\partial y}\left(\frac{\partial\Phi}{\partial p}\right) - \frac{\partial\Phi}{\partial y}\frac{\partial}{\partial x}\left(\frac{\partial\Phi}{\partial p}\right)\right] \qquad (14\text{-}38)$$

In a uniform lapse rate $(\alpha/\theta)(\partial\theta/\partial p)$ can also be taken as constant, or it can be obtained from the thickness pattern.

The two equations to be used are (14-32) and (14-38). In each of these equations the initial state is given on the right. The equations are solved for the two unknown time variables of Φ and w_p on the left. The initial state is taken from maps of vorticity superimposed on constant-pressure maps. To solve the equations, the boundary condition of $w_p = 0$ at $p = 0$ and at $p = 1000$ mb is introduced. (Actually $w_p = 0$ at $\Phi = 0$, but it is more practical and accurate enough to use 1000 mb as the base.) At the boundaries of the map, arbitrary but reasonable values of the height tendencies $\partial\Phi/\partial t$ and of the vorticity advection are assigned. Any errors introduced in this way are suppressed at distances a few hundred kilometers from the edges of the map. The obvious procedure is to construct the map for an area much larger than the area of interest.

A machine computation for a large number of grid points adds the many height increments $\delta\Phi$ to the initial height field as it changes with the advection and with w_p in short time intervals δt. The increments are accumulated to produce prognostic charts for selected isobaric surfaces. By suitable programming the machine can produce the prognosticated vertical velocity at each point—the "omega equation."[1]

Research meteorologists and advanced students can learn much about the atmosphere by working with these two quasigeostrophic vorticity-advection equations. There are several variations of this approach—special multilevel numerical schemes, barotropic and baroclinic variations, and programs for various scales, from the mesoscale to the general circulation. These must be left for more specialized treatises.[2]

EXERCISES

1 Calculate the displacement of isobars past your station in 3 hr when the barograph trace shows a tendency of -1 mb in 3 hr and the isobars are spaced 3 mb in 210 km with lower pressure moving in.

2 Find the wavelength of a stationary wave in the upper westerlies blowing at 60 m sec^{-1} at lat 50°.

3 Use the graph in Fig. 14-1 to find the wave speed C for a wavelength of 60° longitude at lat 45°, with a westerly wind speed U of 12° longitude per day.

4 In Eqs. (14-28) and (14-30) the geostrophic velocities are given in their x and y components. Why is it not suitable to express the geostrophic velocity in the natural coordinate system?

---

[1] Leading theoretical meteorologists in the United States use the Greek omega ω as the symbol for vertical velocity dp/dt because it looks similar to w. Omega was introduced by the Greeks to represent an o sound different from omicron. This abuse of the Greek alphabet seems firmly established in meteorology.

[2] G. J. Haltiner, "Numerical Weather Prediction," John Wiley & Sons, Inc., New York, 1971.

15

CONDENSATION, PRECIPITATION, AND ATMOSPHERIC ELECTRICITY

With the development after World War II of techniques that showed the way to cloud management and weather modification, the study of the physics of clouds and precipitation has become one of the leading subfields of meteorology. While the news media and popular scientific writings have dwelt upon the weather modification or "rainmaking" aspects, atmospheric scientists have intensively explored the microphysics of the transformation of water vapor into fog and cloud droplets, the development of these into raindrops, the formation of snow crystals and hail, the role of electrical phenomena, and a number of related processes. At the same time progress has been made in numerical modeling of internal motions of clouds and the microphysical chain of reactions resulting from these motions. In this chapter the fundamentals of cloud physics and a summary of the distribution and transfer of electrical charge in the atmosphere will be presented. More extensive treatments of these subjects may be found in textbooks[1] and references cited in this chapter.

[1] For example, H. R. Byers, "Elements of Cloud Physics," The University of Chicago Press, Chicago, 1965. N. H. Fletcher, "The Physics of Rainclouds," Cambridge University Press, Cambridge, 1962. B. J. Mason, "The Physics of Clouds," 2d ed., Oxford University Press, New York, 1971.

As demonstrated by several experimenters toward the end of the last century, condensation of water vapor to the liquid phase occurs on certain available nuclei. The nuclei are the myriads of small particles floating in the atmosphere called *aerosols*.   The true aerosol particles are those smaller than about 3 μm radius;[1] those larger than that settle out, but their presence is quite noticeable in dusty or polluted air and detectable in " clean " air while slowly falling.   Those having a radius smaller than about 0.1 μm serve as condensation nuclei only if a considerable percentage of supersaturation occurs.   Some particles are composed of *hygroscopic* substances, that is, those substances that have a chemical affinity for water.   Such a particle starts a water droplet growing around it before saturation is reached.   Other particles may be *hydrophobic*; that is, they resist the spread of a film of water over them.   Given a variety of particles in an atmosphere containing water vapor, one finds that, as saturation is approached, condensation will occur first on those nuclei that are large and *hygroscopic*, or at least not hydrophobic.   It is an observed fact that in the atmosphere with temperatures favorable for liquid condensation appreciable supersaturation is not necessary for condensation and that not infrequently it takes place with less than 100 percent relative humidity.   This fact indicates that there is usually an ample supply of large and chemically favorable condensation nuclei.

The nucleation process for the passage from vapor to the crystalline phase, or crystallization, is not the same as that for condensation; at low temperatures a different set of circumstances prevails, and supersaturation with respect to ice is noted.   This indicates that suitable crystallization nuclei are scarce in the atmosphere and that those suitable for condensation are not very suitable for crystallization.

Saturation, or 100 percent relative humidity, in its ordinary sense and in the sense that it is used in this discussion refers to a plane surface of pure water and means that if the liquid water and the vapor in the space above it are at the same temperature and in equilibrium, they will have the same (saturation) vapor pressure.   Or we may say that then the vapor pressure in the space is the same as the vapor tension of a pure, flat water surface at that temperature.   If supersaturation is an equilibrium condition over any other kind of surface, then that surface must exert a greater vapor tension than does a plane surface of pure water at the same temperature.

As tiny fog or cloud droplets begin to form on soluble nuclei in the atmosphere, they are not at exact saturation according to this standard, because they neither consist of pure water nor have a flat surface.   While the curvature causes the vapor tension at their surfaces to be higher than standard saturation, at the small sizes where this effect is appreciable, the impurity introduced in the form of a soluble nucleus tends to counteract it.   The solution effect has been demonstrated extensively in the laboratory for bulk solutions.

---

[1] When the term "radius" or "diameter" is used for particles, it refers to the equivalent spherical radius or diameter, that is, to a radius that satisfies the relation $M = \frac{4}{3}\pi r^3 \rho_p$, where ρ_p is the density of the particle and M is its mass.

SIZE SPECTRUM OF AEROSOL PARTICLES

Of meteorological interest in the atmosphere are particles as small as molecules and on up in size through clouds and precipitation elements to giant hailstones. The principal particles with which we shall be concerned are listed in Table 15-1.

First on the list are the small ions. An ion is a particle that has one extra elementary charge so that it is not in neutral balance, or it may lack one such charge. An ion can be a molecule or a large particle, as indicated by the range of sizes in the table. The small ions are constantly being formed by the ionizing action of cosmic rays entering the atmosphere and from radioactive gases escaping from the soil and rocks over continents. Ions also are produced through manmade nuclear reactions. The creation of small ions is balanced by recombination among themselves and by their attachment to larger particles to produce large ions. In fact, the medium and large ions apparently are formed only from capture of the small ones by particles of large-ion size.

From the column listing the ranges of radii, it appears that large ions and Aitken nuclei are the same, and other facts about them bear out this identity. Beginning around 1880, the British physicist John Aitken developed an instrument for producing condensation in a chamber through expansion and for counting the condensed droplets. He recognized that each droplet was forming on a nucleus. The nuclei around which droplets condense in an Aitken counter are called *Aitken nuclei.* They may or may not be the nuclei around which natural clouds are formed, since the atmospheric supersaturations may be less than those created by the expansion in Aitken's chamber. The Aitken nuclei are an abundant class of particles, and in addition to their possibilities as natural condensation nuclei, they play an important role in the ionization

Table 15-1 SIZES OF PARTICLES AND SUPERSATURATIONS REQUIRED FOR CONDENSATION ON THEM

Class of particles	Approx. equivalent spherical radius, cm	Supersaturation required, percent	
		Hygroscopic	Nonhygroscopic
Small ions	$<10^{-7}$	.....	400–100
Medium ions	$10^{-7}–2 \times 10^{-6}$	0.50–1.8	100–5
Large ions	$2 \times 10^{-6}–10^{-5}$	1.8–0.4	5–1.2
Aitken nuclei	$2 \times 10^{-6}–10^{-5}$	1.8–0.4	5–1.2
Smoke, dust, haze	$10^{-5}–10^{-4}$	0.85 – 0	2.4–0.001
Large condensation nuclei	$10^{-5}–3 \times 10^{-4}$	0.4 – 0	1.2–0.0003
Giant condensation nuclei	$3 \times 10^{-4}–10^{-3}$	0	0.0003–0.001
Cloud or fog droplets	$5 \times 10^{-5}–5 \times 10^{-3}$		
Drizzle drops	$5 \times 10^{-3}–5 \times 10^{-2}$		
Raindrops	$5 \times 10^{-2}–5 \times 10^{-1}$		

balance of the lower atmosphere. Combustion products contribute greatly to the number of Aitken nuclei; hence the largest counts are found in industrial cities, running as high as millions per cubic centimeter. Over the oceans, sea-salt particles may be abundant as a result of spray and foam in stormy areas, but the numbers seldom exceed a few tens of thousands per cubic centimeter. On the average, the counts are lowest over the oceans. The numbers decrease rapidly with height, dropping off by a factor of 10 every 2 km or so. However, over calm ocean areas where the counts can be extremely low, an increase through the first kilometer or two may be observed, indicating upper transport from a disturbed area.

Smoke, dust, and haze are classed together in the table, since no clear distinction can be made between them. The smaller particles in this category could be Aitken nuclei but are put in a different classification because they have been collected in a different manner. The term "dust" is occasionally used to denote all types of dry material in the atmosphere. Dust blown into the atmosphere in dust storms usually has a yellowish color while smoke tends to be more bluish, but this difference must be due to different particle sizes, the dust-storm particles being fairly large. Haze is often composed of particles to which considerable quantities of water are attached. The hygroscopic nuclei may be emitted in smoke or may be detached from the ocean or land by the air motion. Many of the smoke or "dust" particles are active condensation nuclei.

Most of the atmospheric condensation appears to occur on what are classed as "large" condensation nuclei in Table 15-1. Although less abundant than the Aitken nuclei, they occur in the atmosphere in sufficient numbers to facilitate cloud and fog formation without appreciable supersaturation.

It should be noted that there is considerable overlapping in the size ranges of smoke, haze, large condensation nuclei, and cloud or fog droplets. This circumstance is in agreement with the fact, to be discussed in more detail in this chapter, that condensation is a process of continuous growth of particles.

The giant nuclei, although relatively few in number, play an important part in condensation and precipitation, because as the relative humidity reaches 100 percent, those that are hygroscopic will have already attained the size of drizzle drops. The last two columns in Table 15-1 show the supersaturations required to produce condensation on particles of the size indicated in the preceding column. It is seen that hygroscopic particles of the given size require much less supersaturation than do nonhygroscopic nuclei of the same dimensions, and they can nucleate at subsaturation (relative humidity less than 100 percent) if they are both large and hygroscopic. Evidence indicates that in the natural atmosphere supersaturations greater than 0.1 percent with respect to pure liquid water are not necessary for condensation to occur. Thus hygroscopic particles of a dry equivalent spherical radius greater than about 10^{-5} cm must be the prevailing nuclei. In the lower troposphere where most of the above-freezing condensation occurs, there appears to be an ample supply of such nuclei. In the middle troposphere where summer or tropical temperatures may be high enough to permit liquid condensation, particles in the middle of the Aitken-nuclei spectrum may be utilized owing to a shortage of larger ones. However, the same upward currents that carry water vapor through the troposphere also carry large nuclei, so that regions of shortage of nuclei are likely to be regions of subsidence and dry air, therefore of no consequence as far as condensation is concerned.

The extreme supersaturations required in perfectly "clean" air in which there exist only the ever-present small ions were first demonstrated by C. T. R. Wilson in 1897. His investigations should have dispelled the erroneous notion that small ions serve as condensation nuclei for atmospheric clouds, but this idea is sometimes revived even today.

PHYSICS OF CONDENSATION

To study the physics of condensation one needs to understand the concept of *free energy*. This form of energy is exhibited in the action of a substance or arrangement of molecules against the environment, such as the escaping tendency of gas molecules, including evaporation, and the reduced binding force between molecules at the surface of a liquid as contrasted with the interior. The arrangement of molecules in a solution determines actions such as osmotic pressure and modified surface phenomena. These effects are recognized as *potentials* of a substance or aggregation, expressed as thermo-dynamic-chemical potentials. The free energy, introduced by the American physicist J. W. Gibbs in 1875, is given by the potential, per mole, as

$$G = E - T\varphi + pv \qquad (15\text{-}1)$$

where the symbols represent the thermodynamic quantities introduced in Chap. 5. The differential of G is

$$dG = dE - Td\varphi - \varphi dT + pdv + vdp \qquad (15\text{-}2)$$

but for a gas

$$dE = Td\varphi - pdv \qquad (15\text{-}3)$$

as in Eq. (5-39) combined with (5-36). The expression reduces to

$$dG = vdp - \varphi dT \qquad (15\text{-}4)$$

For an ideal gas at constant temperature the second term is zero, and with a substitution for v from the equation of state we have, again per mole,

$$dG = RT\frac{dp}{p} = RT\,d(\ln p) \qquad (15\text{-}5)$$

In addition to this thermodynamic potential of a pure substance, a chemical potential arising from the degree of concentration of a solution is added. Another form of free energy is brought about by the exposure of molecules at the surface of a liquid or solid. In that position they have less cohesive binding force than interior molecules because they do not have neighbors on all sides. A mass of ice in the shape of a disk would have a higher surface free energy than the same mass in spherical form, because as a disk it would have more exposed area. The specific (per unit area) surface free energy σ is called also the *surface tension* for a liquid, since it represents the force holding the liquid mass together such as in a bubble or cloud droplet. The surface tension acts to hold a free liquid mass in a spherical form, which has the lowest total surface free energy (smallest area per unit volume). This fact illustrates the rule that the most stable state is one in which the free energy is at a minimum. The surface free energy also decreases with increasing temperature.

For pure water vapor at constant temperature, considered as an ideal gas, Eq. (15-5) becomes

$$dG = RT\, d(\ln e) \qquad (15\text{-}6)$$

As the vapor pressure e increases, so does the free energy. Let us take the free energy at saturation as the zero state in which the saturation ratio $e/e_s = 1$ and therefore $\ln(e/e_s) = 0$. Then we may consider the free energy above that base in the case of supersaturation, when $e/e_s > 1$. At constant temperature the difference in free energy between the supersaturated and the saturated state is given by

$$G - G_s = \int_{e_s}^{e} d(\ln e) = RT \ln \frac{e}{e_s} \qquad (15\text{-}7)$$

which, when designated per gram in a droplet of mass $M = \frac{4}{3}\pi r^3 \rho_L$, becomes

$$G - G_s = \frac{4}{3}\pi r^3 \rho_L \frac{RT}{m_w} \ln \frac{e}{e_s} \qquad (15\text{-}8)$$

where ρ_L is the density of liquid water.

Consider the formation of a droplet in condensation from pure vapor in the absence of a foreign nucleating substance. We need to take into account the surface free energy. This would elevate the free energy to a value higher than that for a flat surface, which is the reference for e/e_s. Since we start from zero liquid, the change in area would be $4\pi r^2$, which, multiplied by the surface free energy per unit area σ, would give the amount of free energy developed in the droplet as it grew from zero to the radius r. The elevation of the free energy with respect to saturation ratio, due to curvature of the surface, would be

$$\Delta G = 4\pi r^2 \sigma - \frac{4}{3}\pi r^3 \rho_L \frac{RT}{m_w} \ln \frac{e}{e_s} \qquad (15\text{-}9)$$

For a given value of the saturation ratio at supersaturation it is found that ΔG has a maximum at a certain radius r^*.   From the definition of a maximum, we note that for a given saturation ratio $\partial(\Delta G)/\partial r$ must be zero at this point and the derivative with respect to r of the surface-free-energy term must equal that of the second term.   This maximum represents an equilibrium between the liquid and vapor phases at the surface of the droplet for the given saturation ratio.   At a size less than this equilibrium the droplet does not have a high enough surface free energy to attain a balance with the supersaturation.   For it to reach equilibrium, either the free-energy maximum must be lowered by an increase in the saturation ratio or the embryo droplet must somehow grow against a saturation ratio that is less than equilibrium for its size, and that would therefore cause it to evaporate instead of growing.   If the radius is larger than required for equilibrium at the given supersaturation, the droplet will grow.   The equilibrium radius r^* is therefore called the *critical radius*.   Differentiating to find the maximum, that is, the point of zero slope of the ΔG, r curve, we have

$$\frac{\partial(\Delta G)}{\partial r} = 8\pi r\sigma - 4\pi r^2 \rho_L \frac{RT}{m_w} \ln \frac{e}{e_s} \qquad (15\text{-}10)$$

Where this derivative is zero, we have the critical radius

$$r^* = \frac{2\sigma m_w}{\rho_L RT \ln(e/e_s)} \qquad (15\text{-}11)$$

After the critical radius is surpassed, the embryo droplet grows in microseconds to a fog or cloud-droplet size as ΔG approaches zero (100 percent relative humidity).   A substitution of numbers into Eq. (15-9) shows that this is essentially achieved before the droplet reaches micrometer sizes.

The upper curve in Fig. 15-1 shows the relationship between the critical radius and the saturation ratio.   Since we are dealing with pure water, there is no soluble nucleus involved.   The droplet achieves the critical radius either by (1) growing through accidental aggregations of molecules of pure water in what is known as *homogeneous* nucleation or by (2) growth around an insoluble but wettable nucleus that is already so large that a water film around it presents a spherical droplet large enough to have a critical radius at a low supersaturation.   Growth through (1) can occur only in laboratory cloud chambers where the air is freed of all possible nucleating particles, including ions, and a high supersaturation is obtained.   Laboratory measurements show that saturation ratios of 5 or 6 (500 to 600 percent relative humidity) are needed for homogeneous nucleation by molecular aggregations as in (1).   The lower curves in Fig. 15-1 are concerned with *heterogeneous* nucleation on soluble particles, which we shall now consider.

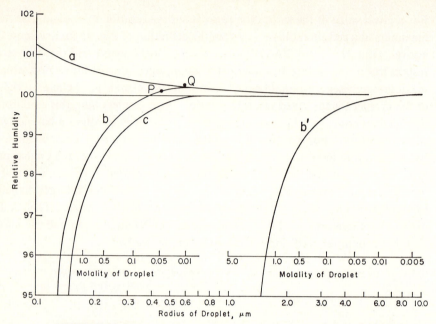

FIGURE 15-1

Equilibrium relative humidity over (*a*) a pure droplet, (*b, b'*) solution droplets of 0.1 μm and 1.0 μm initial radii, and (*c*) bulk solution of the same molality as (*b*).

SOLUTION EFFECT

The surface of liquid around a soluble nucleus growing into a droplet is not flat and the water is not pure. The equilibrium vapor pressure over such a surface will be affected by the curvature and the concentration or molality of the solution thus formed. We shall start by considering the equilibrium over a flat surface of a solution, then combine the two effects—solution and curvature—as they apply to a solution droplet.

The vapor tension of a solution is lower than that of pure water. The fact that the vapor tension diminishes with increasing concentration and varies among the different solutes may be accepted as an observed fact. It is logical to think that water molecules entangled with molecules of another substance which has an extremely low vapor tension would not leave the surface as freely as they would in the absence of these other molecules. The exact arrangement of the molecules and the cohesive forces between them are not well known.

We shall concern ourselves only with solutions of electrolytes, that is, of substances which dissociate into ions when they go into solution. Consider a solution consisting of n moles of electrolyte in an aqueous solution. The molality M is defined as the number of moles of solute in 1000 g of solvent, or $M = 1000 \times n/(V_L \rho_L)$, where the subscript refers to the water solvent. It has been shown experimentally that the water-vapor saturation ratio over the solution—the ratio of the equilibrium vapor pressure e'_s over the solution to that of pure water e_s—is given by

$$\ln \frac{e'_s}{e_s} = - \frac{i m_w M}{1000} \qquad (15\text{-}12)$$

The quantity i itself depends on the molality. The negative sign in the expression shows that the saturation ratio must always be less than unity, that is, that the equilibrium vapor pressure over the solution is less than over pure water. Values of i for a variety of soluble salts are given in the chemical literature[1] as a function of molality. For two of the most common substances that serve as nuclei in the atmosphere, namely, sodium chloride and ammonium sulfate, i has a value close to 2, which number may be used at all molalities without appreciable error in computations relating to cloud droplets.

In computing the equilibrium over a solution droplet, it would be convenient if it could be assumed that the vapor pressure over the curved solution bore the same relationship to the vapor pressure over a flat surface of the solution as in the case of a pure droplet with respect to a pure, flat surface. If that were true, then for the solution droplet

$$\ln \frac{e'_r}{e'_s} = \frac{2\sigma' m_w}{\rho_L R T r^*} \qquad (15\text{-}13)$$

where e'_r is the vapor pressure over the solution droplet and e'_s is that over bulk solution of the same molality. The difference between this equation and the pure-water equation (15-10) is that the density ρ'_L and the surface tension σ' of the solution vary with the molality, and of course, for a given mass of nuclear material the molality varies with the radius. It can be shown that this relationship is valid if terms of higher order are neglected.

To show the validity of Eq. (15-13), we follow a procedure similar to that used in the case of a pure droplet: we balance the derivatives with respect to r of the two free-energy terms, but take into account in this case the change of density and surface tension with r. We have

$$\frac{d}{dr}(4\pi r^2 \sigma') = \frac{d}{dr}\left(\frac{4}{3}\pi r^3 \rho'_L \frac{RT}{m_w} \ln \frac{e'_r}{e'_s}\right) \qquad (15\text{-}14)$$

---

[1] For example, R. A. Robinson and R. H. Stokes, "Electrolytic Solutions," Butterworth & Co. (Publishers), Ltd., London, 1959.

resulting in

$$\left(4\pi r^2 \rho'_L + \tfrac{4}{3}\pi r^3 \frac{d\rho'_L}{dr}\right)\frac{RT}{m_w}\ln\frac{e'_r}{e'_s} = 8\pi r\sigma' + 4\pi r^2 \frac{d\sigma'}{dr} \quad (15\text{-}15)$$

which, divided through by $4\pi r$, becomes

$$\left(r\rho'_L + \tfrac{1}{3}r^2 \frac{d\rho'_L}{dr}\right)\frac{RT}{m_w}\ln\frac{e'_r}{e'_s} = 2\sigma' + r\frac{d\sigma'}{dr} \quad (15\text{-}16)$$

In the ranges of droplet sizes ($r = 10^{-6}$ to 10^{-3} cm) the second term on each side of (15-16) is at least four orders of magnitude smaller than the first term. Neglecting these terms, we find

$$r\rho'_L \frac{RT}{m_w}\ln\frac{e'_r}{e'_s} = 2\sigma' \quad (15\text{-}17)$$

or

$$\ln\frac{e'_r}{e'_s} = \frac{2m_w\sigma'}{\rho'_L RT r^*} \quad (15\text{-}18)$$

which is our Eq. (15-13).

The two expressions (15-12) and (15-18) can be combined to give the equilibrium over a solution droplet in terms of the saturation ratio referred to a pure, flat surface and taking into account the effect of molality change with change of radius. What is wanted is e'_r/e_s. We note that

$$\frac{e'_r}{e_s} = \frac{e'_s}{e_s}\frac{e'_r}{e'_s}$$

or

$$\ln\frac{e'_r}{e_s} = \ln\frac{e'_s}{e_s} + \ln\frac{e'_r}{e'_s} \quad (15\text{-}19)$$

The two terms on the right have been obtained in Eqs. (15-12) and (15-18), respectively. Substitution gives

$$\ln\frac{e'_r}{e_s} = \frac{2m_w\sigma'}{\rho'_L RTr} - \frac{im_w M}{1000} \quad (15\text{-}20)$$

Equation (15-20) will be greater or less than 1 depending on whether the solution effect or the curvature term predominates. In other words, solution droplets of high concentration can be in equilibrium at subsaturation. For a given drop the molality varies as r^{-3}, as shown by the relation

$$M = \frac{1000\, n_p}{V_L \rho_L} = \frac{1000\, n_p}{\rho_L \tfrac{4}{3}\pi r^3 - n_p m_p} \quad (15\text{-}21)$$

where n_p is the number of moles and m_p the molecular weight of the dissolved salt, both of which are constant in a given drop as long as it remains free of collisions with other particles.

Equilibrium values for a bulk salt (NaCl) solution of a given initial molality, and for a droplet of the same molality, are plotted against size in Fig. 15-1. The bulk solution is considered to be diluted with water in the same way as a growing droplet; so the scale of droplet size is related to the diluteness of the solution. Another curve, labeled b', represents growth with increasing humidity of a salt nucleus having an initial dry radius of 0.5 μm, thus falling in the class of large or cloud nuclei. To magnify the effects, only the portions of the curves near saturation are plotted.

If the solution droplet b were in an environment having a saturation of 1.001 (relative humidity 100.1 percent), it would grow to the size represented by the point P and would be in equilibrium at that size. A saturation ratio at Q of 1.003 would be higher than the peak equilibrium value for this droplet. Under this condition it would not be in equilibrium and would keep right on growing. As it became larger and purer, its equilibrium vapor pressure would decrease; so in this supersaturated environment it would be growing at an accelerated rate. Actually this growth does not go on indefinitely; the droplet must compete with other droplets and nuclei in a fog or cloud for the available water. From the maximum point of the curve it is determined that the critical saturation ratio for the growth of droplet b to the usual cloud droplet size of from 10 to 20 μm diameter would be 1.0023. By the time it reaches cloud droplet size its dilution is so great that it may be regarded as a pure water surface. The curvature effect on vapor equilibrium also becomes negligible at that size. The growth of the droplet beyond the peak takes place by diffusion of water vapor to it, and this process will be taken up next.

GROWTH AND EVAPORATION OF DROPS

As pointed out in Chap. 11, the flux per unit area of water vapor in the direction x by diffusion in air is given by the gradient of vapor density multiplied by a diffusion coefficient, or

$$\frac{F}{A} = -D\frac{\partial \rho_w}{\partial x} \qquad (15\text{-}22)$$

where D is the diffusivity of water vapor in air. From the area $4\pi r^2$ of the surface of a drop the flux (evaporation) would be given by

$$F = -4\pi r^2 D\frac{\partial \rho_w}{\partial r} \qquad (15\text{-}23)$$

or, in terms of mass growth, by condensation,

$$\frac{1}{4\pi r^2}\frac{dM}{dt} = D\frac{\partial \rho_w}{\partial r} \qquad (15\text{-}24)$$

The positive sign indicates that the mass of the drop increases when the vapor density increases radially outward from the drop.

It is reasonable to assume that the vapor gradient is symmetrical around the drop, and we can then integrate the expression by setting $r = r$ and $\rho_w = \rho_{w(r)}$ at the boundary of the drop. In the ambient space we may assume that $r = \infty$, since the distance between drops is hundreds or thousands of drop radii in clouds, and the infinity where $\rho_w = \rho_w$ would be somewhere between neighboring drops. In the integration it is proper to assume that the mass accumulation on the drop is at a constant rate over a short period of time, so that dM/dt may be placed outside the integral. Since $(\partial \rho_w/\partial r)\,dr = d\rho_w$, we form the integrals

$$\frac{dM}{dt}\int_r^\infty \frac{dr}{r^2} = 4\pi D \int_{\rho_{w(r)}}^{\rho_w} d\rho_w \qquad (15\text{-}25)$$

which results in

$$\frac{dM}{dt} = 4\pi Dr(\rho_w - \rho_{w(r)}) \qquad (15\text{-}26)$$

The rate of growth of the spherical drop by the shell thickness dr is

$$\frac{dM}{dt} = \rho_L 4\pi r^2 \frac{dr}{dt} \qquad (15\text{-}27)$$

Equating the last two equations, we obtain the growth in terms of radius as

$$r\frac{dr}{dt} = \frac{D}{\rho_L}(\rho_w - \rho_{w(r)}) \qquad (15\text{-}28)$$

By substitution from the equation of state the expression can be written in terms of vapor pressure in the form

$$r\frac{dr}{dt} = \frac{Dm_w}{\rho_L RT}(e - e_r) \qquad (15\text{-}29)$$

under the assumption that the drop is at the same temperature as the ambient air. This assumption is not entirely correct, but with the usual temperature differences of less than 1 percent the result is not affected within its order of magnitude. The e_r will be the vapor tension of the drop, which is equivalent to the saturation vapor pressure at the temperature of the drop.

In condensation and evaporation there is a heat exchange which arises from the latent heat released and leads to a steady state of heat exchange with the air through diffusion.

The heat added to the drop by condensation is expressed as $L \, dM/dt$. Substituting for dM/dt from Eq. (15-26) and transforming vapor density to vapor pressure, we find

$$L \frac{dM}{dt} = \frac{4\pi r L \, Dm_w}{RT} (e - e_r) \qquad (15\text{-}30)$$

The diffusion of heat away from the drop is, as in the case of vapor diffusion, proportional to the thermal diffusivity κ (also cm^2 sec^{-1}) multiplied by the difference in heat content at constant pressure between the droplet and the ambient air,

$$\frac{dQ}{dt} = 4\pi r \kappa \rho c_p (T_r - T) \qquad (15\text{-}31)$$

or, in terms of the more commonly used thermal conductivity $K = \kappa \rho c_p$,

$$\frac{dQ}{dt} = 4\pi r K (T_r - T) \qquad (15\text{-}32)$$

In the balanced state $dQ/dt = L \, dM/dt$; so

$$\frac{e - e_r}{T(T_r - T)} = \frac{RK}{m_w \, DL} \qquad (15\text{-}33)$$

Values for the coefficients on the right at 10°C are $K = 2500$ ergs cm^{-1} sec^{-1} °C^{-1} and $D = 0.241$ cm^2 sec^{-1}. They vary slightly with temperature, and D also varies in direct inverse proportion to the pressure, $1000/p$ mb.

Since the quantities on the right in Eq. (15-33) are always positive, the numerator and denominator on the left must both be of the same sign. Thus, when the droplet is growing, $e > e_r$, and therefore the droplet must be warmer than the ambient air, and during evaporation it must be colder. The gradients of temperature and vapor pressure are always opposed. Note that this equilibrium is similar to that over a wet bulb, given in Chap. 6, Eq. (6-31). In the wet-bulb thermometer a high ventilation rate is applied; so D and κ are assumed to cancel each other.

In the computation of growth or evaporation one can maintain this balance by fitting it in steps or in a computer program in the application of Eq. (15-29). It is, desirable to obtain an expression combining the vapor and heat fluxes. It is also desirable to express the relationship in terms of the saturation ratio of the environment including with it e_s, the saturation vapor pressure at the ambient-air temperature. Realizing that e_r is also a saturation value at the temperature of the drop T_r, we can use the temperature-pressure relationship of the phase-change curve shown in Chap. 6, Fig. 6-2.

First, we divide Eq. (15-29) by e_s to write

$$\frac{e - e_r}{e_s} = \frac{\rho_L R T}{D m_w e_s} r \frac{dr}{dt} \qquad (15\text{-}34)$$

Over the limited temperature and vapor-pressure range dealt with here, the slope of the saturation vapor-pressure curve of Fig. 6-2 may be represented by the Clapeyron-Clausius equation[1] as follows:

$$\frac{d(\ln e)}{dT} = \frac{m_w L}{R T^2} \qquad (15\text{-}35)$$

This expression is integrated between the ambient and droplet values e_s, e_r, T, T_r,

$$\int_{e_s}^{e_r} d(\ln e) = \frac{m_w L}{R} \int_T^{T_r} T^{-2} \, dT$$

$$\ln \frac{e_r}{e_s} = \frac{m_w L}{R} \left(\frac{1}{T} - \frac{1}{T_r} \right) = \frac{m_w L}{R T_r T}(T_r - T) \qquad (15\text{-}36)$$

For the small temperature differences with which we are dealing around the individual droplets, we may substitute T^2 for $T_r T$ in this equation and write

$$\ln \frac{e_r}{e_s} = \frac{m_w L}{R T^2}(T_r - T) \qquad (15\text{-}37)$$

Next we insert the required balance of Eq. (15-33) by substituting for $T_r - T$ in the above expression, obtaining

$$\ln \frac{e_r}{e_s} = \frac{m_w^2 L^2 D}{K R^2 T^3}(e - e_r) \qquad (15\text{-}38)$$

followed by a substitution for $e - e_r$ from Eq. (15-29)

$$\ln \frac{e_r}{e_s} = \frac{\rho_L m_w L^2}{K R T^2} r \frac{dr}{dt} \qquad (15\text{-}39)$$

To obtain an expression in terms of the ambient saturation ratio e/e_s, we separate Eq. (15-34) by writing

$$\frac{e}{e_s} = \frac{e_r}{e_s} + \frac{\rho_L R T}{D m_w e_s} r \frac{dr}{dt} \qquad (15\text{-}40)$$

---

[1] For a development of this equation see physical chemistry or cloud physics texts such as H. R. Byers, "Elements of Cloud Physics," The University of Chicago Press, Chicago, 1965.

then substituting from Eq. (15-39) to obtain

$$\frac{e}{e_s} = \exp\left(\frac{\rho_L m_w L^2}{KRT^2} r \frac{dr}{dt}\right) + \frac{\rho_L RT}{e_s Dm_w} r \frac{dr}{dt} \qquad (15\text{-}41)$$

The relation is now expressed in terms of the ambient saturation ratio S.   In symbolic form the expression is

$$S = e^{ax} + bx \qquad (15\text{-}42)$$

where x represents $r\, dr/dt$.   When ax is much less than 1, as in this case, an expansion of e^{ax} shows that it may be simplified to $1 + ax$.   Thus we have

$$S = 1 + ax + bx$$

and solving for x we obtain the final form of the growth equation as

$$r \frac{dr}{dt} = \frac{S-1}{a+b} = \frac{S-1}{(\rho_L m_w L^2/KRT^2) + (\rho_L RT/e_s Dm_w)} \qquad (15\text{-}43)$$

The numerator is the supersaturation.   In the case of evaporation it would be negative (subsaturation).   In the denominator the two temperature terms and the ambient saturation vapor pressure, which is quite sensitive to temperature, are the important variables.   In many cloud situations the various terms would change with the decreasing temperature and total pressure in the saturation-adiabatic ascent of the air.   When r is moved to the right-hand side of the equation as r^{-1}, it is seen that for a given temperature, pressure, and supersaturation, the growth rate is faster the smaller the droplet, but a small droplet does not overtake an initially larger one.   In terms of mass, the accumulation of water is faster the larger the droplet.

To express the growth in terms of mass, Eq. (15-43) may be written with the aid of Eq. (15-27) as

$$\frac{dM}{dt} = \rho_L 4\pi r \left(r \frac{dr}{dt}\right) = \frac{4\pi r(S-1)}{a' + b'} \qquad (15\text{-}44)$$

where $a' + b' = (a+b)/\rho_L$.   This eliminates ρ_L from consideration, since $a + b$ contains a factor ρ_L.   At atmospheric temperatures the density of water is so close to 1 that the factor has no effect on the computation.

In applying Eq. (15-43) to the evaporation of drops, a ventilation factor must be included.   The ventilation effect is proportional to the speed of fall, therefore the radius.   It has no effect on droplets of radii less than about 40 μm, that is, for cloud droplets, but for falling raindrops the rates of evaporation in Eq. (15-43) must be multiplied by about 1.3 for 60 μm radius, 1.9 for 100 μm (0.1 mm), 3.2 for 200 μm (0.2 mm), and 8.26 for 1 mm radius.   In terms of mass change as given in Eq. (15-44), the evaporation rate is greater for large drops than for small ones, even without ventilation, because dM/dt is proportional to the radius.   In terms of dr/dt, however, the rate is inversely proportional to the radius.

GROWTH OF A POPULATION OF DROPLETS

The growth process in the preceding discussion has been limited to a single droplet. Several factors inhibit the growth of droplets to very large sizes by diffusion. It is generally recognized that the growth to raindrop sizes is accomplished by other means, especially by collision and coalescence, which will be discussed on subsequent pages. Equation (15-27) expresses the fact that for any sphere the ratio of dr/dt to dM/dt is proportional to r^2; so as droplets grow, an increasingly large amount of water is required to produce a given increment of size. Furthermore, the rate of transfer of latent heat to the drop depends on dM/dt, so that for a given increment of radius more heat is added as the droplet grows larger, thus causing a greater reduction of the vapor gradient. Finally, unless there is rapid adiabatic expansion in a cloud updraft, the excess water vapor is used up and the cloud approaches exact saturation equilibrium.

As a result of these limiting factors most cloud and fog droplets have radii less than about 20 μm, with a peak frequency of droplets somewhere between 1 and 10 μm radius. As shown farther on in this chapter, droplets of that size have a negligible fall speed; so it should not be surprising to find that the vast majority of clouds do not produce rain.

Several authors have published results of computations of droplet-size distributions resulting from condensation in natural clouds. Because a number of variables are changing together and mutually affect each other, the computation is done numerically by machine. The computation starts with an assumed distribution of condensation nuclei of uniform composition but different sizes. Some of the larger giant nuclei fall out and evaporate early if the updraft is weak, while most Aitken nuclei are never activated. The strength of the updraft is, of course, very important in carrying the process forward and producing the required supersaturations.

Two sets of results, one by Mordy[1] and the other by Neiburger and Chien,[2] are shown in Fig. 15-2. The initial nuclei distributions are represented in the two left-hand curves and the final droplet distributions on the right. A striking feature is the narrowing of the spectrum of droplets, showing how the small droplets grow fastest, resulting in a tendency toward uniformity of size. These computations show that condensation alone cannot produce raindrop sizes.

The two cases represented here are for slow rates of cooling and weak ascending motions such as one might find in stratified clouds. When faster cooling, such as in cumulus clouds, is imposed, the droplet-size distributions are not greatly different.

---

[1] W. Mordy, Computation of the Growth by Condensation of a Population of Droplets, *Tellus*, vol. 11, pp. 16–44, 1959.

[2] M. Neiburger and C. W. Chien, in Physics of Precipitation, *American Geophysical Union Monograph*, no. 5, p. 191, 1960.

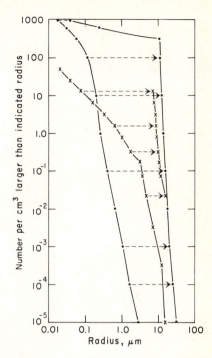

FIGURE 15–2
Initial and final distributions of particle and droplet sizes as computed by Mordy (points marked x) and by Neiburger and Chen (points with heavy dots). The dashed arrows connect initial and final sizes.

Computations of the kind described pose the main problem of cloud physics. Since condensation produces a narrow spectrum of droplets of radii mainly of the order of 5 to 10 μm, which are too small to fall, how does a cloud produce rain? Part of the answer is found in the giant nuclei which, while not numerous, could in some cases be supported in a strong enough updraft to form enough drizzle drops, which in turn could grow to raindrops by collision and coalescence with the main droplet population to produce appreciable rain. Other possible processes will be examined in a later section of this chapter.

GROWTH AND SUBLIMATION OF ICE CRYSTALS IN VAPOR

The growth of ice crystals and their sublimation to vapor are treated in essentially the same way as growth and evaporation of liquid droplets. Since crystals are not spherical but rather in the form of hexagonal plates, stars, needles, columns, and related shapes, a radius cannot be assigned to them. The diffusion of vapor to the crystal is handled in a manner derived from the analogous situation in electricity—that of a current flowing to an object in an electrical field spherically symmetric at infinity.

The surface of the body is considered to be at a uniform potential V_0. The *capacity* C of the body enters into the problem. In the electrical analogy the current to the body is given by an application of Gauss' law as

$$i = \lambda \int_s \frac{\partial V}{\partial n} \, ds = 4\pi C \lambda (V_\infty - V_0) \qquad (15\text{-}45)$$

where the integral is taken over the surface of the sphere. The constant λ is the electrical conductivity of the medium, $\partial V/\partial n$ is the potential gradient normal to the surface, and ds is an element of surface area. The current is the flux of charge in the same sense that dM/dt is the flux of water vapor, and λ is analogous to D, the diffusivity of water vapor. The analogous expression in terms of vapor diffusion is

$$\frac{dM}{dt} = D \int_s \frac{\partial \rho_w}{\partial n} \, ds = 4\pi C D (\rho_w - \rho_{w(s)}) \qquad (15\text{-}46)$$

This expression implies the existence of a density potential with a constant vapor density $\rho_{w(s)}$ at all points at the surface of the body. The capacity C is a geometrical factor which in the electrostatic units used here has the dimensions of length.[1] For a sphere, $C = r$, and for a circular disk, $C = 2r/\pi$. Hexagonal plates have about the same capacity as disks of equal area. Branching at the corners of the hexagons has little effect on the capacity. Needle-like crystals may be represented approximately by $C = a/\ln(2a/b)$, where a is the length and b the cross-sectional diameter.

It is to be noted that Eq. (15-46) is the same as Eq. (15-26) except that C replaces r. In deriving the growth-sublimation equation for ice crystals, it is not necessary to retrace the steps of Eqs. (15-24) through (15-44), since C will everywhere replace r, L will become L_s for sublimation, and e_{0i} for ice will replace e_s for liquid. The saturation ratio $S = e/e_s$ becomes $S_i = e/e_{0i}$. These alterations may be applied directly to Eqs. (15-43) and (15-44) so that for an ice crystal,

$$\frac{dM}{dt} = \frac{4\pi C(S_i - 1)}{(L_s^2 m_w / KRT^2) + (RT/Dm_w e_{0i})} = \frac{4\pi C(S_i - 1)}{A + B} \qquad (15\text{-}47)$$

A unique relationship between the shape of snow crystals and the temperature at which they were formed at normal supersaturations was established by Nakaya[2] in a long series of experiments in the late 1930s and early 1940s. At temperatures down to about $-7°C$ needles predominate. From -7 to about $-10°C$ columns and scrolls or cups are characteristic. From -10 to -13 and again from -17 to about -19

---

[1] In the rationalized mks units C is in farads and, for an isolated sphere, $C = 4\pi\varepsilon r$ F, where ε is the permittivity in farads per meter of the medium, and r is in meters. For any medium $\varepsilon = \kappa\varepsilon_0$, where κ is the dielectric constant (approximately 1 for air) and ε_0 is the permittivity in vacuo.

[2] Summarized in U. Nakaya, " Snow Crystals," Harvard University Press, Cambridge, Mass., 1954.

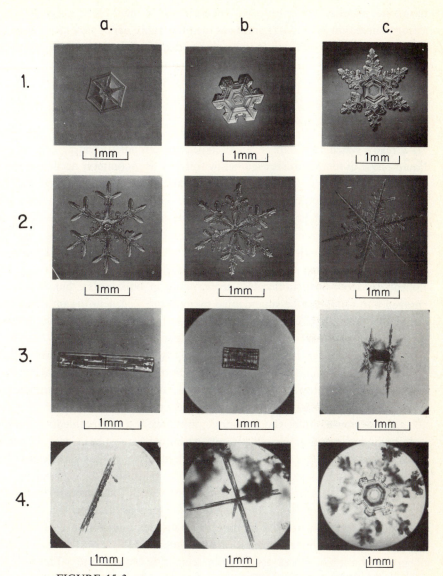

FIGURE 15–3

Microphotographs of some of the elementary forms of crystals of ice or snow: 1a, b, c, hexagonal plates showing growth at corners in b and c; 2a, b, c, dendritic forms; 3a, b, columns; 3c, capped column, shown in end view in 4c; 4a, b, needles. (*Photographs 1a through 3a by Kazuhiko Itagaki; 3b through 4c by Ukichiro Nakaya.*)

hexagonal plates and at -13 to -17 dendritic (branched) forms develop. It is interesting to note that Eq. (15-47) produces the most rapid deposition at $-14°$ at 1000 mb and $-17°$ at 500 mb, which is in the region of dendrite formation. Microphotographs of some of the principal forms are reproduced in Fig. 15-3.

GROWTH IN MIXED CLOUDS

As already stated in Chap. 7, supercooled (undercooled) water, i.e., water in the liquid phase at temperatures well below freezing, is a common occurrence in clouds. Supercooling to -5 or $-10°C$ is to be expected in nearly all clouds within that temperature range. Liquid droplets at -20 to $-30°C$ are not uncommon, and cases of natural clouds of liquid droplets at temperatures approaching $-40°$ have been reported. The supercooling involves not only droplets condensed at above-freezing temperatures and carried to the colder altitudes but also new condensation in the liquid form occurring at subfreezing temperatures. In the laboratory it is very difficult to accomplish noticeable supercooling in bulk water, but the same water sprayed into drops can be supercooled many degrees, depending on its purity. The larger the drop, the higher the temperature at which it will freeze. Small cloud droplets may not freeze until $-40°C$ is reached.

The freezing can be started immediately by mechanically disturbing the water. This effect is important in the formation of ice on airplanes or on stationary objects on windy mountaintops. As the supercooled droplets strike the airplane or other object, they freeze, usually as rime,[1] into an icy mass.

Let us now examine the humidity conditions in a supercooled water cloud. If such a cloud is to maintain itself, the relative humidity must be at least 100 percent or a fraction of a percentage higher. This relative humidity is that with respect to water. With respect to ice this cloud would be at supersaturation, as a glance at Table 6-1 shows, for the saturation vapor pressure over water at subfreezing temperatures is greater than that over ice. The difference is greatest at a temperature of about $-11.5°C$, but since the relative humidity with respect to ice would be given by e_s/e_{0i}, where the subscripts refer to saturation over water and ice, respectively, the percentage of supersaturation increases steadily with decreasing temperature. Figure 15-4 shows curves of the vapor-pressure difference and of the relative humidity with respect to ice corresponding to 100 percent with respect to water, plotted as a function of temperature.

It is significant that ice crystals do not usually form naturally in clouds unless the supersaturation with respect to ice is of the order of 10 percent. This behavior is

---

[1] See definition, Appendix B.

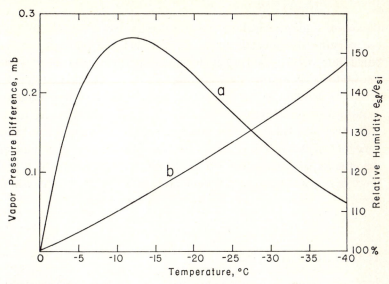

FIGURE 15–4
Difference in vapor tension, water minus ice (curve *a*), and relative humidity at water saturation (curve *b*) measured with respect to ice saturation.

quite different from that of water condensation where supersaturations of less than 1 percent suffice. It is evident that the atmosphere does not contain nucleating agents as suitable for ice as for liquid. The subject of natural and artificial ice-nucleating substances will be treated in later paragraphs.

In cloudy air in a typical supercooled condition, let us say at a temperature of -10 to $-20°C$, the first ice crystals to form would be in an environment with a relative humidity of 100 ± 1 percent with respect to liquid water. At these temperatures this would mean a relative humidity with respect to ice of 110.0 to 121.5 percent; so the crystals would be in a strongly supersaturated environment. Their rate of growth would accordingly be fast. As vapor diffused to the crystals, particularly as they became very numerous, the ambient vapor concentration would diminish to less than the equilibrium value for the droplets. The latter would then start evaporating and we would have a steady state for a while in which water would evaporate from the droplets, move in the vapor form to the crystals, and condense or crystallize there. As long as the crystals remained in a cloud which was predominantly water, they would grow, at least until all the water was used up. Meanwhile some would have attained a size large enough to fall out as snow or, on reaching warm levels, would melt into raindrops.

The process of the formation of snow and rain in mixed liquid and ice clouds is sometimes called the Bergeron process or Bergeron-Findeisen process after its discoverer and principal investigator, respectively.[1] Meteorologists accepted the theory enthusiastically and for a few years believed that this was the only way rain could form outside the tropics, but the coalescence process in warm clouds is now known to be of almost equal importance. Once the rain has started to fall, coalescence, or *accretion* as it is sometimes called, is the only important way in which the raindrops or snowflakes can grow. In the case of snowflakes the collisions can be with other snowflakes to form a clumped mass or with undercooled droplets to form rimed snowflakes or snow pellets. In most situations the great bulk of water reaching the ground has been picked up by the falling precipitation as it passes through the lower portions of clouds or through lower cloud layers not attached to the generating layer.

ICE NUCLEATION

From the preceding discussion it is apparent that the nucleating agents for ice crystals require a considerable degree of supersaturation of vapor with respect to ice before they can become effective. This fact is borne out by laboratory experiments. Tests on undercooled clouds made by blowing steam or one's breath into a frozen-food chest were conducted extensively by Schaefer,[2] who found that the temperature "threshold" for the first appearance of ice crystals in the undercooled clouds depends upon the kind of particles available in the air for nucleating. In fairly clean laboratory air it was discovered that no ice crystals would form until the temperature reached -39 or $-40°C$. This was subsequently called the *Schaefer point* or *temperature of spontaneous nucleation*, suggesting that at this temperature crystallization will occur regardless of nuclei. From a painstaking analysis of the nucleating substances in natural snow crystals, Kumai[3] found that clay minerals, especially kaolinite, were responsible for the great majority of ice-nucleating events in natural clouds.

A variety of cold boxes for measuring the ice-nucleating temperature thresholds in various parts of the natural atmosphere, at the surface and aloft, have been devised. The measurements show that there is a variation from day to day and from place to place. In one situation ice crystals will be seen glittering in the box after a cooling to only about $-5°C$ and on other occasions the temperature may have to reach $-30°C$ before the first crystals appear.

[1] T. Bergeron, On the Physics of Cloud and Precipitation, *Memoirs de l'Union Géodésique et Géophysique Internationale*, Lisbon, 1933. W. Findeisen, Die kolloid-meteorologischen Vorgänge der Niederschlagsbildung, *Meteorologische Zeitschrift*, vol. 55, p. 121, 1938.

[2] V. J. Schaefer, *Science*, vol. 104, p. 457, 1946.

[3] M. Kumai, Snow Crystals and the Identification of the Nuclei in Northern U.S.A., *Journal of Meteorology*, vol. 18, pp. 139–150, 1961.

Impressed by the seeming inadequacy of ice nucleation in natural clouds, Langmuir and Schaefer[1] considered the possibility of hastening the process artificially. They recognized that this would also induce precipitation by the Bergeron process in undercooled clouds. They successfully demonstrated two ways of doing this. One way is to drop very cold particles, such as pellets of solid CO_2, into the clouds from an airplane. Along the path of fall of the pellets a momentary cooling to the Schaefer point or lower occurs, and narrow streaks of ice crystals are created in the cloud. The spread of ice crystals through the cloud is extremely rapid, once a few are created. When a dry-ice pellet is dropped into a fogged cold chamber of about ten cubic feet in size, crystals are found throughout the volume in about three minutes. In atmospheric clouds the process is helped by eddy diffusion. Crystals once produced act as nucleating agents themselves to infect other parts of the cloud.

The crystal-to-crystal nucleation process is well known in laboratory and industrial chemistry. In fact, the effect can be created by crystals that differ slightly from those to be nucleated. Langmuir, Schaefer, et al.,[2] recognized the similarity between water and silver iodide in the crystalline state. The two principal lattice constants of ice, in angstrom units, are 4.53 and 7.41, while the corresponding ones for silver iodide are 4.58 and 7.49. Both crystals have hexagonal symmetry. Other substances such as lead iodide, cadmium iodide, and quartz have similar crystal configurations but not resembling ice as closely as in the case of silver iodide. By test in the cold box, silver iodide was found to have a nucleating threshold temperature of $-4°C$ and to be completely active at $-10°C$. This is the highest threshold temperature of any of the nucleating agents tested, except, of course, ice crystals themselves.

Since it is not practical to create and handle real ice crystals for artificially stimulated nucleation, silver iodide makes a good substitute. It can be spread into the atmosphere as an aerosol by one of several types of generators. Its advantage over the dry-ice method lies in the fact that it can be spread upward by the air currents, and under favorable conditions clouds may be nucleated from generators located on the ground.

With the emphasis on rainmaking less attention has been attracted to another aspect of artificial nucleation, namely, the artificial dissipation of stratified cloud layers. By dispensing dry ice or silver iodide smoke into an undercooled, stable, stratified cloud layer, the droplets can be changed to ice crystals which will settle out and leave a clear area. A rift can be made in the clouds that will grow to a width of a mile or more in 20 or 30 min, remaining that way for a similar period of time. The usefulness of this technique in clearing airport approach areas in certain winter conditions is obvious.

[1] I. Langmuir, V. J. Schaefer, et al., General Electric Research Laboratory, *Project Cirrus Progress Reports*, Schenectady, New York, 1947–1951.
[2] *Ibid.*

Various types of pyrotechnic devices for dispensing silver iodide are in use. These can be fired either from the ground or from airplanes. Rockets, flares, anti-aircraft shells, and similar pyrotechnics can be armed to explode or burn silver iodide dissolved in a flammable solvent or mixed in an explosive. Silver iodide is insoluble in most liquids, but it can be treated to form a solution in acetone, a flammable organic liquid.

In warm clouds, such as warm-weather cumulus with updrafts in them, the rain mechanism can be artificially initiated by spraying water in the cloud from an airplane flying through it. Spray drops from 50 to 100 μm in diameter coalesce with the cloud droplets to form raindrops. By the time these drops reach a diameter of 5 to 7 mm they split into several drops, all of which, in turn, coalesce with cloud droplets to multiply the process several times. An updraft in the clouds is necessary in order to give the water drops a long enough time in the cloud to acquire an appreciable amount of water by accretion. Common salt nuclei of giant size (5 to 20 μm radius) dispensed into a cloud also are effective in producing drizzle drops to start the coalescence mechanism. Other kinds of hygroscopic nuclei have been used for this purpose, notably urea crystals.

GROWTH BY COALESCENCE

Collision and coalescence could conceivably be caused by irregular relative motions similar to Brownian movements or very small-scale turbulence. Knowledge and experience of the mid-twentieth century has indicated no appreciable effect of this kind. Another possibility is the effect of electrical attraction between droplets. It can be demonstrated in the laboratory that colliding drops have a better chance of coalescing when they are in an appreciable electric field than when there is no space charge, but the collisions themselves depend on relative motions among the droplets, which would not be found in clouds consisting of uniformly small droplets The only important way in which collisions can occur is through the relative motions of the droplets in the gravitational field of the earth. The velocity of fall of the drops, the so-called *terminal* velocity, depends on the drop size, as will now be shown

In a medium having a certain viscosity, such as the air, motion under the acceleration of gravity reaches a fixed terminal velocity defined as that velocity attained when the inertial forces are equal to the viscous or resistance forces $F = F_r$. Under the acceleration of gravity, the inertial force is the buoyancy force, or for spherical drops,

$$F = \tfrac{4}{3}\pi \rho_L r^3 g - \tfrac{4}{3}\pi \rho r^3 g \qquad (15\text{-}48)$$

where the first term is the force or weight of the sphere in a vacuum and the second term is the weight of the displaced air, or the air buoyancy. Separating the common factors, we obtain

$$F = \tfrac{4}{3}\pi(\rho_L - \rho)r^3 g \qquad (15\text{-}49)$$

From fluid mechanics the resistance force on a sphere can be calculated, but the result only will be given here; it is as follows:

$$F_r = 6\pi\eta r v N \qquad (15\text{-}50)$$

where η is the dynamic viscosity of the medium (air) and v is the relative speed of the sphere and the medium. N is a quantity involving two parameters—the *drag coefficient* C_d and a nondimensional number Re, much used in hydrodynamics, called the *Reynolds number*—such that

$$N = \frac{C_d\, Re}{24} \qquad (15\text{-}51)$$

Since the viscosity η has the dimensions g cm^{-1} sec^{-1}, it is apparent from Eq. (15-50) that N is nondimensional, and, in fact, Re and C_d are both nondimensional.

For droplets of radius up to 40 μm, $Re = 24/C_d$ and $N = 1$; so in that size range

$$F_r = 6\pi\eta r v \qquad (15\text{-}52)$$

The size range through which this relation applies is known as the Stokes range, after its discoverer.

To obtain the terminal velocity v_T we equate the inertial and resistance forces and obtain for the Stokes range

$$\tfrac{4}{3}\pi(\rho_L - \rho)r^3 g = 6\pi\eta r v_T$$

$$v_T = \frac{2}{9}\frac{\rho_L - \rho}{\eta}\, g r^2 \qquad (15\text{-}53)$$

Outside this range the factor $1/N$ appears on the right, but in practice it is better to use experimentally determined values for drops with radii greater than 40 μm such as the values obtained by Gunn and Kinzer.[1]

Terminal velocities as a function of drop diameters are plotted in Fig. 15-5. In (a) the droplets fall in accordance with Stokes' law. The values in (b) and (c) were determined by Gunn and Kinzer[1] in still air near normal temperature and pressure. The sizes of the larger drops were computed from the mass to give an equivalent diameter, assuming a spherical mass. The larger drops are known to be nonspherical, shaped somewhat like round loaves of bread with the flat side downward. It is seen that in the size range of cloud droplets, from 2 to 40μm in diameter, the fall velocities are negligible and the droplets are for all practical purposes suspended in the air.

[1] R. Gunn and G. D. Kinzer, *Journal of Meteorology*, vol. 6, p. 243, 1949.

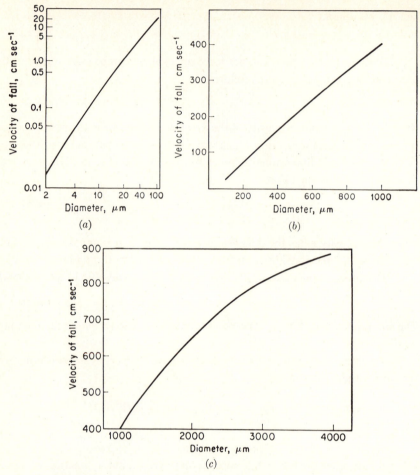

FIGURE 15–5

Velocity of fall of droplets of various diameters. In (a) the droplets fall in accordance with Stokes' law. The values in (b) and (c) were determined experimentally by Gunn and Kinzer. The sizes are computed from the mass in the experiment to give an equivalent diameter, assuming a spherical shape. At sizes in the upper ranges of curve c it is known that the drops are markedly nonspherical.

The significance of clouds as suspended small droplets is emphasized when the terminal velocities and the rates of evaporation are considered together. In Table 15-2, computed by Findeisen,[1] the distance of fall before evaporation of drops of various sizes is given. The relative humidity is assumed constant at 90 percent, a reasonable value under cloud bases, and the pressure is taken as 900 mb with temperature of

[1] W. Findeisen, *Meteorologische Zeitschrift*, vol. 56, p. 453, 1939.

5°C.   It is apparent that there is more reason for differentiating between raindrops and cloud droplets than the mere fact of difference in size.   It is also apparent that any cloud with the usual atmospheric turbulence in it must consist of droplets continually being formed in one part and evaporating in another.   In a small cumulus cloud it is doubtful that a collection of water molecules can remain in the form of a droplet for more than a few minutes.   In a stable atmosphere with stratus, individual droplets may exist as such for many minutes or possibly for one or more hours.

From the data on terminal velocities it is possible to obtain an impression of the spectrum of sizes that would be required in a cloud for collisions to occur.   It is seen that in a cloud with droplets mainly of 10 μm diameter, a drop measuring somewhere between 40 and 100 μm in diameter would have a fair chance of colliding with some of the smaller ones if the latter were of the order of 1 cm apart.   The computation of the collision-coalescence effects on drop growth will now be considered.

Let r_i be the radii of cloud droplets of various small sizes.   The mass of liquid water in a unit volume of air containing n_i droplets is

$$\chi = \tfrac{4}{3}\pi\rho_L \sum_i n_i r_i^3 \qquad (15\text{-}54)$$

A larger drop of radius r falling distance dz through this collection of droplets sweeps out a cylindrical volume

$$dV = \pi r^2\,dz = \pi r^2 v_T\,dt \qquad (15\text{-}55)$$

and sweeps out a mass of water $\chi\,dV$, or

$$dM' = \chi\,dV = \chi\pi r^2 v_T\,dt \qquad (15\text{-}56)$$

where dM' is the mass of water contained in the droplets of radius r_i occupying the volume dV.   If all these droplets struck the larger falling drop and coalesced with it, dM' would also represent the increase in mass of the drop.   However, some of the droplets in the cylindrical volume are carried around the falling drop by the air stream and others bounce off after striking.   Only a fraction E of the droplets is collected; so

Table 15-2   **DISTANCE OF FALL BEFORE EVAPORATION**
(Computed by Findeisen)

Radius of drop, cm	Distance of fall before evaporation	
0.0001	3.3×10^{-4} cm	} Cloud particles
0.001	3.3 cm	
0.01	150 m	} Raindrops
0.10	42 km	
0.25	280 km	

we multiply by this fraction, calling it the *collection efficiency*, to obtain the growth of the drop by accretion in the form:

$$dM = \chi E \pi r^2 v_T \, dt \qquad (15\text{-}57)$$

This produces an additional spherical shell of growth having thickness dr on the drop such that

$$dM = 4\pi \rho_L r^2 \, dr \qquad (15\text{-}58)$$

This value of dM is substituted on the left in Eq. (17-56) and the expression is solved for dr/dt, resulting in

$$\frac{dr}{dt} = \frac{E \chi v_T}{4 \rho_L} \qquad (15\text{-}59)$$

This equation is deceiving in its simplicity. A major effort in cloud physics is devoted to the determination of the collection efficiency E. Analytical and experimental approaches have both been used, but there are some conflicting data. The efficiency is greatly increased by the presence of an electric field of a strength slightly greater than that occurring in fair weather. Analytically the collection efficiency can be shown to depend on a factor involving the ratio of the square of the radius of the small droplets to the first power of the radius of the larger drop, indicating that the size of the collected droplets is more critical than the size of the falling drop. Of course, v_T is dependent on the radius of the falling drop; so the growth rate is not necessarily the fastest where the collection efficiency is the greatest.

The development of rain in clouds appears to result from accretion in many cases. If the size spectrum in a cloud is broadened by the development of a few drops per cubic meter approaching 100 μm diameter, with fall speeds of tens of centimeters per second, collision-coalescence will occur with the small cloud droplets to create growing raindrops. The process is favored by thick clouds and sustained updrafts, so that the drop can grow quite large before falling out of the bottom of the cloud.

ATMOSPHERIC ELECTRICITY

General Geophysical Aspects

The surface of the earth and the conducting layers of the atmosphere above about 50 km, referred to as the *electrosphere*, may be regarded as the plates of a spherical condenser. The outer plate has a net positive charge and the inner one a net negative charge. The condenser leaks because the atmosphere conducts electricity between the two plates; therefore, they have to be recharged frequently or continuously.   It is estimated that the leakage current is about 1800 amperes (A) and that the atmosphere has an effective resistance of 200 ohms (Ω), thus giving a potential of 360,000 volts(V). Measurements show that these values are not uniformly distributed throughout the atmosphere.   At sea level the electrical potential gradient is of the order of 100 V per m, decreasing with height.

The air conductivity is brought about by the presence of ions.   The conductivity is proportional to the number and *mobility* of the ions, the mobility being defined as the velocity an ion would have in a field of 1 V per cm.   Only the small ions, having mobilities of the order of 1 to 2 cm per sec/volt per cm are of importance as conductors.   Large ions, generally considered to be charged Aitken nuclei or condensation nuclei, have such low mobilities (about $\frac{1}{500}$ of the average for small ions) that they are not effective conductors.

The small ions are a result of ionization by cosmic rays and emanations of radioactive gases from the solid earth.   Ionization proceeds when an electron is stripped from a molecule, leaving a positively charged ion.   The electron then attaches itself to a neutral molecule to form a net negative charge on that molecule and thus to create a negative ion.   In this way ions are formed in pairs of opposite sign.   At the surface of the earth between 10 and 50 ion pairs are produced per cubic centimeter per second. The rate of formation does not decrease with height as one leaves the source of radioactive gases, because the cosmic-ray activity increases with height; at 12 km the rate of production is usually greater than at the surface of the earth.   Manmade releases of radioactive materials increase the ionization rate, at least temporarily.

Ion formation is balanced by processes of small-ion destruction—recombination between small ions of opposite signs, combination with large ions of opposite signs, and coalescence with neutral condensation nuclei.   A balance exists between the rate of ion formation q and the rate of destruction, of the form

$$q = \alpha n_+ n_- + \eta_{+-} N_- n_+ + \eta_{+0} N_0 n_+ \qquad (15\text{-}60a)$$

for positive small ions and

$$q = \alpha n_+ n_- + \eta_{-+} N_+ n_- + \eta_{-0} N_0 n_- \qquad (15\text{-}60b)$$

for negative small ions.   Here n_+ and n_- are the number of positive and negative small ions per cubic centimeter; the N's similarly signify the large ions, with N_0 meaning neutral particles of large-ion size.   The α is the recombination coefficient for small ions and η_{+-}, η_{+0}, η_{-+}, and η_{-0} are the combination coefficients of small ions of the sign represented by the first part of the subscript with large ions of the character represented by the second part of the subscript.   Over most regions of the earth, at least in the lower layers, the second term—combination of small ions with large ones of opposite signs—is the most important, so that if q remains the same, n_+ decreases as N_- increases and n_- goes down as N_+ goes up.

The electrical conductivity is given by

$$\lambda = e \sum k_i n_i$$

where e is the charge on an electron and n_i represents the various kinds of ions of mobilities k_i. If an ion balance exists, n does not vary appreciably, except locally with the variations in large ions produced by smoke pollution. The atmosphere therefore readily conducts a current. If no recharging occurs, this air-to-earth or "leakage" current is estimated as being sufficient to completely discharge the earth's condenser shell to a negligible value in an hour or so.

To preserve the electrical balance, some process must supply a replenishing negative charge to the earth. In seeking this "supply current" we look to regions of disturbed weather. Charges are exchanged with the earth's surface by three processes acting in disturbed weather, namely: charges brought down by precipitation, by lightning, and by point discharges. Except for some snow situations which bring down negative charge, most precipitation particles reach the earth with a net positive charge, thus operating in the same sense as the fair-weather conduction current toward destroying the potential gradient. Lightning strokes predominantly transfer negative charge to the surface and help replenish the charge lost by atmospheric conduction. Perhaps more important are the point discharges which occur at the surface under strong electric fields such as occur in the vicinity of thunderstorms or other situations of strong potential gradient.

Around a pointed conductor extending upward from the earth the electrical lines of force are concentrated so that the surface charge density at the point and the potential gradient near it are greater than in the general surroundings. If the field is of sufficient strength around the point, an electron formed by initial ionization can acquire enough energy in the interval between its detachment and its first collision with a molecule to ionize the molecule instead of combining with it, thus creating a fresh ion pair. The multiplication of ions in this way about the point establishes a movement of ions of one sign into the point and those of opposite sign away from it. When the potential gradient is the reverse of that found in fair weather, that is, with positive charge at the ground and negative at a cloud base, electrons move toward the point and positive ions move away, so that negative electricity is conducted to the earth. The movement of charge occurs in a pulsating manner.

For pointed conductors at roughly the height of trees, point discharges occur when the potential gradient at the ground is of the order of 600 to 1000 V per m, but a higher threshold value is necessary for trees themselves. In disturbed weather, especially thunderstorms, the gradient can reach several times this value. From measurements in various parts of the world the ratio of negative to positive charge supplied to the surface by point discharges ranges between 1.5 and 3.0. For the worldwide balance, Israel[1] estimates that to compensate for the $+90$ coulombs (C) km^{-2} $year^{-1}$ conducted to the earth in fair weather and $+30$ C km^{-2} $year^{-1}$ from precipitation, negative charge in the amount of -100 C km^{-2} $year^{-1}$ from point discharge and -20 C km^{-2} $year^{-1}$ from lightning is transported to the earth.

For point discharges to carry negative charge to the earth the potential gradient must be the reverse of that observed in fair weather. The earth's surface and the point must have a net positive charge with respect to the space above. Thunderstorms are predominantly negatively charged at their bases and therefore provide the required direction of the field and, in addition, strengthen it greatly. Conditions for both lightning and point discharges to operate in bringing down negative electricity are therefore to be found in thunderstorms. Climatological data indicate 1800 thundery situations, on an average, more or less continuously over the earth.

An indication of a link between thunderstorm activity and the daily regeneration of the earth's electric field was given by Whipple,[2] whose curve of the diurnal variation of the area covered by thunderstorms on land areas of the earth, reproduced in Fig. 15-6, matches the diurnal variation of potential gradient on the oceans. The oceans are used because land areas have a diurnal effect caused by the creation of an excessive number of immobile large ions by smoke pollution and a tendency to concentrate particles under nocturnal inversions. Figure 15-6 shows the values of the potential gradient over the oceans plotted in terms of absolute time during the course of the day. In each case a maximum is shown around 18 hr GMT. It seems that at the same hours of absolute time on the average day the atmospheric condenser is recharged more vigorously than at other hours. This is at about the time of the maximum number of thunderstorms over land areas of the earth. The preponderance of thunderstorms at 14 to 20 hr is due to the great effect of the afternoon and early evening thunderstorms over equatorial Africa and South America, summer and winter.

Cloud Electricity

Detectable charge centers form in convective clouds in the early stages of their development and reach the spectacular magnitudes leading to lightning discharges in thunderstorms. There are a number of plausible theories of cloud-charge generation, but rather than advocate one of them it is better to start out by noting some of the observed and measured conditions.

[1] H. Israel, Bemerkung zum Energeiumsatz im Gewitter (Remark on the Energy Return in the Thunderstorm), *Geofisica Pura e Applicata*, vol. 24, pp. 3–11, 1953.

[2] F. J. W. Whipple, Modern Views on Atmospheric Electricity, *Quarterly Journal of the Royal Meteorological Society*, vol. 64, pp. 199–213, 1938.

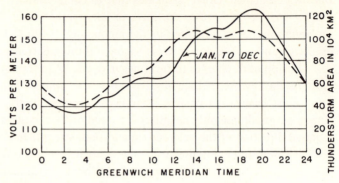

FIGURE 15-6
Potential gradient and thunderstorm areas of the globe, dashed line, as functions
of absolute time during the day.

From airborne and mountain electrical measurements in clouds it is found that
the building cumulus containing no radar echo is typically charged negatively with
respect to its surroundings, with horizontal potential gradients of up to 20 V per cm in
and near the cloud.

In warm clouds studied in the Caribbean area, not much more happens electrically even after warm rain has started. In some cases a positive center forms
around the rain echo but it does not strengthen appreciably. In the clouds of the
continental United States where the first precipitation echoes are forming at or near
the freezing level at a time when the cloud may be 15,000 ft tall from base to top,
rapid electrical developments take place after the first echo appears. The first precipitation echo usually represents a center of charge that is positive both with respect to the
rest of the cloud and with respect to the outside environment. If frozen precipitation
elements, such as snow pellets, are present, this positive center increases in magnitude
very fast so that within 10 min the potential gradient around it increases from less than
50 to perhaps 1000 V per cm or more. After a lightning discharge the thunderstorm
can completely recharge itself in about two minutes. The cloud-to-ground lightning
flashes are between the negatively charged lower part of the cloud and the earth, which
locally has an induced positive charge. The top parts of the cloud, in the ice areas,
carry a positive charge, and locally, in the heaviest downpour, also at the base of the
cloud, there is a core of positive charge. Some of the cloud-to-ground lightning
discharges transfer this positive charge downward, but the vast majority of them
carry down negative electricity. Normally the in-cloud or cloud-to-cloud discharges
outnumber the cloud-to-ground lightning strokes by about two to one.

With this very general description of the electrical conditions in thunderstorms, the subject will be left at this point while in the next chapter the structure and dynamics of thunderstorms will be considered.   In this way the relationship between the water circulation and the electrical aspects can be approached more meaningfully.

EXERCISES

1 A cloud contains 1 g of liquid water per cubic meter consisting of spherical droplets uniformly of a diameter of 50 μm.   (*a*) How many droplets are there in a cubic centimeter?   (*b*) In this cloud raindrops, also spherical, are formed with diameters of 2 mm. These are suspended in the updraft in the cloud such that there are 500 per cubic meter of cloud space.   If there still is 1 g per m³ in the 50-μm size, what is the new liquid-water content including raindrops and droplets?   (*c*) How many droplets of diameter 20 μm would have to be coalesced to make the 500 raindrops per cubic meter described in (*b*)?

2 Determine the liquid-water contents of the following clouds, considered to have in each droplets of uniform size, as indicated: (*a*) fair-weather cumulus, droplet radius 10 μm, droplet concentration 314 cm^{-3}; (*b*) cumulus congestus, droplet radius 20 μm, droplet concentration 63 cm^{-3}.

3 Substitute r^* from Eq. (15-11) into Eq. (15-10) and show that the free-energy peak

$$\Delta G^* = \frac{16\pi\sigma^3\, m_w}{3[\rho_L\, RT \ln\,(e/e_s)]^2}$$

Then substitute for ln (e/e_s) from Eq. (15-11) into Eq. (15-10) and show that

$$\Delta G^* = \tfrac{1}{3}\pi r^* \sigma$$

4 In Eq. (15-33) consider a droplet at 9.5°C in a cloud at air temperature 10°C.   From values for the constants given following the equation and for L as given in Appendix C, find the difference in vapor pressure over the droplet and in the ambient cloud air.   Is the droplet growing or evaporating?

5 Find the rate of growth of a cloud droplet with an initial radius of 5 μm in a cloud at a saturation ratio of 1.05 at a temperature of 10°C and pressure of 850 mb.

6 Obtain the rate of growth of a raindrop of diameter 1 mm falling through a cloud containing 5 g of liquid water per m³.   Assume a collection efficiency of 1.

16

THUNDERSTORMS AND RELATED PHENOMENA

More than 200 years ago, Benjamin Franklin conducted his experiments to show that lightning was a form of what was then known as electricity, and then proceeded with his invention of the lightning rod. Since that time thunderstorms have been studied seriously as scientifically explainable phenomena rather than as supernatural manifestations. The fact that several investigators have lost their lives in exploring the electrical processes and violent air currents accounts in part for the slow progress in understanding thunderstorms and attests to the violence of the phenomena. Thunderstorms and their related weather are seen as the most violent of all storms, and since they occur on a scale too small both in space and in time for recognition in the synoptic network, only the *probability* of development of severe local storms can be predicted. Warnings of an individual storm can be issued only in the hour or minutes that a severe thunderstorm or its associated tornado or hailstorm can be traced locally by radar.

DEFINITION AND RELATION TO CONVECTION

By agreement in the World Meteorological Organization, a thunderstorm is reported if thunder is heard at the station.   The thunder is the noise of the lightning discharge and it is noticeably delayed if the lightning is at a considerable distance, owing to the great difference between the speed of light and the speed of sound.   Thus thunderstorms are defined in terms of their electrical manifestations.   From the more general meteorological point of view, this definition may be regarded as based only on a rough measure of the size and intensity of a cumulonimbus cloud system.   In studying thunderstorms we are therefore concerned with intense convection in damp air. Most theories of charge generation and separation in thunderstorms relate the electrical development to the air circulation and to hydrometeors in the clouds.

Recognizable units of cumulonimbus convective systems can range from the turrets only a mile or so in diameter, which bulge upward in a growing part of the cloud and which often contain a separate radar echo and their own electrical generating unit, to great connected masses or lines of thunderstorms extending for 50 miles or more.   The synoptic analysis on the *meso*scale of the large thunderstorm systems and squall lines has been discussed briefly in Chap. 12.   A unit of convection, or "cell," having a diameter of the order of 5 miles has been identified as characteristic of thunderstorms by a number of investigations, particularly by the 1946–1947 U.S. Thunderstorm Project.[1]   An isolated cell forms from several growing cumulus clouds or from active turrets in a less well-defined cloud mass.   It is not often that a single-celled thunderstorm is seen very long, except possibly in arid, mountainous regions, because there is a tendency for the development and joining together of adjacent cells.   Although the cells are connected by cloud structure, they can be distinguished in airplane flights by a less turbulent zone in the connecting part, and they can usually be recognized through separate patterns of radar echoes and of precipitation on the ground.

STRUCTURE AND LIFE CYCLE OF A CELL

The life cycle of the thunderstorm cell naturally divides itself into three stages determined by the magnitude and direction (upward or downward) of the predominating vertical motions.   These stages are:

1   The cumulus stage: a cell formed by a collection of cumulus clouds, characterized by an updraft throughout the cell.

[1] H. R. Byers and R. R. Braham, Jr., Thunderstorm Structure and Circulation, *Journal of Meteorology*, vol. 5, pp. 71–86, 1948.   U.S. Weather Bureau, "The Thunderstorm," Government Printing Office, Washington, D.C., 1949.

2 The mature stage: characterized by the existence of both updrafts and downdrafts, at least in the lower half of the cell.

3 The dissipating stage: characterized by weak downdrafts throughout.

The cumulus-stage cell in reality develops from a cluster of cumulus clouds, and since it is defined by hindsight as a system destined to become a true thunderstorm cell, it is bigger and has a stronger and more uniform updraft than the usual cumulus. The updraft is strongest at the higher altitudes and increases in magnitude toward the end of the stage. Converging air feeds the updraft not only from the surface but also from the unsaturated environment at all levels penetrated by the cell. Thus, air is entrained into the cloud system and is accommodated by the evaporation of some of the liquid water carried in the updraft. This entraining continues throughout all the stages.

In-cloud temperatures in a strongly developing cell are higher than those of the environment at corresponding altitudes. It is noted that the hydrometeors are not reaching the ground. Although pilots and observers flying through the clouds in this stage report rain or snow (flakes or pellets), particularly near the end of the stage, these condensation products may be suspended by the updraft. The first radar echo appears in this stage, often extending downward at a speed greater than the rate of fall of the precipitation particles, suggesting an almost simultaneous growth of precipitation through a considerable thickness of the cloud. In most sections of the United States in summer the first echo appears not far from the freezing level. Depending on location and the conditions characteristic of a given day, the first echo may form by the coalescence of liquid drops or around ice particles. Although concentrations of electrical space charge develop rapidly in this stage, no lightning occurs. (Isolated reports of lightning out of small clouds or "out of the blue," while authenticated, will not be considered here.)

The mature stage begins when rain first falls distinctly out of the bottom of the cloud. Except under arid conditions, the rain reaches the ground. The weight and drag of the precipitation helps to change the updraft into a downdraft which, once started, can continue without this frictional drive, as will be demonstrated in the next section of this chapter. The beginning of the rain at the surface and the initial appearance of the downdraft there are nearly simultaneous. The downdraft starts at the level of rain initiation, above or below the freezing level, later growing in vertical as well as in horizontal extent (Fig. 16-1).

The updraft also continues and often reaches the greatest strength in the early mature stage in the upper part of the cloud system. The updraft speeds may locally exceed 80 ft (about 25 m) per sec. The downdraft is usually not as strong as the updraft and is most pronounced in the lower part of the cloud, although naturally weakening and spreading laterally near the ground. Areas of rain, downdraft, and horizontal divergence are found together at the surface.

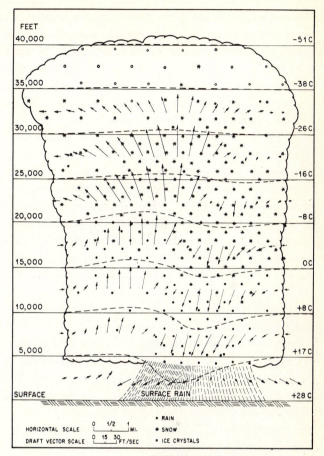

FIGURE 16–1
Thunderstorm cell in mature stage.

Temperatures are low in the downdraft, compared with the environment, and contrast especially with the updraft temperatures. The greatest negative temperature anomalies are found in the lower levels. As might be expected, there is a close association between updraft and high temperatures and between downdraft and low temperatures.

In the updraft, mixing of entrained air causes evaporation of some of the liquid water, thus removing some of the heat gained from condensation. The updraft air has its temperature reduced at an entrainment wet-adiabatic rate after the manner described in Chap. 6. Despite this effect the updraft is enough warmer than the environment air to be strongly buoyant. The downdraft in many cases seems to be characterized by reversible wet-adiabatic temperature increases in which evaporation counteracts to some extent the compression effects. Since the downdraft starts at a temperature very near that of the environment, its wet-adiabatic descent assures that it will be colder than the environment, which has a lapse rate greater than the wet-adiabatic. The cold downdraft spreads out at the surface as a cold air mass to form the pseudo-cold front advancing against the warmer surrounding surface air, a type of front often dealt with in the mesosynoptic scale.

The mature stage represents the most intense period of the thunderstorm in all its aspects, including electrical activity. At the ground, heavy rain and strong winds are observed while in the clouds the airplanes encounter at this stage the most severe turbulence, including in addition to the drafts the short, intense accelerations known in aeronautics as "gusts." Hail, if present, is most often found in this stage. The cloud may extend to more than 60,000 ft, penetrating the tropopause, although more often the maximum height reached is from 40,000 to 45,000 ft. With very strong updrafts it is possible for liquid water to be carried well above the freezing level. On the Thunderstorm Project in Ohio one case of heavy rain at 26,000 ft, nearly 10,000 ft above the 0°C line of the environment, was reported.

When the updraft disappears and when the downdraft has spread over the entire area of the cell, the dissipating stage begins. Dissipation results from the fact that there is now no longer the updraft source of condensing water. As the updraft is cut off, the mass of water available to accelerate the descending air diminishes; so the downdraft also weakens. The entire cell is colder than the environment as long as the downdraft and the rain persist. As the downdrafts give out, the temperature within the cell is restored to a value approximately equal to that of the surroundings. The complete dissipation occurs or only stratified clouds remain. All surface signs of the thunderstorm and the downdraft ultimately disappear.

THERMODYNAMICS OF ENTRAINMENT AND OF THE DOWNDRAFT

Computed and observed inflow rates in American thunderstorms show that the cell of cumulus clouds that develops into a thunderstorm entrains environment air at a rate of approximately 100 percent per 500 mb of ascent; i.e., it doubles its mass as it rises through a pressure decrease of 500 mb. This is a lower rate of entrainment than that for individual small cumulus clouds that may be less than 15,000 ft in depth. With very much higher rates of entrainment the updraft could not be maintained, since it would then become colder than the environment or would evaporate all its

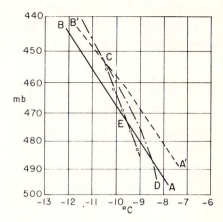

FIGURE 16–2
Thermodynamics of entrainment in thunderstorm cell.

liquid water to dissipate the cloud. For different air-mass conditions different critical rates of entrainment, i.e., rates which would either use up all the water or deprive the updraft of its buoyancy, can be calculated. In summer tropical air in the Eastern United States the critical rate would be 100 percent in 250 or 300 mb. In Europe the entrainment rates permissible for sustained convention would have to be less.

The updraft appears to follow the required thermodynamic pattern of entrainment. The downdraft is a special case, however. In Fig. 16-2 the updraft-entrainment wet-adiabatic rate in a typical, well-developed growing cumulus cell is represented by line $A'B'$ and a typical environment lapse rate for American tropical air is given by line AB. A saturated parcel displaced downward from C would follow the wet-adiabatic CD, if no environment air were entrained into it. With entrainment it would warm at some other, less rapid, rate such as CE. If the parcel is forced downward beyond D or E, it will become colder than the environment and sink. The frictional drag of the mass of liquid water provides the means whereby a parcel in a thunderstorm can thus be brought below point D or E, whence it continues as the thunderstorm downdraft. With a large quantity of liquid water available for evaporation, saturation can be maintained in spite of the increasing temperatures during descent, and the parcels will reach the ground, arriving there with a temperature several degrees lower than the surface environment wet-bulb temperature.

Another way for the downdraft air to be cooled is from the presence of hail. If hail occurs at a level where the downdraft is generated or at a place through which the downdraft passes, the hailstones extract the heat of fusion (melting) from the air as they melt.   This process may, in fact, start the downdraft.   The hailstones remain at a temperature of 0°C while they are melting.   The melting of a hailstone 1 cm in radius can cool a cubic meter of air by nearly 0.3°C at the melting level.

THUNDERSTORM WEATHER NEAR THE SURFACE

Figure 16-3 shows the course of events as a cell passes a station just after reaching the mature stage.   The data are from recording instruments of the U.S. Thunderstorm Project in Ohio.   Not very many thunderstorm occurrences are of this extreme nature; if the cell has been in the mature stage for some time, the pseudo-cold front has usually spread ahead of the rain and has weakened.   The complicating factors will become more obvious as we take up the details of the different weather elements.

Rainfall

The rainfall pattern follows closely the arrangement of the cells and reflects to a considerable extent their stages of development.   Along with the downdraft and the area of horizontal divergence, the rain from a newly developed cell first covers a very small area and then gradually spreads.   However, the cold air of the downdraft is able to spread laterally from the cell while the rain falls directly to the ground, so that an expanding outer area of cold air without rain develops.   In the dissipating stage this cold-air area continues to expand while the rain area contracts.

If the rainfall is considered with respect to the moving cell, it is found that the duration of moderate to heavy rain from a single cell may vary from a few minutes in the case of a weak, short-lived cell to almost an hour in a large, active one.   At a fixed point on the ground the duration of the rain depends upon such factors as the number, size, and longevity of the cells passing over the point, the position of each point with respect to each passing cell, and the rate of translation of the cells.   In the Eastern and Southern United States the average duration of thunderstorm rain at a given station is about 25 min, although it is highly variable from case to case.

The most intense rain occurs under the core of the cell within 2 or 3 min after the first measurable rain from that cell reaches the ground, and the rain usually remains heavy for a period of from 5 to 15 min.   The rainfall rate then decreases, but much more slowly than it first increased.   Around the edges of the cell, lesser rainfall rates occur.

Wind Field

Early in the cumulus stage there is a gentle inward turning of the surface wind, form-ing an area of weak lateral convergence under the updraft.   As the cell grows and a

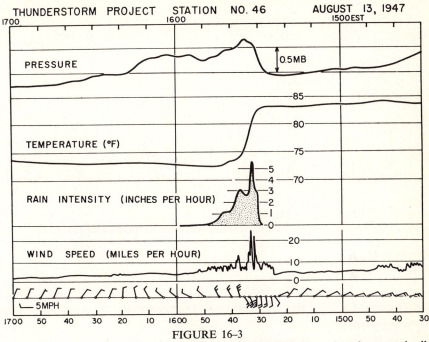

FIGURE 16–3

Autographic records for passage of newly matured cell. Time runs from right to left.

downdraft develops, the surface winds become strong and gusty as they flow outward from the downdraft region. The outward-flowing cold air underruns the warmer air which it displaces and a sudden change in the wind and temperature fields is established (the pseudo-cold front). The windshift line moves outward, pushed by the downdraft, resulting in strong horizontal divergence behind it. Divergence values of almost 10^{-2} per sec over an area of 50 sq miles have been measured.

The outflow is radial in the slowest-moving storms, but in most cases the wind field is asymmetrical with considerably more movement on the downwind side. The prevailing air movement of the lower layers nullifies the radial flow on the upwind side. With respect to the moving cell the outflow may still be radial in character, although not with respect to the ground. Thus the wind discontinuity in most cases is easily detected only in the forward portions of the storm, where it appears as the pseudo-cold front.

The cold dome of outflowing downdraft air has a form illustrated in Fig. 16-4. In this sketch the cell is considered to be in the mature stage and is moving from left to right. The cold air is represented as having spread out considerably farther on

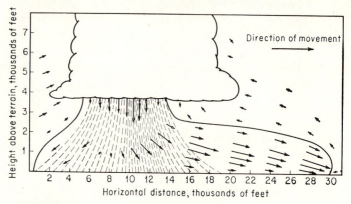

FIGURE 16–4
Vertical cross section through cold dome of outflowing downdraft air.

the downwind side of the cell than on the upwind side, as would be expected in a moving system.

In from 15 to 20 min after the outflowing starts, the pseudo-front zone has traveled about 5 or 6 miles from the cell center. The surface winds near the pseudo-front are still strong and gusty, but well within the cold-air dome wind speeds have decreased so that the strongest winds are no longer underneath the cell itself. A continued settling of the outflow air, transporting momentum downward, causes the wind speed to increase as one approaches the discontinuity zone from within the cold air. Pushed by the air somewhat above the ground, the pseudo-front often travels at a speed greater than any sustained speed observed at the ground.

With the pseudo-front, the wind shows clockwise shifts in most cases. This is especially true in American tropical-air currents in middle latitudes where the winds are usually from southwest or south and shift to west or northwest at the discontinuity.

Temperature

The "first gust" and the "temperature break," marked by the point on the thermograph where the temperature suddenly starts its drop, are two of the most pronounced features observed at the surface, and they occur essentially together. At the time of the formation of summer afternoon thunderstorms in the United States the temperature is usually above 85°F, often in the 90s. As a result of the rain and the downdraft, the temperature may reach a value as low as 65°F without change in air mass in the usual sense.   As the thunderstorm activity dies down after sunset, the temperature has usually recovered to an intermediate value representing a mixture of strongly cooled and uncooled portions of the air mass.

The area affected by the cooling is many times greater than that over which rain falls, but the temperature change is, of course, most marked in the rain core (downdraft center).   Cooling may be detected to as much as 15 to 20 miles downstream.   Near the center of a mature cell the temperature reaches a minimum 10 to 15 min after the temperature break; farther from the cell the temperature drop is much slower.   The amount of the temperature decrease observed in any given storm varies inversely with the distance of the observation point from a cell core.   Since the downdraft is only a few miles in diameter, the area over which the first and most rapid temperature fall occurs is relatively small.   As a result, strong horizontal temperature gradients are created after the downdraft reaches the ground.   Gradients exceeding 20°F per mile have been observed.   As the storm ages, the cold air spreads out and the magnitude of the gradient decreases.   Regardless of the spread of the cold air, the area of minimum temperature remains in the general location where the cold downdraft made its first appearance at the surface, except in cases where a new cell with its own downdraft and rain core develops over another part of the cold dome.

Pressure

Early in the cumulus stage a fall in surface pressure almost invariably occurs.   This fall is observed before the radar echo forms, and is recorded over an area several times the maximum horizontal extent of the echo.   When the radar echo appears, the pressure trace levels off in the region directly underneath it, but continues to fall, frequently at a more rapid rate, in the surrounding areas.   The pressure drops in the cumulus stage are usually small in magnitude—less than 0.7 mb below the diurnal change of the particular time of day—and take place over a period of 5 to 15 min. Following the fall, the pressure trace remains steady for as long as 30 min.

The pressure falls appear to be caused by the combined effects of vertically accelerated air motions, the expansion of the air due to the release of the latent heat of condensation, and the failure of the convergence near the surface to compensate fully for the expansion or divergence aloft.   Wind patterns in the vicinity of these cells show velocity convergence below 20,000 ft and divergence above, both of considerable magnitude; so if the pressure changes at the surface are due to a divergence imbalance, we are dealing with, as usual, a small difference between two large quantities.

In the mature stage, two features of the pressure trace—the "dome" and the "nose"—are recognized. The dome is registered at all stations to which the cold-air outflow penetrates. The pressure nose, the abrupt, sensational rise that some meteorologists regard as typical of the thunderstorm, really occurs only at stations that happen to be passed by the main rain and downdraft just after they have first reached the earth in the beginning of the mature stage. It is superimposed upon or may mark the start of the pressure dome.

The displacement of the warmer air by the cold outflowing air from the downdraft results in the pressure rise, initiating the pressure dome. A study of 206 thunderstorm-pressure records from the U.S. Thunderstorm Project surface micronetwork showed that in 182 of the traces there was a pressure rise associated with the arrival of the cold outflowing air. Since the rate and total amount of pressure rise depend on the slope of the cold-air mass, the temperature difference between the cold air and the displaced warm air, the depth of the cold air itself, and the speed with which the system travels, the most marked pressure changes are found near the cell core and they decrease with distance from it. The areal extent of the pressure dome is similar to that covered by the cold air. Therefore, the pressure remains high for a period of from one-half hour to several hours, depending on the amount of cold air (number of cells) involved.

The distinction between the pressure nose and the pressure dome can be made only with difficulty on the conventional week-long barograph traces in most weather stations; a twelve-hour recording drum is necessary to show the effect.

After the brief pressure nose, the pressure remains at the value of the pressure dome which prevails for the particular thunderstorm. The dome persists through the dissipating stage of the cells, after which the pressure returns to the trend prevailing before the passage of the storm. In the case of a thunderstorm associated with a cold front or a fast-moving squall line, the pressure remains high or even continues to rise as a result of cold-air advection or the passage of a wave in the pressure and wind fields.

DRY THUNDERSTORMS

In cases where the rain does not reach the ground, as is often the case in arid regions, the downdraft and outflow may still be felt. The pseudo-cold front with the usual manifestations of first gust, temperature break, and pressure nose or dome can be quite well developed even if the rain does not reach the ground. Dust storms are frequently stirred up by the first-gust line. Cloud-to-ground lightning strokes from such thunderstorms are treacherous as producers of forest fires because of the absence of rain to quench the flames.

NIGHT THUNDERSTORMS

In large areas of the middle western United States thunderstorms occur predominantly at night. These are not to be confused with evening thunderstorms left over from the daytime convection; in fact they show a peak of occurrence between midnight and 4 A.M. Studies have shown that they are caused by a diurnal variation in the large-scale wind system over the continent which is favorable for producing convergence in the low levels in the regions concerned at night.

The night thunderstorms usually fail to exhibit the same phenomena at the surface as those described on the preceding pages. The downdraft has difficulty penetrating to the surface because of the nocturnal temperature inversion there; so the wind and temperature features usually noted at the surface are either missing or very restricted in extent. The pressure and precipitation patterns are very nearly the same as those of daytime thunderstorms.

DEVELOPMENT OF NEW CELLS

From a study of numerous cases of new cell development, the outflowing air appears unmistakably to be important in contributing to the new growth. When two cells in the mature stage are within a few miles of each other, the cold outflows collide and ascent of the displaced warm air triggers the building of new cells. The greatest frequency of new development is in the area between two existing cells whose edges are three or less than three miles apart. A three-mile band downwind is next in importance, then the lateral edges, and least frequently the upwind or rear side.

In many cases the time interval between the beginning of the outflow and the appearance of the new cell on the radarscope is too short to permit explanation of the new one as a result of the underrunning cold air or a similar time-consuming process. There are cases, as indicated by the radar echoes, in which one new cell or a cluster comes into existence almost simultaneously with the initial or parent cell; this suggests that a preferred region of convergence and ascent favors the development of several cells.

Most thunderstorms consist of several cells representing all stages of development. The cells may have been separated at one time, but have come together. The cloud mass connecting the cells one with the other is usually inactive as far as

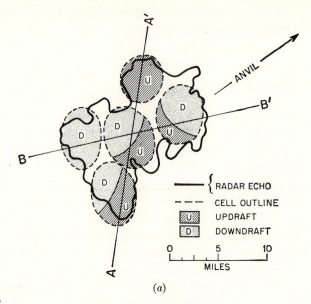

(a)

FIGURE 16-5
Three-dimensional structure of a thunderstorm containing several cells. In (a) the thunderstorm and its cells are represented in map plan. In (b) (next page) a vertical cross section along the line AA' of (a) is represented, while (c) (page 388) shows a vertical cross section along the line BB'. U stands for updraft, D for downdraft. Other symbols are the conventional ones used in meteorology, such as rain, snow, and ice symbols.

vertical motions and precipitation are concerned, but generally contains enough liquid in raindrop form to return a weak radar echo. A typical thunderstorm is represented in map plan and in two vertical cross sections in Fig. 16-5. In a the collection of cells that make up the storm are seen as they would appear in outline on the PPI-scope of a radar. From flight measurements of gusts and vertical motions the cell boundaries and the areas of updraft and downdraft are delineated. The lines AA' and BB' along which the two cross sections in b and c are constructed are shown in the map in a.

Figure 16-5 is a slightly simplified and idealized version of a thunderstorm studied by radar and flight measurements of the U,S. Thunderstorm Project. It should be noted that while the two lines of flight penetration corresponding to the cross sections were through the center of the storm, the traverses were not ideally suited to obtaining complete representation of the features of each cell. Obviously a single penetration cannot reveal the distribution of the cells and their various stages of development.

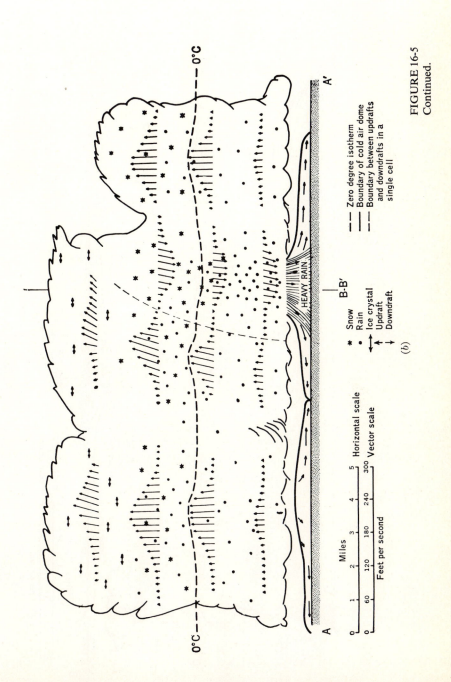

FIGURE 16-5
Continued.

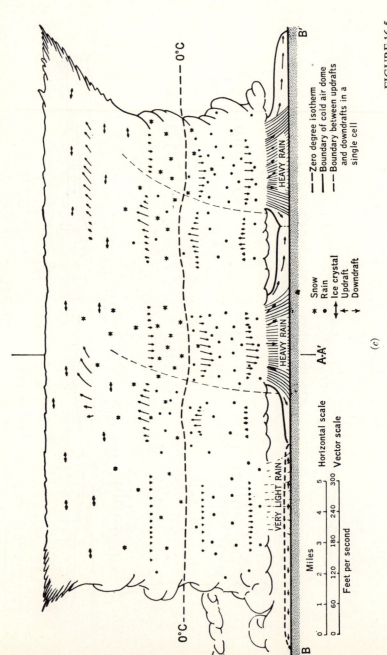

FIGURE 16-5
Continued.

THUNDERSTORM ELECTRICITY

The predominant charge distribution in a thundercloud is that of an upper positive and lower negative charge, but often there is a concentration of positive charge in a limited region at the cloud base. Most of the modern theories relate the separation of charges in a thunderstorm cell to the distribution of the hydrometeors—rain, snow, and ice particles. The upper positive charge is in most cases in the snow region with a negative charge centered a kilometer or so below it, and extending to the cloud base. The small area of positive charge also found at the base is in the region of heavy rain and strong downdraft. Kuettner,[1] from observations on a high mountain, associated this positive area with the occurrence of ice pellets. This shaft of positive charge extending through the heaviest rain downward from the ice level and surrounded by the negative charge provides a picture in which both vertical and horizontal gradients occur. This distribution helps explain the observation that in-cloud lightning discharges outnumber cloud-to-ground discharges by nearly three to one and that individual lightning flashes within clouds are likely to be more nearly horizontal than vertical.

Radar returns from thunderstorms in the United States in summer show that, with few exceptions, the first lightning discharge—usually in-cloud—occurs after the echo-producing part of the cloud has pushed upward to a level where the temperature is at about $-20°C$ or colder. As adjoining cells develop, the horizontal and vertical gradients become more complicated and lightning flashes continue into the dissipating stage after the radar echoes have subsided so that their tops are warmer than is required for the initial lightning strokes. In a thunderstorm cell the main electrical activity starts only in the most active updraft region, which normally occupies only about 15 percent of the cell area at that height.

Modern theories of charge separation in the upper dipole are concerned with processes involving ice. In laboratory experiments of Workman and Reynolds reported in 1948[2] and later by Reynolds, Brook, et al.,[3] various phenomena associated with the freezing of water were emphasized. It was found that when an ice crystal struck a riming ice pellet, charging of the pellet occurred. The sign of the charge was dependent on the temperature difference between the pellet and the crystal.

[1] J. Kuettner, The Electrical and Meteorological Conditions inside Thunderclouds, *Journal of Meteorology*, vol. 7, pp. 322–332, 1950.

[2] E. J. Workman and S. E. Reynolds, A Suggested Mechanism for the Generation of Thunderstorm Electricity, *Physical Review*, vol. 74, p. 709, 1948.

[3] S. E. Reynolds, "Compendium of Thunderstorm Electricity," Socorro, New Mexico, 1954 (New Mexico Institute of Mining and Technology), summarized by M. Brook, Thunderstorm Electrification, in S. C. Coroniti (ed.), "Problems of Atmospheric and Space Electricity," Elsevier Publishing Company, Amsterdam, 1965. Also: S. E. Reynolds, M. Brook, and M. F. Gourley, Thunderstorm Charge Separation, *Journal of Meteorology*, vol. 14, pp. 426–436, 1957.

A slightly different effect was produced in the laboratory by Aufdermaur and Johnson.[1] They performed an experiment in a cold (−5 to −15°C) wind tunnel with supercooled droplets striking riming pellets of ice in an electric field of 50 to 150 kV m⁻¹, characteristic of clouds of less than thunderstorm magnitude. They found that a certain percentage of the droplets collided with the pellet at grazing incidence and bounced off. From this encounter the droplets picked up some of the charge existing at the point of contact on the surface of the pellet. Translated into thunderstorm charge separation, this experiment suggests a process as follows:

1 In the prethunderstorm cloud, ice pellets or hailstones are found which fall faster, or are carried upward less rapidly by updrafts, than the cloud droplets.
2 In the cloud an electric field with positive charge above negative charge causes an induced positive charge on the undersurface of each pellet of ice.
3 The droplets moving upward with respect to the pellet collide with the positively charged undersurface of the pellet. Those that hit at grazing incidence, perhaps 1 to 10 percent of them, break away from the pellet and carry upward and away a positive charge picked up from the pellet.
4 The pellet is left with a net negative charge because of the stripping off of positive charges by the droplets.
5 In the gravitational field the negative pellets fall downward with respect to the droplets, resulting in increased negative charge in the lower part of the cloud and an increased positive charge in the upper part.

Mason[2] calculated that a process of this kind could cause the electric field to increase from 0.5 to 300 kV m⁻¹ (the breakdown potential for lightning in dry air) in 5 to 10 min, depending on the number concentrations and sizes of the droplets and the percentage of grazing collisions. He calculated that the same result could be accomplished by ice crystals colliding with and rebounding from polarized ice spheres.

Some of the proponents of ice theories have prescribed that hail should be present or forming in thunderclouds, but the Thunderstorm Project observations described at the beginning of this chapter showed that only about 20 percent of the cells had hail in them. Small, irregularly shaped ice particles 1 to 5 mm in size are found at −5 to −10° in growing cumulus clouds. Evidence indicates that they are formed by the freezing of small drizzle or raindrops swept up from below in the

[1] A. N. Aufdermaur and D. A. Johnson, Charge Separation Due to Riming in an Electric Field, *Quarterly Journal of the Royal Meteorological Society*, vol. 98, pp. 369–382, 1972.
[2] B. J. Mason, The Physics of the Thunderstorm, *Proceedings of the Royal Society* (*London*), Series A, vol. 327, pp. 433–466, 1972.

updraft, as reported by Braham.[1] Evidence also supports the idea that there is a multiplication of ice crystals at these temperatures in natural clouds.[2]

There has been no general acceptance of a theory for the lower positive charge, small though it may be in terms of the general charge distribution in a thundercloud. In his 1967 edition of "Atmospheric Electricity" (Pergamon Press, New York), J. A. Chalmers discussed some fifteen theories of charge separation, all based on some experimental evidence. A possible explanation is derived from the breaking-drop theory of Simpson,[3] who found that as large drops break, the residual drop carries a positive charge. Chalmers offers the conjecture that if ice pellets grow to large sizes, they would form water drops on melting which would be too large to be stable and would break up.

Once the electric fields build up in a thunderstorm cell and the water and ice particles carry charges, separations of charge can occur from various interactions. Sartor[4] has shown that when precipitation particles collide in an electric field or with charge, there is a charge transfer. Other experiments show that in an electric field there is an enhancement of coalescence when drops collide. This effect has the additional implication that rain intensification by increased drop coalescence should be expected in thunderstorms.

To further confuse the problem, cases of lightning from clouds in which no icing has been achieved and which are not influenced by neighboring thunderstorms have been reported. Vonnegut[5] has investigated and attempted to explain this phenomenon. It is of very rare occurrence.

HAIL

Two problems enter into a study of hail formation—the cloud physics problem of the growth of hail*stones* and the synoptic-thermodynamic investigation of the conditions

[1] R. R. Braham, Jr., What Is the Role of Ice in Summer Rain Showers? *Journal of Atmospheric Sciences*, vol. 21, pp. 640–645, 1964.

[2] L. R. Koenig, Some Observations Suggesting Ice Multiplication in the Atmosphere, *Journal of Atmospheric Sciences*, vol. 25, pp. 460–463, 1968.

[3] G. C. Simpson, On the Electricity of Rain and Its Origin in Thunderstorms, *Philosophical Transactions of the Royal Society of London*, Series A, vol. 209, pp. 379–413, 1909. The Mechanism of a Thunderstorm, *Proceedings of the Royal Society (London)*, Series A, vol. 114, pp. 376–401, 1927.

[4] J. D. Sartor, Calculations of Cloud Electrification on a General Charge-separating Mechanism, *J. Geophys. Res.*, vol. 66, pp. 831–838, 1961. Induction Charging Thunderstorm Mechanism, in S. C. Coroniti (ed.), "Problems of Atmospheric and Space Electricity," pp. 307–310, Elsevier Publishing Company, Amsterdam, 1965.

[5] B. Vonnegut, Thunderstorm Theory, in S. C. Coroniti (ed.), "Problems of Atmospheric and Space Electricity," pp. 285–292, Elsevier Publishing Company, Amsterdam, 1965.

FIGURE 16–6
Hailstones compared with golf ball, some showing protuberances occasionally found in hailstones. (*NOAA National Weather Service.*)

that produce hail*storms*. Few thunderstorms have hail reaching the ground and not many of them have it even in the most suitable parts of the clouds. Furthermore, there is an unusual geographic distribution of hail frequency; in the United States, for example, it is from four to eight times more frequent in the Great Plains and adjacent Rocky Mountain areas than in the Middle West, South, and East, even though the latter regions have up to two or three times as many thunderstorms as the former.

Physically, a hailstone appears to be formed by collision and coalescence of undercooled water drops with some kind of ice pellet. It is not difficult to understand how a coating of ice can accumulate around such a globule if it remains for some time in an undercooled cloud; we see the same thing happening on other objects such as airplanes and on mountain peaks. It is reasonable that the cloud updraft could keep the stones at levels of subfreezing temperatures for a considerable period of time. How the initial pellet forms and becomes a sizable stone is not entirely clear, however. Snow pellets are commonly observed inside cumulus congestus clouds in the precipitation-initiation and later stages at or somewhat above the freezing level. In most cases they fall below the freezing level and melt before they have grown appreciably.

To form hailstones the pellets apparently must either originate at a much greater height than commonly observed or be supported at the subfreezing levels by unusually strong updrafts. Both these circumstances would ensure a long sojourn in undercooled portions of the cloud to permit a substantial accumulation of ice. It is probable that mammoth hailstones such as those shown in Fig. 16-6 represent severe updraft conditions at the subfreezing heights underlain by a downdraft which lets the stones fall rapidly to earth before appreciable melting.

A characteristic often observed in hailstones, when one collects and dissects them, is a system of layers or successive shells of ice, suggesting that the stones have gone through a series of icing periods, possibly interspersed with melting. Oscillations of the stones up and down through the freezing level have been surmised. It is possible that the laminations represent the difference in texture between relatively clear ice from large, runny drops and the less-dense rime ice from smaller cloud droplets, all encountered above the freezing level. Recent findings suggest that uninterrupted descent from very high parts of the giant cloud through an updraft characterized by stratifications in the water content is probably sufficient to account for the laminated structure.

Fragments of apparently spherical stones are sometimes observed. Occasionally these are almost perfectly shaped angular sectors of spheres of uniform size. One explanation of this condition is that the ice formed on the outside of the stone while there was still some liquid water in the core. The expansion inside as the core freezes causes the sphere to explode or fracture from inside pressure.

Spherical hailstones of 1 to 4 cm diameter and specific gravity of 0.8 have terminal velocities of from about 10 to 20 m sec^{-1}. Updrafts of this speed have been measured in some thunderstorms.

Recent studies have suggested that hailstorms as well as tornado-producing storms are sustained in stronger vertical wind shears than are ordinary thunderstorms. A steady-state circulation is suggested, in which the shearing wind carries the top part of the updraft away laterally instead of allowing it to sink back down in pulsating turrets. Extensive downwind cirrus anvils support the idea of efficient convective pumping.

TORNADOES

Tornadoes are the most violent disturbances of the atmosphere, yet they are so small in their horizontal extent (usually 1 mile or less) that they never appear on the synoptic chart. Because of their highly localized nature and random distribution it is impossible to forecast the spot where they will strike. The best the meteorologist can do is to forecast the conditions under which tornadoes are likely or possible in a certain large region. Once a tornado has been spotted and the meteorologist has been informed of it by telephone, he can follow the thunderstorm echo associated with it on his radarscope and get police and local radio and television stations to warn of its approach and set in motion organized disaster plans.

Out of a great mass of literature on tornadoes, only general theories and data have come to explain their cause or describe their mode of formation. Meteorologists are practically unanimous in agreeing that they are a result of excessive instab-

FIGURE 16–7
Tornado at Dallas, Texas, April 2, 1957. It is estimated that the larger pieces of
debris are the major parts of houses exploded and carried upward. (*NOAA
National Weather Service.*)

ility and therefore steep lapse rates in the atmosphere. It is known that tornadoes
are associated with intense thunderstorms.

A tornado is easily recognizable in the visual range as a pendant funnel cloud
reaching downward from the clouds, as shown in Fig. 16-7. After it strikes the
ground the funnel sucks up dust and debris, but the main part of the funnel is considered to consist of condensed water as in a cloud. The tornado often comes with
the first gust of the squall before appreciable rain has fallen, but instances of occurrence in virtually all parts of a major thunderstorm area have been reported.

Barograph, wind, and radar data show that the tornado is inside what is known
as the *tornado cyclone*, a low-pressure area perhaps 10 miles in diameter with winds

of 50 knots, more or less.   The tornado seems to be between the center and right-hand side of this cyclone, that is, to the right of the all-over upper wind and therefore to the right of the direction of displacement of the system.   As might be expected, the horizontal wind speeds are greatest on that side.   The tornado itself has exceptionally low pressure at its center, although no barograph has survived to record it.   The pressure is so low that much of the damage to buildings comes from explosion resulting from the sudden drop in pressure outside as the tornado hits.   Winds up to 500 mph are indicated by the patterns of damage.

Of the nearly 200 tornadoes per year occurring on the average in the United States, about 90 percent of them are ahead of cold fronts, the average distance ahead being something like 150 miles.   This is the same distribution as that observed for squall lines, particularly in the central part of the country.   Tornadoes have also occurred in advance of warm fronts and behind cold fronts, breaking down through the lower layer of cold air from their origin in the unstable tropical air above.   Their paths range from a few miles to some hundred or so.   However, in the case of long paths there are indications of dissipation, reforming, and skipping, with new funnels replacing the old.

In some instances a radar with antenna sweeping the horizon receives echo patterns by means of which it is possible to recognize the circulation of the tornado cyclone in which the tornado is embedded.   At first a hook-shaped extension from the thunderstorm echo is observed; then as the circulation carries the echo-producing cloud elements in a cyclonically curved path the hook closes into a pattern looking like a 6, and finally like a doughnut.

A synoptic situation that is known to have occurred in tornado conditions in the United States is one in which strong cold-air advection is found aloft.   Under certain conditions, cold air masses, usually of Pacific origin, become heated up so much in the low levels over the western plateaus that they ride up over the maritime-tropical air of the Mississippi Valley.   The aridity and clear skies of the high-level heat sources of the West result in unusually high surface temperatures in air that was originally quite cold.   This air preserves its low temperature at greater heights, and thus a steep lapse rate is developed.   Apparently, as the air moves aloft over the maritime-tropical air, nothing much happens at first, because of the slight temperature inversion or stable layer separating it from the air below.   Presently, however, some cloud formation in the maritime-tropical air may release enough heat through condensation so that it will be able to penetrate into the upper layer.   Here the lapse rate is so steep that an ascending parcel or cloud mass would be greatly accelerated upward, resulting in the almost explosive type of convection necessary for the development of severe local storms.   The condition is illustrated in Fig. 16-8 on the pseudo-adiabatic diagram.   Apparently it is not until the overrunning air has reached to about the 100th meridian that in many cases the moisture content of the lower air is

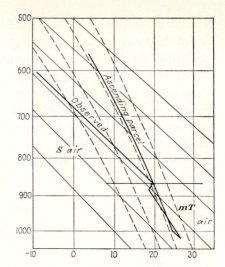

FIGURE 16–8
Hypothetical conditions for formation
of tornadoes in unstable dry air (*S*) overrunning maritime-tropical (*mT*) air.

high enough to release the heat of condensation (to follow a pseudoadiabatic line of sufficiently high value) necessary to overcome the stability at the air-mass boundary.

A well-developed cyclone with open warm sector is a fairly common breeder of tornadoes in the central United States in spring. Figure 16-9 illustrates the synoptic situation as described by Newton.[1] A stream of moist air pushes northward from the Gulf of Mexico over the Great Plains under the influence of an approaching cold front to form a warm-sector cyclone. Cold air in the middle and upper troposphere under a cold jet stream increases the instability in the Gulf air. The latter exhibits a low-level jet stream, a condition that is characteristic of the Plains under strong tropical inflows. This warm air ascends at the warm front. The combination of low-level convergence, frontal lifting, cold-air advection aloft, warm-air advection below, and in some instances, daytime heating of the air, results in marked instability. The parallelogram in the figure delineates an area where tornadoes are most likely to occur.

Photographs, eyewitness reports, and surveys of paths of destruction indicate a wide variation in the horizontal dimension of an individual tornado. Funnel clouds up to a mile in diameter at the ground are observed. Others have an appearance more like an elephant's trunk, while others are described as "ropes." The pendant tornado does not always touch the ground; sometimes it skims the treetops, twisting off only the top branches. It has been observed to extend itself intermittently, then recede, and thus to create an interrupted damage path. In some cases the

[1] C. W. Newton, Charakteristische Merkmale von nordamerikanischen Tornados (Characteristic Features of North America Tornadoes), *Die Umschau in Wissenschaft und Technik*, vol. 8, pp. 234–237, 1963.

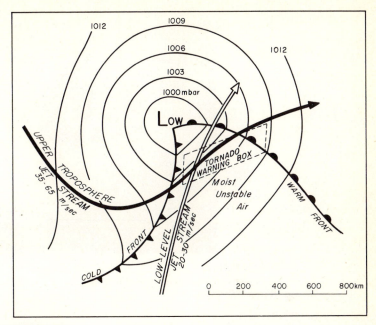

FIGURE 16–9
Typical synoptic situation for tornadoes in the central United States. (*After Newton.*)

skipping of areas in the path may be the result of a second tornado replacing the first from a different part of the cloud.   Multiple funnels are not unusual.   In the famous Palm Sunday 1965 series of tornadoes a pair of very large funnels moved abreast at a separation of less than a mile.

Paths of individual tornadoes probably seldom exceed about 50 miles in length, but series of tornadoes associated with the same squall or frontal system can cover hundreds of miles in a scattered and unpredictable manner.   Winds of several hundred miles per hour, pressure drops of 50 mb or more, and tons of debris swirling through the air combine to cause houses to shake, explode, or blow into pieces. Although the rotating winds are extreme, the displacement of a tornado is relatively slow, being that of the mother cloud—50 mph and less.

WATERSPOUTS

Tornadoes generated over bodies of water are called waterspouts.   Over near-shore waters such as the Great Lakes or the continental shelf of the Atlantic and Gulf, they may be part of a system that has also produced tornadoes or might have had the

FIGURE 16–10
Tornadoes or waterspouts over the ocean near the Bahamas. (*NOAA National Weather Service.*)

potentiality of producing them over land.   In certain favored areas of warm waters, waterspouts are observed in relatively undisturbed weather.   Such an area is that surrounding the Florida Keys, where waterspouts are of such frequency that a project has been carried out there to investigate them.

Published photographs of waterspouts show them to be relatively narrow pendants from the cloud, such as the one shown in Fig. 16-10, although some are quite broad at the top.   Cases have been reported of spouts unaccompanied by a cloud.

OBSERVATIONS AND STATION INSTRUMENTS

Meteorological observations may, for convenience, be considered as of two types—surface observations and upper-air observations. The latter have been made systematically only since the 1920s, but some form of weather observation by man from his position on the surface of the earth has been made since earliest times. As crude methods of measurement were developed, people became interested in knowing the depth of rain falling during a storm; the degree of heat or cold, i.e., the temperature; the strength of the wind, or wind velocity; the degree of dampness in the air, or humidity and possibly the intensity of the sunshine. These are all quantities which can be judged, or even measured in a rough way by our own perceptions. With the development of barometry, a very important meteorological measurement, that of pressure of the atmosphere, a quantity of which we were almost wholly unaware, could be made.

MEASUREMENT OF TEMPERATURE

Modern thermometry has been developed to the point where temperatures can be measured to within a thousandth of a degree if desired. If an instrument measuring with such accuracy were exposed to the air under average conditions, it would show rapid and large fluctuations in the temperature which would have no practical meaning. The air, at least near the ground, does not ordinarily maintain a constant enough temperature to warrant reading to within less than a tenth of a degree.

Of perhaps greater importance are the variations in the temperature read as the thermometer is moved from place to place within a very short distance, especially over rough or rolling terrain, or as it is placed at different heights above the ground. On clear, calm nights, the net loss of heat by radiation from the surface of the earth is sufficient to make the temperature several degrees lower next to the ground than at a height, let us say, of 2 or 3 m. On hot summer days the temperature at the ground may be 2°C higher than at a height of 1 m. Obviously, under such conditions, very different temperature readings are going to be obtained depending upon the height at which the thermometer is placed. On cold nights, when there is not much wind to keep the air mixed, the coldest air, because of its greater density, settles into the hollows. We have all experienced this perhaps in traveling through the country at night.

Liquid-in-Glass Thermometers

Thermometers based on the expansion of liquids with increasing temperature are the most convenient for temperature measurements of the accuracy required in meteorology. They are easy to manufacture, easy to read, and require no maintenance. The mercury thermometer is the most convenient and the most reliable, but in localities where the temperature is likely to reach the freezing point of mercury, $-38.9°C$, alcohol thermometers are used. The liquid expands into a glass tube out of a glass bulb at the bottom, and lengths in the tube are scaled off in terms of temperature to be read at the top point to which the liquid has expanded in the tube. For accuracy, the degree markings should be etched on the glass so that if the glass changes its position with respect to the frame or expands at a different rate, the readings are still reliable.

The thermometer used for official observations in the United States is a glass tube about 10 in. long, mounted on a strip of metal for support and graduated in some 150 Fahrenheit degrees. With a little practice, temperatures in tenths of a degree can be interpolated with good approximation. As stated above, there is no need for such close reading of the temperature, but it should be pointed out that in certain types of humidity determination involving wet- and dry-bulb thermometers, readings to this accuracy are important.

It is interesting to know the highest and lowest temperatures reached during any specified period. Weather reports usually give the maximum and minimum temperature occurring each day. Except during rapid changes in the weather, the minimum temperature usually occurs in the early morning and the maximum in the middle or late afternoon. Self-registering thermometers that indicate the highest or lowest reading reached between times in which they are reset are used for this purpose. They look very much like ordinary thermometers. The maximum thermometer has a constriction in the glass tube which permits the mercury to pass when forced up the tube by temperature expansion, but gravity is insufficient to let the mercury down into the bulb as it cools; hence it stays at its highest reading until the thermometer is shaken or whirled. To ensure proper performance, the thermometer is mounted

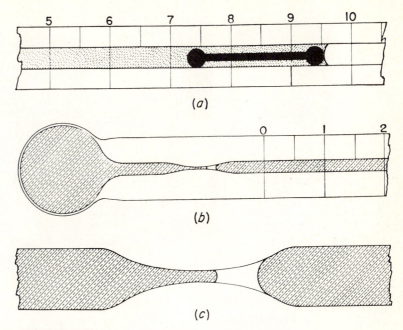

FIGURE A–1
(*a*) Detail of minimum thermometer. (*b* and *c*) Details of maximum thermometer. (*NOAA National Weather Service.*)

nearly horizontal but is read in the vertical position. The details of a maximum thermometer are shown in Fig. A-1 (*b* and *c*).

The minimum thermometer is an alcohol thermometer with a small glass index having the appearance of a pin with a head at both ends. The index is in the bore inside the alcohol column, kept just below the meniscus by the surface tension there (Fig. A-1*a*). The thermometer is mounted in a horizontal position, and as the meniscus retreats toward the bulb, it drags the index with it. As the temperature rises, the liquid leaves the index, which remains with its outer end still showing the lowest temperature reached. To reset the thermometer, the bulb end is tilted upward, and this is sufficient to cause the index to float to the end of the alcohol column. There, again, the surface tension of the meniscus prevents the index from breaking out of the liquid column into the empty part of the bore.

Deformation Thermometers

When two strips of metal having different coefficients of thermal expansion are welded and rolled together, temperature changes will cause the bimetal strip made in this fashion to become distorted by the shearing stress between the two faces of the strip. With most metals at ordinary temperatures this bending can be reproduced time and again in exactly the same

FIGURE A–2

Hygrothermograph, combining Bourdon-tube-type thermograph and hair hygrograph.

amount and thus can be calibrated in terms of temperature.   Invar, a metal having an extremely low coefficient of expansion, is commonly used for the inner portion of the arc-shaped strip, with brass or steel on the outer face.

The Bourdon thermometer consists of a curved tube, usually of elliptical cross section, made of metal and filled with some organic liquid.   Expansion of the liquid with increasing temperatures forces the tube into a shape of less curvature.

Deformation thermometers are used not for direct measurements of the temperature, but rather for writing a record of it.   An instrument designed to make a temperature record in this way is called a *thermograph*.   In the usual arrangement, the bimetal or Bourdon thermometer moves a pen arm that writes with ink or scratches on a lampblacked surface on a drum rotated by clockwork (Fig. A-2).   These thermometers can also be made to operate a rheostat, make and break electric circuits, control radio sets, etc., to make remote records of temperature.   Their most common nonmeteorological use is in thermostatic controls, such as in house-heating installations, hot-water heaters, stoves, and innumerable manufacturing processes.

Electric Thermometers

For measuring and recording temperature accurately, with great sensitivity or in a very small space, electrical methods are best.   Two general types of electrical thermometers are in use: (1) the thermocouple or thermopile and (2) the resistance thermometer.   A discussion of these can be found in physics textbooks.   The thermocouple is based on the fact that a current flows between two junctions of two metals if one junction is considerably hotter or colder than the other.   The junctions, made of such pairs of metals as copper and constantan, can be made so small that they can be fitted or fused into a small space, and being almost massless, their sensitivity is high and time lag very small.   In its use in meteorology the thermocouple is often arranged with the cold junction in a bath of dry ice and acetone or of liquid nitrogen and the other junction in the place where the measurement is wanted.   The current may be amplified and recorded on any one of a number of suitable recorders either by direct wire or by a radio-transmitted signal.   A thermopile consists of a large number of thermocouple pairs exposed in the same surface.   If connected in series, the output voltage of a single thermocouple is multiplied by the number of pairs; if in parallel, the current is thus multiplied.   They are useful in radiation measurements where the output from a single pair of thermocouples may be too small.

Resistance thermometers are based on the principle that the electrical resistance of a conductor varies with its temperature.   The platinum-wire thermometer is of this type. High sensitivity and highly satisfactory results can be obtained from ceramic resistors.   These have negative resistance coefficients; that is, the resistance decreases with increasing temperature instead of increasing as it does in metals.   One type of ceramic resistor in use, called the *thermistor*, has a very high coefficient, varying from approximately 20,000 to 1,500,000 Ω for the temperature range from $+60$ to $-90°C$.   This makes it very suitable for meteorological measurements.   It is made about 0.5 mm in diameter and 2 to 5 cm in length and is coated with a pigment which makes it more than 90 percent reflective to the solar radiation. Its small mass assures good sensitivity.   It is widely used for upper-air measurements, with the readings telemetered to a ground recorder.   Thermistors are available in a variety of sizes and shapes, including a bead thermistor, shaped as the name implies.

Exposure of Thermometers

One often hears people talking about the temperature "in the shade" as against that "in the sun."   Actually, the temperature we measure is that of the air, and it is about the same whether it is over a shady spot or a sunny one.   The air is free to move about to equalize, by stirring, any purely local horizontal heat differences that may develop.   We feel cooler when we are in the shade because our bodies are not receiving direct rays of the sun or radiation and convection from sun-heated surfaces.

A thermometer is also capable of absorbing heat if exposed to the sun or to radiation from any other surface.   Since the air absorbs an almost negligible amount of the solar radiation, a thermometer in the sun will be warmer than the air, and the temperature it records will be not that of the air but that of its radiation-warmed bulb.   The temperature thus recorded would be different for every design of thermometer, depending to a large extent on its mass and absorbing properties.   By means of thermocouples or other suitable apparatus we can measure the temperature of such things as solid objects exposed to the sun and find that they are much warmer than those in the shade, but this will not be found to be true of the air.

FIGURE A–3
A simple type of shelter showing maximum and minimum thermometers through open door. (*NOAA National Weather Service.*)

In order for a thermometer to have the same temperature as the air, it must be protected from all kinds of radiation that it can absorb but which the air cannot. This can be accomplished by placing a lightweight highly polished tube around the thermometer, opened at both ends and with plenty of air space for ventilation. It is well to ensure ventilation in a stationary installation by drawing air through the shielding tube with a fan. As an extra precaution, the protection may consist of two concentric cylindrical tubes with air space between them. If properly shielded and ventilated in this way, the instrument should give the same readings whether mounted in the sun or in the shade.

For convenience when several instruments have to be used, a special shelter to house all the instruments, such as thermometers (wet and dry bulb, maximum and minimum) and humidity recorders, is installed. This instrument shelter, or thermometer screen, as it is sometimes called, is constructed as protection from the weather as well as from the sun or other radiating surfaces.   It is usually made of wood and is shaped like a box with louvered sides and bottom and double roof or top having a 2- or 3-in. air space.   The shelter is painted white and mounted on an open framework, usually of such a height that the thermometers are about 5 ft above the ground.   The shelter is set up with the door on the north side so that the sun cannot enter while the observer is reading the instruments.   For winter observations or in northerly latitudes, the north side of the shelter may be left open, provided that there is no nearby building or other surface radiating from that direction. A standard United States shelter is shown in Fig. A-3.

In spite of careful attention paid to details in the design of instrument shelters, on calm, clear days of intense sunlight the air in the shelter may acquire a temperature as much as 4°F higher than that in the surroundings, owing to radiation absorbed by the shelter.   On clear, calm nights, the shelter may lose enough heat by net radiation to make its temperature lower than that of the air.   These radiation effects can be minimized by proper ventilation.

HUMIDITY MEASUREMENTS

Humidity and water-vapor measures are defined and discussed in Chap. 6.   A general class of instruments sensitive to changes in humidity or water-vapor content are called *hygrometers.*

Absorption Hygrometers

The most obvious way to measure the water-vapor content is to pass a sample of known volume or mass of the atmosphere through chemicals that absorb all the water vapor.   By weighing the absorbing material before and after the measurement, the mass of water vapor in the sample can be determined.   Sulfuric acid is one of the better-known chemicals having this property, but it is difficult to use because of its corrosive nature.   Certain salts, such as $CaCl_2$, P_2O_5, NH_4Cl, $Ca(NO_3)_2$, and $MgCl_2$, are well suited to this type of measurement.

The chemical-absorption method of measuring humidity, while supposedly quite accurate, is difficult to practice.   It takes so much time to take a reading that it can be used accurately only to obtain a mean value over an hour or perhaps a whole day.   In evaporation studies, in which an integrated value of the humidity over a considerable period of time is desired, chemical-absorption hygrometers are useful.

Another type of absorption hygrometer is based on electrical changes produced by the absorption of moisture on surfaces coated with absorbing materials. Lithium chloride, a highly hygroscopic substance, is sometimes used.   The electrical conductivity across the surface varies with the amount of water absorbed from the vapor.   The resistance also depends on the temperature, so that a correction of the readings due to temperature is necessary.

The Hair Hygrometer

Anyone living in the variable climates of middle latitudes is familiar with the nuisance of doors and windows sticking in warm, moist weather and being rather loose in dry conditions.

This is caused chiefly by the swelling of the wood with increasing humidity.   Other normally dry organic tissues, such as skin and hair, exhibit similar changes in dimensions with changing humidity.   Animal hairs, partly because of their fine texture, are the most sensitive, and in this respect human head hairs are best.

The human head hair increases its length by about $2\frac{1}{2}$ percent as it is brought from 0 to 100 percent relative humidity.   Different hair specimens have different ranges of total expansion, but there is fair uniformity in the relation between the relative humidity and the fraction of the total elongation.   This relation is very closely logarithmic.

Like most materials, the hair also increases its length with increasing temperature. This requires a temperature correction that can be obtained only through calibration.   The correction is not a constant one, being greater when the humidity is high.   The temperature effect is quite small compared with the humidity effect on the hair, but it cannot be neglected.

The hair hygrometer is utilized where a written or transmitted record of humidity is required.   Usually a single hair is not strong enough to support or drive a pen arm or switch. For this purpose a bundle of hairs is used.   Much of the sensitivity of the hairs is lost in this way, and it has been argued that a single, coarser and stronger hair, such as a horsehair, might be better than a bundle of human hairs.

The hair hygrometer or hygrograph is far from being a satisfactory instrument.   It is reliable only if calibrated frequently and carefully.   Its chief disadvantage, especially for use in ascending airplanes or balloons, is that it has a very large time lag.   Vertical gradients of humidity are often quite large in the atmosphere, and this instrumental lag is serious. The electrical-absorption hygrometer has much less lag and is therefore preferred.   The greatest difficulty is encountered at low temperatures, since the hair lag increases as the temperature decreases, becoming approximately infinite at a temperature of $-40°$.   This means that at high levels or in polar regions the instrument is practically worthless.   A hair type of hygrograph is shown as part of Fig. A-2.

The Psychrometer

If an ordinary liquid-in-glass thermometer bulb is covered with a piece of tight-fitting muslin cloth, it becomes a wet-bulb thermometer; for, if the cloth-covered bulb is wetted with pure water and the thermometer is properly ventilated, the temperature reading will decrease to a certain point at which it will remain until all the water is evaporated.   This final temperature is called the *wet-bulb temperature*.   (See Chap. 6 for details.)

The humidity values obtained from the measurement depend on the difference between the dry- and wet-bulb readings, or what is called the *depression* of the wet bulb, and on the actual temperature.   The relationship is determined semiempirically and is represented in various convenient tables, such as those of the Smithsonian Institution (Washington, D.C., 1952).

It is seen, then, that the determination depends on the reading of two temperatures, the wet and the dry.   A wet-bulb thermometer and an ordinary thermometer mounted side by side make up what, curiously, is regarded as a separate individual instrument called by the high-sounding name *psychrometer*.

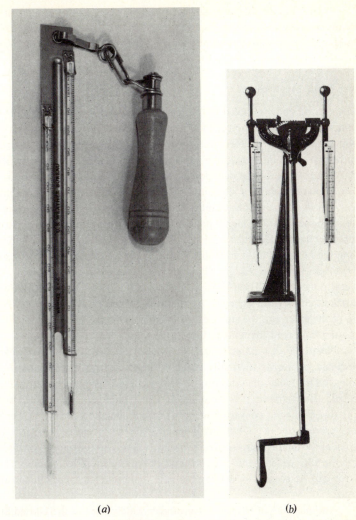

FIGURE A–4
(*a*) Sling psychrometer. (*b*) Whirling psychrometer.

Most psychrometric tables are computed on the basis of a ventilation of from 4 to 10 m per sec. Several methods of providing this ventilation are used. In the United States the most common method is to whirl the two thermometers. When the thermometers are mounted in an instrument shelter, they are whirled by turning a crank, geared for proper speed, or about 4 revolutions per sec. For measurements in the open, a crank handle is attached to the upper end of the thermometer mountings. This arrangement is known as a *sling psy-chrometer*. Two types of psychrometers are shown in Fig. A-4.

The aspiration psychrometer is considered standard for general use. It ventilates the thermometers by drawing air past them by means of a fan through double-walled metal tubes in which the thermometers are mounted. The metal tubes also serve as radiation shields.

Dew-Point Hygrometer

When a solid body is cooled to the dew point, dew begins to form on it. If the temperature of the surface of such a body can be measured, it will provide the means of obtaining the dew-point temperature. This quantity, once obtained, together with the temperature and pressure, can be converted into any other expression of the humidity that may be desired.

Some forms of the dew-point apparatus are crude and simple. A small reservoir of ether is placed behind a silvered disk or in a silvered tube. By bubbling air through the ether, it is made to evaporate and cool, and the silvered surface is supposed to cool at the same rate. The surface is so highly polished that one notices immediately the dulling of reflections caused by the condensation of dew. The temperature of the ether is read on a thermometer immersed in it, and if the instrument is properly constructed and the thermometer is read just at the right time, the reading will be the temperature of the dew point. The instrument has several difficulties. The condensation depends on condensation nuclei, the nature of the particles that inevitably cling to the surface, and the close relationship between the temperature of the ether and the surface. It is difficult for the observer to note the instant of condensation and read the temperature at the same time.

Electronic dew-point sensors have been devised. One used widely in the U.S. Air Force has the mirror thermally bonded to but electrically insulated from a thermoelectric cooler. The cooler operates on the Peltier effect.[1] When condensate forms on the mirror, the reflective characteristics from a light source change. As detected by a pair of photoresistors, this change is converted to an electrical signal to drive a power supply producing a direct-current signal. The current actuates the thermoelectric cooler. The system holds the mirror at the temperature of the dew point. It is not necessary to reverse the cycle, since heat from the hot side leaks into the cooler when the current is decreased. The temperature of the mirror can be sensed by a thermocouple or resistance thermometer and continuously recorded.

Another type of electrical hygrometer widely used in the United States is the so-called "dewcel." It is based on the principle that the vapor tension of an electrolyte solution is less than that of pure water, as discussed in Chap. 15. A large thermometer cell is surrounded by a glass-wool wick impregnated with a saturated solution of lithium chloride. Two bare silver wires are wound in a spiral mesh along the full length of the wick-covered bulb. An electric current is impressed in the wires and solution. The flow of current through the solution generates enough heat to raise the temperature of the device until the vapor tension of the solution equals the vapor pressure in the air. Then, as water begins to evaporate from

---

[1] If two dissimilar metals are brought into electrical contact and a current is passed through the junction, this is heated or cooled according to the direction of the current.

the solution, the resistance increases, and the current decreases and with it the heat.   If the device cools again, condensation occurs and the resistance decreases and the current, and therefore the heating, increases again.   A balance is maintained such that the ambient vapor pressure is in equilibrium with the solution.   By definition, the vapor pressure is in equilibrium (at saturation) at its dew-point temperature.   The temperature is recorded.

A carbon humidity element is used in United States radiosonde instruments to record the relative humidity.   A polystyrene slide or strip with two metal electrodes along the long edges is sprayed with a mixture of carbon particles and a hydroxyethyl cellulose binder.   The binder changes its volume with relative humidity in such a way that it separates the carbon particles from each other as the relative humdity increases, thus increasing the resistance between the electrodes.   The relative humidity is calibrated as a function of resistance.

THE MEASUREMENT OF PRESSURE

Torricelli, in 1643, found that when a tube somewhat less than a meter long was filled with mercury, stoppered, inverted, and unstoppered as it was immersed in a container of mercury, a small quantity of mercury would run out of the tube until the mercury column stood at a height of about 76 cm, thus leaving a vacuum above.   This experiment amounted to measuring the pressure of the atmosphere by balancing it against the weight of a column of mercury, and it was found that the pressure of the atmosphere was equivalent to that exerted by approximately a 76-cm depth of mercury.   This type of glass tube filled with mercury and having its lower end open and immersed in a dish or cistern of mercury is called a *Torricellian tube*, and is essentially the standard form of modern mercury barometers.

The balancing of the atmosphere against a mercury column has become so common a procedure that in most practical work in the physical sciences the pressure is expressed in terms of the length of the supported mercury column, such as 760 mm or 29.92 in., instead of the regular pressure units of dynes per square centimeter or newtons per square meter.

In a Torricellian tube the cross-sectional area of the column does not affect the reading unless the tube is of such narrow bore that capillary forces between the glass and the mercury are important.   If the column is 3 cm² in cross section the force, being pressure times area, is three times as great as the force per unit area (the pressure); therefore, the mercury will stand at the same height against a vacuum regardless of the cross-sectional area as long as it is greater than capillary size.

The mercury in the glass also acts as a thermometer, expanding as the temperature rises, contracting as it decreases.   In order to make comparable readings in terms of the length of the mercury column, the values are reduced to that which would be obtained if the mercury were at a temperature of 0°C.   Therefore, when we state that the pressure is 760 mm Hg, we mean that the atmosphere has a pressure equal to that of mercury, which at a temperature of 0°C would have a height of 760 mm.

In the tube the length that is measured is that of the column extending above the level of the mercury in the dish, or *cistern*, as it is called.   The measuring scale must therefore have its zero point at the level of the mercury in the cistern.   In the barometers in general use, the designs are such as to overcome the necessity of moving the scale as this level changes.

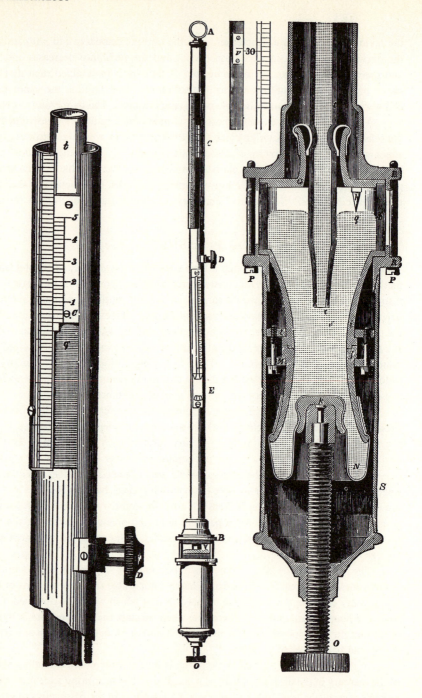

Two kinds of barometer are in general use—the Fortin type and the fixed-cistern type. In the Fortin type the level of the mercury in the cistern is raised to the zero point of the scale while the scale remains fixed. The bottom of the cistern is made of kidskin, which may be raised or lowered by a setscrew to the desired zero point before reading. A small ivory pointer indicates the zero point. The reflection of this pointer is seen in the mercury, and the screw is turned up until the ivory tip just touches its image. Air has access to the instrument through a leather collar or porous gasket between the cover of the cistern and the tube. The Fortin barometer is shown in detail in Fig. A-5.

In the fixed-cistern or Kew type, no attempt is made to have the zero point always at the surface of the mercury in the cistern. Instead, the fall of the mercury in the cistern accompanying a rise in the tube is taken into account by altering the graduations on the scale. A change in pressure of 1 cm will change the height of the mercury in the tube by an amount less than 1 cm, depending on the relative areas in the tube and in the cistern. If the cistern is of large cross section compared with the tube, this discrepancy will be small.

Let us designate volume, area, and displacement in the cistern by primed symbols and for the tube unprimed. Then

$$\Delta V' = \Delta V$$

$$A'\Delta x' = A\,\Delta x$$

$$\Delta x' = \Delta x\,\frac{A}{A'}$$

The displacements with pressure increase are given by

$$\Delta p = \Delta x + \Delta x'$$

$$= \Delta x + \Delta x\,\frac{A}{A'} = \Delta x\left(1 + \frac{A}{A'}\right) = \Delta x\,\frac{A' + A}{A'}$$

A unit of pressure change is produced when this last expression is equal to 1, or when

$$\Delta x = \frac{A'}{A' + A}$$

So the scale is graduated in this unit.

Usually A' is about 15 times A; so the scale is compressed by $\frac{1}{15}$, or $\frac{2}{3}$ mm is subtracted for each centimeter. The details of a fixed cistern are shown in Fig. A-6.

FIGURE A–5
Fortin-type mercurial barometer. *Left*, details of vernier scale: the sliding scale (*C*) is operated by a thumbscrew (*D*) on the housing over the tube (*t*). The slide rests at the apex of the meniscus of the mercury (*q*) for correct reading. *Center*, full view showing attached thermometer (*E*) for temperature correction, suspension ring (*A*), base (*B*), and screw (*O*) for adjusting zero of mercury column. *Right*, details of cistern: mercury (*q*) is adjusted to the level of the index pointer (*h*); the tube (*t*) enters the cistern through the collar (*G*). The portion (*N*) of the bottom of the cistern is kidskin, adjusted by the screw (*O*) through the head (*k*). (*NOAA National Weather Service.*)

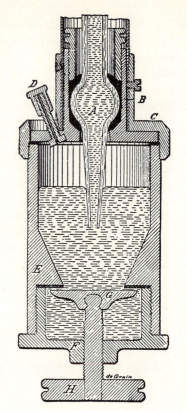

FIGURE A–6
Details of the cistern of a fixed-cistern
barometer. The tube (*A*) rests in a
collar (*B*). The thumbscrew and head
(*H* and *G*) are used in carrying.
(*NOAA Nattional Weather Service.*)

All mercury barometers are constructed for accurate readings.   A sliding vernier
scale gives graduations to tenths of a millimeter or even thousandths of an inch.   The scale
is intended for reading at the highest point of the meniscus of the mercury.   A thermometer
is attached for the purpose of reducing the readings to correspond with a mercury column at
standard temperature, 0°C.   The temperature corrections are to be found in the Smithsonian
Meteorological Tables or other tables.   The tables usually include also a correction for the
temperature effect on the brass scale.

The fixed-cistern barometer has the advantage that it can be read quickly, easily, and
accurately, but has the disadvantage that any loss of mercury during transportation affects
the readings.   The Fortin type is not affected by ordinary small losses of mercury, but the
necessity for setting the cistern to zero before each reading presents a difficulty.   Observers
cannot be consistently sure of getting the exact zero in their settings.   Considerable attention
given to the lighting around the barometer, especially if the mercury has lost some of its
luster, is necessary in order to ensure that the ivory pointer and its image can be matched
exactly be semiskilled observers.

FIGURE A–7
Aneroid barometer.

Aneroid Barometers

The aneroid barometer gets its name from the Greek word for *dry*, since no liquid is used. Just as in weighing machines there are those which balance the object to be weighed against known weights and those which use springs, so in barometry there are two types—those which balance against weights (mercury barometers) and those which operate with springs. The aneroid is of the latter type.

Fundamentally, the aneroid barometer consists of an evacuated chamber and a spring. The spring keeps the chamber from collapsing under the pressure of the atmosphere and restores the chamber to a larger shape when the pressure is reduced. The chamber is of thin metal that can expand or contract the chamber as the pressure changes. The best types are in the form of metal bellows. The spring may be in the form of either a spring arm outside the chamber or a helical spring inside. Some newer forms have the bellows themselves built for resilience like a spring. A common type of aneroid is shown in Fig. A-7.

The deflections of the spring and chamber are multiplied by levers so that they may operate a pointer on a dial marked with pressure readings. The aneroid has a temperature correction due to temperature effects on the characteristics of the spring and on the small amount of gas contained in the chamber. Modern designs have been fairly successful in providing compensating mechanisms for these temperature effects.

FIGURE A–8
Aneroid barograph.

A common use of the aneroid is for providing a written record of the pressure, i.e., *barographs*. Instead of a pointer, a pen arm is moved which writes on a drum rotated by clockwork. Barograph traces are of considerable importance in weather forecasting. A barograph is pictured in Fig. A-8.

The great advantage of the aneroid barometer is its portability. It is especially suitable for use on shipboard, because it is not affected appreciably by swaying, can be hung on any convenient wall, and is as easy to read as a compass. The disadvantage is that its pointer must be reset at fairly frequent intervals by comparison with the mercury barometer. Unfortunately there are many inferior aneroid barometers on the market and only a few excellent ones. No general statement about the performance or reliability of aneroids can be made because of the wide variation in design and workmanship. The weather services however, use aneroid barometers with great success at hundreds of airports.

Altimeters are aneroid barometers made with a scale of altitude instead of pressure. This can be done because of the close relationship between height and pressure. Under average conditions, the pressure decreases with height in the lower part of the atmosphere at a rate of about 10 mb per 100 m or about 1 in. of mercury per 1000 ft. The rate of decrease is not always the same, varying with temperature, pressure, temperature lapse rate, and humidity, stated in the order of their importance. The altimeter, if it is accurate, measures the pressure correctly. By setting the zero of the scale to the pressure at sea level or at the ground, altitudes for a standard atmosphere can be read off directly.   If the atmosphere is colder than the standard, the altimeter will give too high an altitude and if it is warmer, the reading will be too low.   The readings probably never vary more than 5 percent from true values. The effects of other meteorological elements besides temperature can be neglected in practical air navigation.   Tables or slide rules can be made up for applying temperature corrections to altimeter readings.   It should be emphasized again, however, that these corrections have nothing to do with the instrument itself but apply to the pressure-height relationship that the graduations on the scale assume.   The correction should be applied in terms of the mean temperature of the atmosphere intervening between the altitude in question and the ground.

Weather reconnaissance and some commercial and military airplanes carry radio altimeters which give geometric height very accurately from reflected signals.   *D*-values as described in Chap. 12 can then be read directly as the difference between the readings of the two forms of altimeter.

"Exposure" of Barometers

Most meteorological instruments have peculiar problems of exposure to the air in order to obtain representative readings.   This is not true of barometers, because they measure not the properties of a "sample" of air, as do thermometers, hygrometers, etc., but simply the force exerted at the height of measurement by the overlying atmosphere.   Barometers do not have to be exposed in the open; in fact, because of obvious convenience, they are always mounted indoors.   The pressure inside a building is always the same as that at the same level outside, except for momentary differences in severe wind storms.   In mounting a mercurial barometer, care must be taken not to place it where the temperature is likely to fluctuate appreciably, because the temperature correction is based on the assumption that the mercury in the barometer has the same temperature as that in the attached thermometer.   The latter is more sensitive to temperature changes, and a wrong correction value may be obtained under fluctuating temperature conditions.   For this reason it is best to keep the barometer on the inside wall of a room that is kept at a fairly uniform temperature.   This is not much of a problem in most government or airport buildings in which meteorological offices are housed.

The problem of exposure of aneroids in airplanes is more difficult, since airplanes are especially built to create local pressure differences.   The pressure in the vicinity of the instrument panel on most airplanes is far different from the static pressure of the free air.   Therefore, airplane altimeters are connected to the outside air by means of a tube with its opening on some part of the airplane where no dynamic pressure effects are present.   Such precautions are never necessary in the vicinity of ordinary structures, such as buildings.   This emphasizes the very special nature of airplane design, for it is doubtful that any other structure could be found that would produce systematic dynamic pressure effects of this nature, even in the strongest winds.   There is a nonsystematic dynamic pressure effect noticeable around many buildings in strong winds which causes the barometric pressure to oscillate very rapidly through a small range.   This effect is called *pumping* and is especially marked when the winds are strong and gusty, particularly at observatories on mountain peaks.   There is no satisfactory way to overcome or correct pumping.

Sea-Level and Gravity Reduction

In charting atmospheric pressures over the earth, meteorologists are principally interested in pressure differences between one place and another, or the gradient of pressure. They are not interested in the small differences that may be due to differences in the acceleration of gravity at various locations. The acceleration of gravity varies with latitude and with elevation and is also affected by mountains. By international agreement, meteorologists reduce all their pressure observations to standard gravity, the value which prevails at sea level in lat 45° being specified. The value of 980.665 cm per sec^2 is used for this standard. If the local acceleration of gravity cannot be measured, it can be obtained by interpolation between measuring stations or by an empirical formula. Formulas and tables are given in the International Meteorological Tables published by the World Meteorological Organization. At most stations the gravity correction may be considered a constant value to be added to all barometer readings.

In order to be able to compare readings at stations having different elevations, it is necessary to reduce all observations to a common level. Since most of the earth is at or near sea level, it is common practice to reduce all pressures to sea level or heights to 1000 mb for charting or other comparative purposes. In some sections, such as the western part of the United States, where much of the ground is on an elevated plateau, pressures reduced to some higher standard level, such as 5000 ft or 850 mb, are frequently used.

The problem of reduction of pressure to sea level is a very serious yet practical one and could well be made the subject of a complete book. For isolated mountain peaks the problem is not very difficult, being simply that of the airplane altimeter worked in reverse. Over great plateaus and elevated plains, such as in the Western United States, the problem defies exact solution, because there is no atmosphere intervening between the station and sea level about which temperature approximations can be made. In winter, small valleys in the elevated plateaus may be filled with cold air while surrounding stations may be in a place where cold air does not collect. Stations in the cold valley will use the low temperature as representative of the fictitious air column between them and sea level, while the other stations will use a much higher temperature. The colder stations will have a much larger reduction and will therefore have a pressure altogether too high at sea level. It is unfortunate that pressures, which can be measured with great accuracy, should become so inaccurate in certain regions for use on weather charts because of the necessity of reduction by crude methods through some thousands of feet of rock.

THE MEASUREMENT OF WIND

There are many simple ways of determining the speed and direction of the wind. The old-fashioned farm windmill is a crude form of instrument that could be used for this purpose. The wind sock used at small airports is an excellent wind-direction indicator, as is also the wind vane found on many church steeples. The meteorological instrument used officially to measure the wind is a combination of the windmill principle and the old-fashioned wind vane.

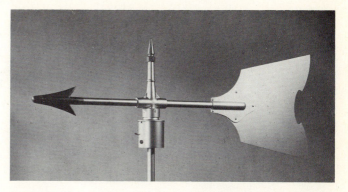

FIGURE A–9
A simple type of wind vane.

A conventional design of wind vane is shown in Fig. A-9. In order to function properly, the vane should be mounted on bearings that will reduce friction to a minimum, and the vane should be balanced at the point of rotation. The arrowhead, often made of lead, not only gives the vane a finished appearance but serves to balance the weight of the tail.

The cup anemometer, pictured in two forms in Fig. A-10, has been used by meteorological services for many years. It has an advantage over the common windmill type in that the vertical axis of rotation obviates the necessity of keeping the instrument facing the wind. The instrument can be calibrated so that the number of revolutions in a given time may be translated into wind speed. Obviously the cup wheel must be carefully balanced and mounted in bearings that can be kept properly lubricated in any kind of weather.

A great deal of work has been done by instrument research workers toward perfecting the cup anemometer, but it still has disadvantages, and there is lack of agreement as to which design is best. One of the chief difficulties is that in a gusty wind the anemometer, because of its inertia, indicates only the mean wind. Furthermore, the cup accelerates more rapidly than it loses speed, so that the mean wind is registered slightly too high. All winds are more or less gusty, and much valuable information concerning the turbulence of the lowest layers of the atmosphere is lost through use of this instrument.

The cup anemometer is usually arranged with a dial that counts the revolutions, translating them into miles of wind passage. This information is not especially useful, since the instantaneous speed or, at least, that averaged over a minute or two is desired. Most modern installations provide for remote recording indoors. The rotating cups can drive a small ac generator for recording on a recording ammeter. A current can be supplied externally to the system so as to include the recording of wind direction through a contactor moved over a variable resistor by the wind vane. A variety of sophisticated electronic transducers have been devised.

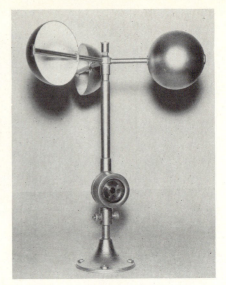

FIGURE A–10
Cup anemometers. (*a*) Simple airway type. (*b*) Totalizing-dial type. (*NOAA National Weather Service.*)

Various propeller or windmill anemometers are on the market. The Friez aerovane is one that is widely used in America. A general external view of the sensing part of the instrument is shown in Fig. A-11. The vertical tail fin serves to keep the head facing into the wind. The wind direction and rotation speed of the propeller are transmitted to recorders or indicators by self-synchronous motors or other electronic devices. Wind-tunnel tests have shown the rotation speed of the propeller to be essentially a linear function of actual wind speed, which is not true of cup anemometers.

Detailed studies of boundary-layer turbulence require special anemometers, including those capable of measuring vertical components. The state of this art is highly advanced, and is better covered in treatises on turbulence.

Exposure of Anemometers

Because the wind increases rapidly with height, one of the first requirements of anemometer installations should be that they all be at the same height. Practical considerations of placing the instruments on city buildings or at airports apparently are so important that little attention is paid to standardizing the height. A height of 9 m has been suggested because it corresponds to the average height above the water at which winds are measured or estimated on shipboard. At major airports the equipment is placed near the runways, where it is desired that the sensors be mounted near the average height of the wings and fuselage of rolling jet transports. The center of a large airfield is an ideal exposure.

FIGURE A–11
Sensing part of aerovane.

Beaufort Scale

Most ships at sea are not equipped with any type of instrument for determining the wind. It then becomes necessary to estimate it. Direction and speed can be estimated from the state of the sea and the heading and speed of the ship. Admiral Beaufort of the British Navy devised in 1805 a scale for estimating the *wind force*, with definitions based on the effect of the wind on the sails carried. The scale is still used today, and since its first use, the wind speeds corresponding to the various forces, which are 12 in all, have been determined experimentally.

The Beaufort wind scale with specifications for land and the open sea as agreed upon by the World Meteorological Organization is given in Table A-1. The equivalent speeds are those used by the British and United States meteorological services. General acceptance of a set of values has been difficult to obtain among the WMO delegates. Note in the specifications that British words and thought dominate ("white horses," "chimney pots," "slates," and failure to specify, on land, winds of 70 to 100 mph, known to Americans).

Table A-1 BEAUFORT WIND SCALE

Beaufort number	Description	Speed, mph	Specifications on land	Specifications at sea
0	Calm	1	Calm; smoke rises vertically	Sea like a mirror
1	Light air	1–3	Direction of wind shown by smoke drift but not by wind vanes	Ripples with the appearance of scales are formed, but without foam crests
2	Light breeze	4–7	Wind felt on face; leaves rustle; ordinary vanes moved by wind	Small wavelets, still short but more pronounced; crests have a glassy appearance and do not break
3	Gentle breeze	8–12	Leaves and small twigs in constant motion; wind extends light flag	Large wavelets; crests begin to break; foam of glassy appearance; perhaps scattered white horses
4	Moderate breeze	13–18	Raises dust and loose paper; small branches are moved	Small waves, becoming longer; fairly frequent white horses
5	Fresh breeze	19–24	Small trees in leaf begin to sway; crested wavelets form on inland waters	Moderate waves, taking a more pronounced long form; many white horses are formed (chance of some spray)
6	Strong breeze	25–31	Large branches in motion; whistling heard in telegraph wires; umbrellas used with difficulty	Large waves begin to form; the white foam crests are more extensive everywhere (probably some spray)
7	Near gale	32–38	Whole trees in motion; inconvenience felt when walking against wind	Sea heaps up and white foam from breaking waves begins to be blown in streaks along the direction of the wind
8	Gale	39–46	Breaks twigs off trees; generally impedes progress	Moderately high waves of greater length; edges of crests begin to break into spindrift; foam is blown in well-marked streaks along the direction of the wind
9	Strong gale	47–54	Slight structural damage occurs (chimney pots and slates removed)	High waves; dense streaks of foam along the direction of the wind; crests of waves begin to topple, tumble, and roll over; spray may affect visibility

Table A-1 (continued)

Beaufort number	Description	Speed, mph	Specifications on land	Specifications at sea
10	Storm	55–63	Seldom experienced inland; trees uprooted; considerable structural damage	Very high waves with long overhanging crests; the resulting foam, in great patches, is blown in dense white streaks along the direction of the wind; on the whole, the surface of the sea takes a white appearance; the tumbling of the sea becomes heavy and shock-like; visibility affected
11	Violent storm	64–72	Very rarely experienced; widespread damage	Exceptionally high waves (small and medium-sized ships might be for a time lost to view behind the waves); the sea is completely covered with long white patches of foam lying along the direction of the wind; everywhere the edges of the wave crests are blown into froth; visibility affected
12	Hurricane	72	Not specified	The air is filled with foam and spray; sea completely white with driving spray; visibility very seriously affected

MEASUREMENT OF PRECIPITATION

Rainfall measurements are extremely easy to make at any given location. All that is needed is a bucket of uniform area and a measuring stick. The depth of the water in the bucket is measured after each rainstorm before the water begins to evaporate, and the bucket is emptied to be ready for the next storm.

It is difficult to say to what extent a network of such observations represents the total precipitation falling over an area. It is known that over level terrain in winter, precipitation amounts are fairly uniform, but summer showers may occur entirely in the space between rain gages even though these may be less than 10 miles apart. In mountainous regions, great batteries of rain gages on all watersheds would be required in order to estimate the total precipitation. A rain gage collects an absurdly small sample of the depth of water over a large area. However, as the number of gages increases from year to year, it is to be expected that we are getting more and more representative values.

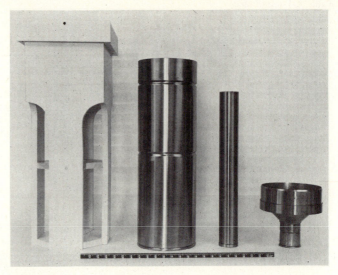

FIGURE A–12
United States standard 8-in. rain gage with, *l* to *r*, wooden housing, large over-flow cylinder, inner measuring tube, and 8-in. mouth funnel. Measuring stick in lower foreground. (*NOAA National Weather Service.*)

Precipitation gages may be classified into two types—recording and direct measuring. For nonrecording purposes, a cylindrical gage 8 in. in diameter is standard in the United States (Fig. A-12). In Canada a cylinder having an area of 10 sq in. (diameter 3.57 in.) is used, and the British standard is 5 in. in diameter. In these gages, the precipitation that enters passes through a funnel into a measuring tube or bottle. In the United States type, the measuring tube has one-tenth of the cross-sectional area of the exposed gage, so that the depth of water in it is ten times that which has fallen. A depth of 10 in. therefore corresponds to 1 in. of rainfall. The water is measured by a measuring stick graduated in terms of a ten-times exaggerated scale so that the nearest hundredth of an inch is easily read. The funnel serves the dual purpose of running the water into the tube and preventing appreciable evapo-ration. The whole tube is enclosed in a large overflow cylinder so that the hole in the funnel is the only egress for evaporated water. The funnel and measuring tube are removed when snow is expected.

For recording purposes, the weighing-type gage developed by Ferguson is now being used in the United States (Fig. A-13). In this instrument, precipitation is also caught in an 8-in. cylinder. It accumulates in a bucket resting on a spring-type weighing platform. The weight, transposed into inches of precipitation by means of proper linkages, is written by a pen on a clock-driven drum. The capacity is 12 in. of precipitation, and the pen is of the dual traverse type, recording up to 6 in. on the upstroke and the remaining 6 on the down-stroke. For detailed recording the mouth area is made larger in order to catch a greater

FIGURE A–13
Weighing rain gage showing housing and 8-in. throat removed.

amount of water. An increase in area of $2\frac{1}{2}$ times results in a diameter approximately that of the outer shell of the gage. This makes the full stroke of the pen correspond to 2.4 in. instead of 6 in. as in the case of the 8-in. mouth. The clock drum can be speeded up by using gears with a smaller reduction factor. When snow is not expected, a funnel is placed inside the gage above the bucket to check evaporation of water after it is collected so it can be measured with a stick as a proof of the accuracy of the recording mechanism.

Rain gages affect the flow of the wind around them. The air flow may act in such a way as to cause some of the precipitation to escape being caught, or it may cause the gage to act as a "trap." For the measurement of snow a shield to smooth out the air flow is positively necessary, since the feathery flakes are carried with the wind current in such a way that the gage is an unlikely place for them to land. A platform and shield for a rain gage, much used in the United States, is shown in Fig. A-14.

FIGURE A–14
Shield around weighing rain gage. (*NOAA National Weather Service.*)

Where a properly shielded and representative gage is not available, it is possible to measure precipitation in the form of snow by carefully selecting a sample, cutting it out like a biscuit, then weighing it or measuring it in a graduate after melting. Care should be taken not to include old snow in the sample. If the snow has drifted badly, it is practically impossible to be assured of a representative sample, and considerable guesswork is involved. For snow surveys in mountain regions, a sampling tube about 3 in. in diameter and having a sharp cutting edge is used, equipped with a spring balance for immediate weighing to determine the water equivalent. For general information, it may be stated that the water equivalent of snow is roughly one-tenth of its unmelted depth. In many cases this is quite far from correct.

Radar has been used with considerable success as a means of obtaining the total rainfall over a specified area, especially under showery conditions where rain-gage networks are wholly inadequate. The pulse transmitted by radars in the 3- to 10-cm wavelengths is scattered back by water and ice particles of raindrop size suspended or falling in the air. As a radar scans the horizon, the rainy portions of all clouds that are producing rain are outlined on the radarscope. The ratio of the power returned to the power transmitted at a given wavelength and at a given distance of the scatterers from the radar is proportional to na^6, where n is the number of spherical scatterers of radius a per unit volume. The volume of water contained in such spherical drops is proportional to na^3. Thus the ratio of the echo intensity to the volume of water is proportional to a^3.

By comparing the volume and intensity of a radar echo with rainfall rate at the same time and place over a dense network of recording rain gages, the radar can be "calibrated" as a means of measuring rain.   In practice, the area covered by the echo is used instead of the volume, since to get the volume requires vertical as well as horizontal scanning.   Instead of measuring the intensity, the shrinkage of the area as the receiver gain is reduced is noted. With these simplifications the method loses little in accuracy.   Since only the more intense rainfall produces an echo at lower receiver gain, the various power settings produce areas on the scope that bear some correspondence to the areas enclosed by lines of equal rainfall (*isohyets*).   A variety of factors limits the accuracy of the radar-rain relationship.   Even with its greatest quantitative inaccuracies the radar will not miss a shower altogether, as the standard rain-gage networks sometimes do.   The duration of rainfall over a given area is one of the most important factors in determining the total that falls.   This factor of time the radar obtains with complete accuracy.

FIGURE A–15
Projector of transmissometer.   The detector is similar in external appearance, but with a narrower optical tube. (*NOAA National Weather Service.*)

REMOTE SENSING INSTRUMENT SYSTEMS

At major airports instruments are mounted near the runways with all the data recorded on dials or charts at a single console in the weather office. Anemometers and wind vanes, thermistors, hygristors, or dew-point hygrometers plus ceiling and visibility sensors are all wired to the observer's console.

The ceiling is measured by a rotating-beam ceilometer. It consists of two parts, a light receiver and a transmitter, located 500 to 1000 ft apart. A light source is transmitted from a pair of parabolic mirrors rotating in a vertical plane directed toward the receiver. The detector points vertically and senses the spot of light on the cloud as the spot arrives directly overhead. From the known base-line distance and the angle of the beam at the moment of detection the height is obtained by solving the right triangle ($h = x \tan \alpha$, where x is the base-line distance). The light source is chopped so that it produces a distinctive pulse which the detector senses even in bright daylight. The photocell is not sensitive to the steady daylight.

For obtaining the visibility, a transmissometer is used. This consists of a light source (Fig. A-15) transmitted to a detector over an optical path long enough to show the depletion effects of light fog or haze. The detector measures the intensity of the light. Different calibration factors have to be applied for different background light intensities, such as between day and night.

Remote sensing instrument packages have been designed for radiotelemetering from remote locations. Buoys at sea and automatic weather stations in the polar regions have been established, using systems with varying degrees of sophistication.

UPPER-AIR AND CLOUD OBSERVATIONS

In order to determine the state of the atmosphere at rest at any given place and time, three quantities must be measured—pressure, temperature, and humidity. These three quantities determine the height or geopotential through integration of the hydrostatic equation. It is sufficient to solve the relationship through 100-mb Δp layers using the mean virtual temperature, $\overline{T}^*$ of each layer.

In the early days of upper-air soundings the quantities were measured from instruments mounted on giant kites, captive balloons, instruments to be recovered from unmanned free balloons, and airplanes. Today, although airplanes can still be used, the most economical and practical way is by means of disposable (sometimes recoverable and reconstructed) instruments, borne on free balloons, and telemetering their measurements back to earth by radio signals. These systems are called *radiosondes*, or if wind-finding equipment is included, *rawinsondes*.

The instrument and radio transmitter must be lightweight and inexpensive. These are compactly housed in a corrugated cardboard or light plastic container about $6 \times 5 \times 8$ in., less antenna, and weigh about 1 kg, including battery. To achieve a good signal from small power output (approximately 400 mW) ultra high frequency is used, commonly 1680 MHz.

In the United States and some other countries an instrument is used in which the audio-output is modulated by the sensing elements. Distinguishing reference signals also are

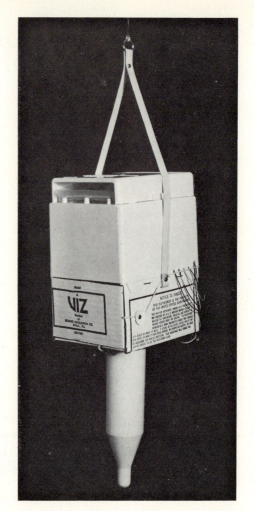

FIGURE B–1
Radiosonde before extending thermistor support frame. The antenna is in the bottle-shaped lower extension. (*NOAA National Weather Service.*)

transmitted. Two systems of modulation are available, one in which the audiofrequency is modulated and the other with amplitude modulation.

A special pressure-measuring and actuating system—a *baroswitch*—is used. An aneroid-barometer cell moves a contact arm over a series of printed conductors. These conducting strips are spaced in accordance with set intervals of pressure so that the pressure-change steps can be added together as the balloon ascends. This commutator system alternately actuates the temperature, humidity, and reference signals.

The temperature sensor is the thermistor and the humidity is sensed by the carbon-element "hygristor," both described in Appendix A.

A modified form of the radiosonde to provide the necesssry precision of pressure measurement at heights where the Δp of the baroswitch is too coarse can give reliable pres-

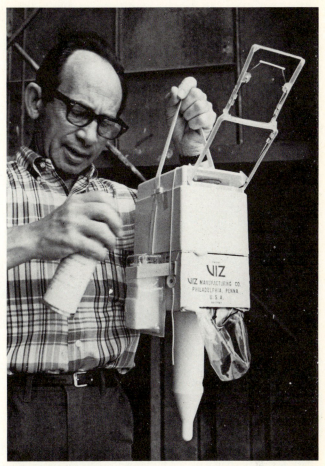

FIGURE B–2
Radiosonde being readied for flight.   The thermistor frame is extended outside
the main housing and Freon is being poured into the hypsometer.   The hygristor
is in the upper ventilated portion of the housing.   A larger ventilation system
was introduced in the 1973 model. (*NOAA National Weather Service.*)

sure readings down to 2 mb or less.   The instrument, called a *hypsometer*, is based on the
principle that a liquid boils when its vapor tension is equal to the total ambient pressure.
At a given pressure the temperature of the boiling point is uniquely determined.   In the
hypsometer the requirement is to keep the liquid boiling and translate its temperature into
pressure from the p, T relationship of the liquid.   Freon 11 is used because it has a low boiling
point and therefore does not require appreciable heating at low temperatures and pressures.

The 1968–1971 hypsometer assembly consists essentially of a bead thermistor and heater
assembly mounted within a small dewar flask.   The flask is filled just prior to launch with a

FIGURE B–3
Launch of radiosonde after removal from inflation shelter, *left*. Note parachute
in the instrument train to permit gentle landing after the balloon bursts. (*NOAA
National Weather Service.*)

small quantity of Freon 11.   The hypsometer is made to operate from 50 to 1.5 mb.   The
flask is contained in a plastic blister with vents for the escaping vapor.   It is mounted under
an elastic strap on the side of the instrument.

 With suitable radio-ranging and direction-finding equipment at the ground and trans-
ponding device in the balloon-borne instrument, the radiosonde system becomes a rawinsonde.
A receiver for a 403-MHz ground-based transmitter is carried.   A 75-kHz sine wave modu-
lates the amplitude of the 403-MHz transmitter.   This signal is directed by an antenna array
to the 403-MHz receiver, where it is detected, amplified, and used in the radiosonde circuit
to frequency-modulate the regular 1680-MHz signal at 75 kHz.   The 1680-MHz frequency-
modulated signal is received and demodulated by the ground receiver.   The incoming 75-
kHz modulation is compared with that generated at the ground set, and the phase difference
between the two signals is a measure of the slant range.   Simultaneously the meteorological
information is transmitted in the usual way on the 1680-MHz carrier.   With altitude from
the radiosonde computation, elevation angle, and azimuth angle, and slant range from the
ground system, the altitude and horizontal drift of the balloon can be traced.   The ground
unit computes and records automatically through suitable electronic circuitry.

EARTH-CIRCLING HIGH-ALTITUDE BALLOONS

Techniques have been developed for floating large free balloons at fixed altitudes which can survive a number of trips around the world.   At the National Center for Atmospheric Research in Boulder, Colorado, a group under Vincent Lally developed a technique known as the GHOST concept, for Global Horizontal Sounding Technique.   A breakthrough making this technique possible was the introduction of a thin plastic film under the trade name Mylar, which is tough and relatively inelastic.   The balloon is filled with gas and sealed at an overpressure so that it expands as it ascends.   At the surface it is shaped like a sausage, but at 15,000 ft it becomes nearly spherical with a diameter of 5 ft, expanding to 10 ft at 55,000 ft and 40 ft at 90,000 ft.   It floats with zero buoyancy when it displaces the mass of atmospheric air equal to its own mass.   As long as there is an overpressure, it maintains its size and shape. There is some diffusion through the balloon material, but the larger the balloon the less the diffusion, which is another way of saying the lower the pressure and the temperature the less the diffusion.   Such balloons have remained at fixed altitudes for months at a time.

The usual meteorological information can be telemetered from the balloons.   The location is revealed by a sensor of the sun angle.   Thin-film electronics has been used in the instruments and circuitry with success.

SOUNDINGS WITH ROCKETS

For soundings in the upper stratosphere and mesosphere beyond the reach of balloons, rocket systems have proved useful.   Rockets going to 160 km and higher, well into the ionospheric layers, have been of importance in aeronomy.   Meteorological experiments have included the detonation of explosive charges at intervals along the rocket trajectory and measurements on free-falling spheres separated at peak altitude.   The explosions produce a measurement at the surface of the pressure perturbations from which the mean temperatures and winds between explosions can be determined.   The falling spheres provide a measure of the density through the effect of changing drag shown on accelerometers inside the spheres.

During the 1960s rockets with specific meteorological applications were developed, capable of carrying the scientific payload to 75 km or higher.   Equipment consisting of a specially adapted and more sophisticated radiosonde system is released with a parachute at peak altitude.   A 403-MHz transmitter, power supply, and associated components weigh a little more than half as much as the conventional radiosonde.

SATELLITE SOUNDINGS

The dream of meteorologists—to see the earth and its swirling cloud systems from space— came true when man-made satellites began orbiting the earth.   Some of these have been synchronous with the earth's turning so that they can remain over one spot to transmit the picture at the same place day after day.   The more or less continuous monitoring of the cloud pictures over the earth is now commonplace, and most of the principal meteorological subcenters have readout equipment.

Besides the photographic and numerous other missions the satellites perform, they can make greatly needed radiation measurements with respect to the earth and the sun. The solar constant, the earth's albedo, the distribution of energy through the spectrum, and scattering, reflection, and absorption in the atmosphere can be measured with the various sensors of radiance.

The Nimbus III and IV satellites, launched April 14, 1969, and April 8, 1970, respectively, were equipped with the Satellite Infrared Spectrometer (SIRS), marking major steps in the application of optics to meteorological requirements. The instrument design is based on the principle of the conventional spectrometer. In the first application of the system the spectral radiance in the 15-μm band of carbon dioxide was measured. This gas was selected because it is known to be almost uniformly mixed at about 320 ppm by volume throughout the appreciable atmosphere. Ozone and water vapor are also of significance in the 15-μm band. Therefore, later versions of the system have included the Backscatter Ultraviolet Spectrometer (BUV), which monitors the spatial distribution of atmospheric ozone by measuring the intensity of ultraviolet radiation back-scattered from the earth's atmosphere, and the Filter Wedge Spectrometer (FWS), which monitors the distribution of water vapor.

Over clear areas of the earth this equipment makes it possible to compute the vertical distribution of temperature and, through the FWS, also water vapor. The soundings thus obtained have been found to compare remarkably closely with those taken from radiosondes in the same vicinity. With a cloud cover the process becomes more complicated, but it can be helped by working from surface observations and knowledge of what the low-level air mass should be like. The Selective Chopper Radiometer (SCR) determines the temperature profile of the atmosphere from the earth's cloud-top level to about 65 km altitude over a strip 11.3 km long in the direction of flight by 112.7 km wide.

In a sun-synchronous nearly circular orbit the Nimbus satellites circle the earth every 107 min, viewing the entire planet twice daily. As of the date of publication of this book the question of the extent to which the satellite soundings will replace the worldwide network of radiosondes is being energetically discussed.

CLOUD CLASSIFICATION

Meteorologists of the world have agreed on a uniform system of cloud classification. The World Meteorological Organization (WMO) fosters adherence to the system and is responsible for any changes or improvements that may be made from time to time. The present classification is an outgrowth of a system published in 1803 in England by Luke Howard, improved upon by the Frenchman Renou and the Swede Hildebrandsson. The WMO publishes the definitions and photographs of the different types in the form of an atlas. A two-volume atlas was published in 1957 in several languages.

The classification does not attempt to consider the processes forming the clouds but rather adheres to distinguishable shapes, shading, general appearances, and optical effects

FIGURE B–4
Cirrus.

which a subprofessional meteorologist can be trained to recognize.   The meteorologist at the analysis center can put together the cloud observations from a number of stations to obtain clues to the processes that are operating.

There are 10 *genera* of clouds, as listed in Table B-1, which may be further subdivided into *species* and *varieties*.

Surface and aircraft observations have shown that clouds are generally encountered over a range of altitudes varying from sea level to the height of the tropopause.   That part of the atmosphere in which clouds are usually present in the troposphere has been divided into three "étages": high, middle, and low.   Cirrus, cirrocumulus, and cirrostratus are in

Table B-1 CLOUD GENERA

Cirrus	Nimbostratus
Cirrocumulus	Stratocumulus
Cirrostratus	Stratus
Altocumulus	Cumulus
Altostratus	Cumulonimbus

the high étage; altocumulus is in the middle étage; and stratus and stratocumulus are in the low étage. Altostratus is usually found in the middle étage, but it often extends higher; nimbostratus, cumulus, and cumulonimbus extend through several levels. The étages overlap and vary with latitude, but their approximate limits are given in Table B-2.

In the following descriptions, the definitions of the 1957 World Meteorological Organization Cloud Atlas are given in italics, and additional information to help the student has been supplied by the author.

Cirrus

Detached clouds in the form of white, delicate filaments or white or mostly white patches or narrow bands. These clouds have a fibrous (hair-like) appearance.

The prefix *cirro-* is applied to forms at the same general level but having somewhat different appearance. Cirrus is the name for detached clouds as defined. A variety of forms is noted, such as tufts, delicate lines across a blue sky, branching plumes, curved lines ending in tufts, and unshaded white smears against a blue sky. Effects of perspective sometimes give the clouds the appearance of vertical extent or converging of bands to a point on the horizon, but this is a false impression.

All the clouds of the cirrus or cirro- type are composed of ice crystals. The sun or moon shining through these ice-crystal clouds produces a halo. In the simple cirrus types, however, the irregular cloud distribution fails to produce this action on the light rays in a noticeable way.

Because of their great height and brightness, cirrus clouds are brilliantly colored well above the horizon at sunset and sunrise, often being of a bright yellow or red color upward from the horizon and almost directly overhead.

Cirrocumulus

Thin, white patch, sheet or layer of cloud without shading, composed of very small elements in the form of grains, ripples, etc., merged or separate, and more or less regularly arranged; most of the elements have an apparent width of less than one degree.

They often look like small flakes or very small globular masses. When well marked in a uniform arrangement they form what the sailors used to call a *mackerel sky*. Cirrocumulus is commonly connected with cirrus or cirrostratus. It occurs less frequently than the forms of cirrus and cirrostratus.

Table B-2 LIMITS OF ÉTAGES

Étages	Polar regions	Temperate regions	Tropical regions
High	3–8 km	5–13 km	6–18 km
Middle	2–4 km	2–7 km	2–8 km
Low	Surface to 2 km	Surface to 2 km	Surface to 2 km

FIGURE B–5
Cirrocumulus, *upper right*, with cirrus and cirrostratus.

Cirrostratus

Transparent, whitish cloud veil of fibrous (hair-like) or smooth appearance, totally or partly covering the sky, and generally producing halo phenomena.

Sometimes the cirrostratus is so thin that it only slightly whitens the blue of the sky. At other times it has the appearance of a heavy white sheet. Occasionally it has irregular filaments. The border of the sheet of cirrostratus is usually indefinite, often ending in patches of cirrus or cirrocumulus. On rare occasions the edge of the sheet is straight and clear-cut. It never obscures the sun to the extent that shadows are not cast by objects on the ground, never at least when the sun is fairly high.

Altocumulus

White or grey, or both white and grey, patch, sheet or layer of cloud, generally with shading, composed of laminae, rounded masses, rolls, etc., which are sometimes partly fibrous or diffuse and which may or may not be merged; most of the regularly arranged small elements usually have an apparent width of between one and five degrees.

Altocumulus clouds have shapes somewhat similar to those of cirrocumulus; however, they are distinguishable from the latter in that they are larger and usually have definite dark shading underneath and in the middle of each cloud element. They do not produce halo phenomena. The edges of the elements are often thin and translucent and exhibit irisations (bright plays of colors) which are found only in this type of cloud.

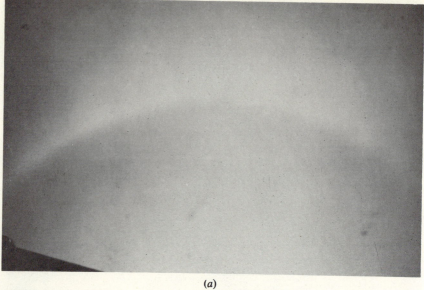

(a)

(b)

FIGURE B–6
Distinction between cirrus with halo (a), and altostratus (b) with sun appearing as through ground glass. (*R. G. Beebe.*)

FIGURE B–7
Altocumulus, high scaly.

Low-level altocumulus may be distinguished from stratocumulus by the size of the elements. Altocumulus may occur at more than one level at the same time. The fact that they do not produce halos should not be taken to mean that they do not occur at temperatures below freezing. As a matter of fact, altocumulus clouds, as well as altostratus, are often composed of liquid droplets undercooled to temperatures well below freezing.

Altostratus

Greyish or bluish cloud sheet or layer of striated, fibrous, or uniform appearance, totally or partly covering the sky, and having parts thin enough to reveal the sun at least vaguely as through ground glass. Altostratus does not show halo phenomena.

FIGURE B–8
Altocumulus, massive, closely packed.

Altostratus is sometimes thin in light patches between very dark parts, but it never shows definite configurations.   Under an altostratus sheet, shadows of objects on the ground are never visible.

Low altostratus may be distinguished from stratus or nimbostratus because of the darker, more uniform gray of the lower forms and the fibrous structure and whitish gleam often visible in the altostratus.   In a very smoky atmosphere it is almost impossible to make this distinction without some direct height measurement.   If the sun or moon is completely hidden at all times, i.e., if not even a light spot shows in the vicinity of these luminaries, then one may conclude that the sheet is stratus or nimbostratus.

Altostratus may result from a transformation of a sheet of altocumulus, and on the other hand altocumulus may represent a dissipating stage of altostratus.

Nimbostratus

Gray cloud layer, often dark, the appearance of which is rendered diffuse by more or less continuously falling rain or snow, which in most cases reaches the ground.   It is thick enough throughout to blot out the sun.

Low, ragged clouds frequently occur below the layer, with which they may or may not merge.

The precipitation does not have to reach the ground.   The rain or snow may evaporate in the air, in which case the base of the cloud usually is ill-defined and has a " wet " look caused by trails of precipitation and low " scud " clouds (fractostratus).

FIGURE B–9
Altocumulus, detached, from convection under a stable layer.

Nimbostratus is most often a low cloud form. It is distinguishable from stratocumulus in that it has no discrete or, at least, regular cloud elements and it is distinguishable from stratus in that it is darker, shows rain conditions, and has a highly variable lower surface.

In the usual storms of middle latitudes, nimbostratus often develops from a thickening and lowering altostratus. Streaks of snow or rain not reaching the ground are called *virga*. The fractostratus or scud clouds, sometimes also fractocumulus, form underneath the nimbostratus to give the sky a dark chaotic system of low clouds that are characteristic of bad weather. Less frequently, nimbostratus may develop from stratocumulus.

Stratocumulus

Gray or whitish, or both gray and whitish, patch, sheet or layer of cloud which almost always has dark parts, composed of tassellations, rounded masses, rolls, etc., which are nonfibrous (except for virga) and which may or may not be merged; most of the regularly arranged small elements have an apparent width of more than five degrees.

The arrangements of the cloud elements are in groups, lines, or waves, aligned in one or in two directions. Rolls often appear, sometimes so close that their edges join, but even when they cover the whole sky they have an undulated appearance on their lower surface.

Stratocumulus clouds often have an appreciable vertical development and are therefore likely to be confused with small cumulus. However, they have a softer, more irregular shape than cumulus and when viewed from above are seen to have their tops at a uniform height and to exhibit a quilted pattern. Sometimes they are confused with large lumps of stratus that have broken off from a uniform stratus sheet or that result from the dissipation or

FIGURE B–10
Altocumulus, tightly packed in waves or rolls.

precede the formation of a stratus sheet. Sometimes stratus changes into stratocumulus, but this does not happen simply by the breaking off of lumps of stratus. There must be some regular pattern and distinct masses.

A transition from stratocumulus to nimbostratus is sometimes seen. The elements of thick stratocumulus fuse together completely, and the cloud is called nimbostratus when, after the disappearance of the cloud elements, falling trails of precipitation destroy the clear-cut boundary of the lower surface.

Stratocumulus *vesperalis* are flat, elongated clouds that are often seen to form about sunset as the final product of the diurnal changes of cumulus. Another type is *cumulogenitus*, formed by the spreading out of the tops of cumulus clouds, which have disappeared.

Stratus

Generally gray cloud layer with a fairly uniform base, which may give drizzle, ice prisms, or snow grains. When the sun is visible through the cloud, its outline is clearly discernible. Stratus does not produce halo phenomena except, possibly, at very low temperatures.

Sometimes stratus appears in the form of ragged patches.

FIGURE B–11
Stratus in the Golden Gate, San Francisco, California.

When viewed from above, stratus always has a uniform top, usually marked by a temperature inversion or another thermally stable layer. It can have the same characteristics as a fog, except that it does not occur on the ground.

Cumulus

Detached clouds, generally dense and with sharp outlines, developing vertically in the form of rising mounds, domes, or towers, of which the bulging upper part often resembles a cauliflower. The sunlit parts of these clouds are mostly brilliant white; their base is relatively dark and nearly horizontal.

Sometimes cumulus is ragged.

Over land areas, cumulus is most often found in the daytime, generally dissipating at night. Cumulus cannot produce more than light precipitation. It often, however, makes the transition to cumulonimbus, which is the heavier shower cloud.

Cumulus clouds represent strong vertical convection currents; hence they occur in air that is heated from below or cooled from above, so that the warm air is displaced by surrounding colder air. The "boiling" of the tops is often noticed, indicating the strong vertical currents.

FIGURE B–12
Nimbostratus above low clouds of bad weather.

Four species of cumulus are given in the International Cloud Atlas. They are cumulus *humilis*, cumulus *mediocris*, cumulus *congestus*, and *fractocumulus*. Cumulus humilis, often referred to as fair-weather cumulus, have little vertical development. Nevertheless they have rounded tops and flat bases. They are often more or less equally spaced but appear more crowded toward the horizon owing to perspective. Also, in looking toward the horizon, one notes the flat bases in a series of steps or in *echelon*. Cumulus congestus have marked vertical development, but not reaching the cumulonimbus stage. The tops are seen to "boil" vigorously. Cumulus mediocris and fractocumulus are obvious in their meaning.

Cumulonimbus

Heavy and dense cloud, with a considerable vertical extent, in the form of a mountain or huge towers. At least part of its upper portion is usually smooth, or fibrous, or striated, and nearly always flattened; this part often spreads out in the shape of an anvil or vast plume.

Under the base of this cloud which is often very dark, there are frequently low ragged clouds either merged with it or not, and precipitation sometimes in the form of virga.

FIGURE B–13
Cumulus humilis over New Jersey and New York City. (*NOAA National Weather Service.*)

At a distance, cumulonimbus is recognizable as the most massive and tallest of clouds Its development from a collection of cumulus congestus indicates the strong vertical currents that are present.   The cloud top often reaches to the cirrus level or bursts through the tropopause.

From underneath, the base of the cloud is like nimbostratus, having low scud clouds of fractocumulus character.   In general, the turbulent motion of these ragged low clouds is much more noticeable than in nimbostratus.   Cumulonimbus are outgrowths of cumulus, whereas nimbostratus occurs from clouds that have little vertical development.

Cumulonimbus is the great thundercloud so familiar in the summer weather of most of the United States, and at other seasons in the tropics.   It always produces at least a pronounced shower.   Its development is cellular, and there is always more or less clear space between two adjacent clouds of this type, except in certain general storm conditions.   When the cloud is overhead and the features recognizable at a distance cannot be seen, the fall of a real shower and sudden darkening of the sky is ample evidence of cumulonimbus.

FIGURE B–14
Cumulus congestus. (*NOAA National Weather Service.*)

Cumulonimbus gives rise to a great variety of associated clouds. The sky takes on a broken-up, menacing appearance in the low levels, and the spreading out or dissipation of the high parts forms layers of cirriform and alto-form clouds. In the forward portion of an intense cumulonimbus, a rolling scud cloud is seen just under the cloud base. Also in the forward portion, *mamma* forms, pendant from middle to low clouds like mammalian teats or udders, are sometimes observed.

METEORS (NONASTRONOMICAL)

The general definition of meteors, given by nearly all dictionaries, is that they are any visible or optical phenomena in the atmosphere. Astronomical meteors entering the atmosphere from space are a special class, but since there is no other common word to describe them they virtually monopolize the term in ordinary usage, at least in the English language.

FIGURE B-15
Cumulonimbus, upper part as seen from the air.

The World Meteorological Organization has agreed on the definitions and descriptions of four classes of meteors: (1) *hydrometeors*, composed of water in various forms; (2) *lithometeors*, consisting of essentially dry particles and their visible manifestations; (3) *photometeors*, that is, optical phenomena; and (4) *electrometeors*, arising from atmospheric electrical phenomena. The definitions below are taken from Volume I of that organization's International Cloud Atlas, published in 1956 from the headquarters in Geneva.

Hydrometeors

Rain Precipitation of liquid water particles, either in the form of drops of more than 0.5 mm diameter or in the form of smaller widely scattered drops. *Freezing rain* is recorded when the drops freeze on impact with the ground, with objects on the earth's surface, or with aircraft in flight.

Drizzle Fairly uniform precipitation composed exclusively of fine drops of water (diameter less than 0.5 mm), very close to one another. *Freezing drizzle* is recorded when the drops freeze on impact with the ground, with objects on the earth's surface, or with aircraft in flight.

Snow   Precipitation of ice crystals, most of which are branched (sometimes star-shaped).

Snow pellets   Precipitation of white and opaque grains of ice.   These grains are spherical or sometimes conical; their diameter is from about 2 to 5 mm.

Snow grains   Precipitation of very small white and opaque grains of ice.   These grains are fairly flat or elongated; their diameter is generally less than 1 mm.

Ice pellets   Precipitation of transparent or translucent pellets of ice, which are spherical or irregular, rarely conical, and which have a diameter of 5 mm or less.

Hail   Precipitation of small balls or pieces of ice (hailstones) with a diameter ranging from 5 to 50 mm or sometimes more, falling either separately or agglomerated into irregular lumps.

Ice prisms   A fall of unbranched ice crystals, in the form of needles, columns, or plates, often so tiny that they seem to be suspended in the air.   These crystals may fall from a cloud or from a cloudless sky.

Fog   A suspension of very small water droplets in the air, generally reducing the horizontal visibility at the earth's surface to less that 1 km.

Ice fog   A suspension of numerous minute ice crystals in the air, reducing the visibility at the earth's surface.

Mist   A suspension in the air of microscopic water droplets or wet hygroscopic particles, reducing the visibility at the earth's surface.   Note that in the International Codes for weather reports, the term "mist" is used when the hydrometeor mist or fog reduces the horizontal visibility at the earth's surface to *not less* than 1 km.

Drifting snow and blowing snow   An ensemble of snow particles raised from the ground by a sufficiently strong and turbulent wind.   *Drifting* means that the particles are raised to small heights and the visibility is not sensibly diminished at eye level.   *Blowing* means that the particles are raised to moderate or great heights and the horizontal visibility at eye level is generally very poor.

Spray   An ensemble of water droplets torn by the wind from the surface of an extensive body of water, generally from the crests of waves, and carried up a short distance into the air.

Dew   A deposit of water drops on objects at or near the ground, produced by the condensation of water vapor from the surrounding clear air.   *White dew* is a deposit of white frozen dew drops.

Hoarfrost   A deposit of ice having a crystalline appearance, generally assuming the form of scales, needles, feathers, or fans.

Rime   A deposit of ice, composed of grains more or less separated by trapped air, sometimes adorned with crystalline branches.

Glaze (clear ice)   A generally homogeneous and transparent deposit of ice formed by the freezing of supercooled drizzle droplets or raindrops on objects the surface temperature of which is below or slightly above 0°C.

Spout   A phenomenon consisting of an often violent whirlwind, revealed by the presence of a cloud column or inverted cloud cone (funnel cloud), protruding from the base of a cumulonimbus, and of a " bush " composed of water droplets raised from the surface of the sea or of dust, sand, or litter, raised from the ground; *tornado*, or *waterspout*.

Lithometeors

Haze   A suspension in the air of extremely small dry particles invisible to the naked eye nda sufficiently numerous to give the air an opalescent appearance.

Dust haze   A suspension in the air of dust or small sand particles, raised from the ground prior to the time of observation by a dust storm or sandstorm.

Smoke   A suspension in the air of small particles produced by combustion.

Drifting and blowing dust or sand   An ensemble of particles of dust or sand raised, at or near the station, from the ground to small or moderate heights by a sufficiently strong and turbulent wind.   *Drifting* means raised to small heights, and the visibility is not sensibly diminished at eye level.   *Blowing* means raised to moderate heights, and the horizontal visibility at eye level is sensibly reduced.

Dust storm or sandstorm   An ensemble of particles of dust or sand energetically lifted to great heights by a strong and turbulent wind.

Dust whirl or sand whirl (dust devil)   An ensemble of particles of dust or sand, sometimes accompanied by small litter, raised from the ground in the form of a whirling column of varying height with a small diameter and an approximately vertical axis.

Photometeors

Halo phenomena   A group of optical phenomena in the form of rings, arcs, pillars, or bright spots, produced by the refraction or reflection of light by ice crystals suspended in the atmosphere (cirriform clouds, ice fog, etc.).

Corona   One or more sequences (seldom more than three) of colored rings of relatively small diameter, centered on the sun or moon.

Irisation Colors appearing on clouds, sometimes mingled, sometimes in the form of bands nearly parallel to the margin of the clouds. Green and pink predominate, often with pastel shades.

Glory One or more sequences of colored rings, seen by an observer around his own shadow on a cloud consisting mainly of numerous small water droplets, on fog or, very rarely, on dew.

Rainbow A group of concentric arcs with colors ranging from violet to red, produced on a "screen" of water drops (raindrops, droplets of drizzle, or fog) in the atmosphere by light from the sun or moon.

UNITS

Although in much of the English-speaking world weather information given to the general public is expressed in British units, meteorologists, like other scientists, calculate in the metric system. The "International System of Units" established by the 1960 General Conference of Weights and Measures, is founded on the six basic units:

The *meter*, for length
The *kilogram*, for mass
The *second*, for time
The *ampere*, for electric current
The (*degree*) *Kelvin*, for temperature
The *candela*, for luminous intensity

This system is coherent in the sense that it produces compatible derived units in mechanics, heat, electricity, and optics. It is sometimes referred to as the *mks*—meter, kilogram, second system.

In the mechanical and thermodynamic applications in the atmosphere the centimeter and gram (*cgs* system) are widely used in place of the meter and kilogram, while some meteorologists and oceanographers prefer to use the metric ton of 1000 kg (*mts* system), but these modifications should not be confusing because of the obvious interrelations through factors of 10. Strict adherence to the *mks* system has the advantage that in electricity and magnetism the units come out as the familiar ampere, volt, ohm, coulomb, etc.

The accepted scale of temperature is the Kelvin scale, which has the same unit degree as the Celsius or centigrade scale. The latter places 0° at the ice point of water and 100° at the boiling point for standard sea-level pressure. In 1954 the Kelvin scale was internationally defined by setting the triple point of water at 273.16 ± 0.01 K.

Special names are given to some of the derived units, for example:

Force (kg m/sec²) $\equiv$ newton
   or (g cm/sec²) $\equiv$ dyne
Work, energy (kg m²/sec² = newton m) = joule *or* dyne cm $\equiv$ ergs
Power (kg m²/sec³ = joule/sec) $\equiv$ watt *or* ergs/sec
Voltage (watt/amp) $\equiv$ volt
Electric charge (amp sec) $\equiv$ coulomb

and other electromagnetic units (ohm, farad, weber, henry, tesla) which do not enter into consideration in this book.

Some convenient units which are multiples of the ordinary derived units are

Pressure (force per unit area)—1 bar = 10^5 newtons/m²
                                  = 10^6 dynes/cm²
Heat energy—1 gram-calorie = 4.186 joules
                                  = 4.186×10^7 ergs

The gram-calorie is the amount of heat required to raise the temperature of 1 g of water 1°C, from 15 to 16°C. The expression of pressure in terms of a length, such as inches or millimeters of mercury, is a direct statement from a reading of a barometer before equating to a pressure unit. The most frequently used pressure unit in meteorology is the millibar (10^{-3} bar). The standard or "normal" pressure supporting 760 mm of mercury at 0°C, referred to as NTP, may be seen as the weight (static force) of a mercury column of unit cross-sectional area:

$$\text{Density} \times \text{height of column} \times 1 \times \text{acceleration of gravity}$$
$$13.595 \times 76 \times 980.665 = 1.0136 \times 10^6 \text{ dynes/cm}^2$$
$$= 1.0136 \times 10^5 \text{ N/m}^2$$

which is 1013.6 mb. This pressure is also taken as the unit of one atmosphere, and it is seen that it is slightly more than 1 bar. The factor to convert from mm Hg to mb is almost exactly $\frac{4}{3}$. From inches Hg to mb the factor is 33.88; 1 mb of pressure supports approximately 0.03 in. Hg.

Units related to the measurement of distance and time should be mentioned here. The number of waves reaching a point in unit time—the frequency (sec^{-1})—is given by the speed of propagation divided by the wavelength. For electromagnetic waves, which travel at the speed of light, the unit of frequency (sec^{-1}) is called the Hertz, abbreviated Hz.

Multiples of measures of this type carry prefixes such as the following:

mega- (abbr. M) for 10^6
kilo- (abbr. k) for 10^3
hecto- (abbr. h) for 10^2
deca- (abbr. da) for 10
deci- (abbr. d) for 10^{-1}

centi- (abbr. c) for 10^{-2}
milli- (abbr. m) for 10^{-3}
micro- (abbr. μ) for 10^{-6}
nano- (abbr. n) for 10^{-9}

For 10^{-6} meters (1 μm) the word *micron* was used, but *micrometer* is now preferred. An area of 1 hm² (10^4 sq m) is called a *hectare*, and one cubic decimeter (dm³) is a *liter*. In some sciences, a cubic centimeter (cm³) is called a milliliter (ml).

The conventional unit for measuring angles should be mentioned also. The circle is divided into 360 parts or degrees, which in turn are divided into 60 minutes, and these into seconds, but for the greatest ease of computation the circle is divided into 2π *radians* (1 rad $= 180/\pi = 57.2956°$). The unit of solid angle is the steradian, 4π of which comprise a complete sphere.

Meteorologists and other scientists hope that the English-speaking public will soon be persuaded to forsake the illogical Fahrenheit temperature scale, which sets 0°C at 32 and 100°C at 212, making 180° between these two phase-transition points of water. The relation is $°F = 32 + (1.8 \times °C)$. In several fields of engineering the British units are firmly entrenched. Practical meteorologists in the English-speaking countries are accustomed to dealing with feet, miles, inches, and Fahrenheit temperatures in addressing the users of their data.

In considering processes over the surface of the earth, it is sometimes convenient to use the nautical mile, which is equal to a minute of latitude or arc on the spherical earth. Speeds are sometimes expressed in knots, which is the term for nautical miles per hour ("knots per hour" is redundant and obviously incorrect).

CONSTANTS

The following values are given for constants of use in meteorology. They have been adopted by the World Meteorological Organization (WMO):

Velocity of light	2.9979×10^8 m sec^{-1}
Avogadro's constant	6.0225×10^{23} mol^{-1}
Boltzmann's constant	1.3805×10^{-23} J K^{-1}
Planck's constant	6.6256×10^{-34} J sec
Stefan-Boltzmann constant	5.6697×10^{-8} J m^{-2} K^{-4} sec^{-1}
Standard molar volume of ideal gas	$22,413.6 \times 10^{-6}$ m³ mol^{-1} or 22.4136 liters mol^{-1}
Molar constant of ideal gas	8.3143 J mol^{-1} K^{-1} or 8.3143×10^7 ergs mol^{-1} K^{-1}
Apparent molecular weight of dry air	28.9644
Molecular weight of water vapor	18.0153
Specific heat capacity, dry air:	
Constant pressure	1005 J kg^{-1} K^{-1} or 1.005×10^7 ergs g^{-1} K^{-1}
Constant volume	718 J kg^{-1} K^{-1}
Specific heat capacity for water vapor:	
Constant pressure	1850 J kg^{-1} K^{-1}
Constant volume	1390 J kg^{-1} K^{-1}

Heats of transformation of phase of
    water:
    Vaporization

2.406×10^6 J kg^{-1} ($40°$C)
2.501×10^6 J kg^{-1} ($0°$C)
2.635×10^6 J kg^{-1} ($-50°$C)

Fusion 0.334×10^6 J kg^{-1} ($0°$C)

Sublimation 2.834×10^6 J kg^{-1} ($0°$C)
2.839×10^6 J kg^{-1} ($-30°$C)

Acceleration of Gravity

The value 980.616 cm sec^{-2} has been chosen by the WMO as the most representative of the acceleration of gravity at latitude $45°$. This is the value generally used by physicists and meteorologists, but in reducing mercurial barometers to standard gravity a conventional value of 980.665 cm sec^{-2} has been adopted. The sea-level value ranges from approximately 978 cm sec^{-2} at the equator to about 982 cm sec^{-2} at the poles. It decreases with height about 0.3 cm sec^{-2} per km in the free atmosphere.

Other Useful Constants

Joule's constant, or mechanical
    equivalent of heat (cal$_{15}$) 4.186×10^7 ergs cal^{-1} or 4.186 J cal^{-1}

Mean radius of the earth   6371.2 km

Mean solar distance 1.4968×10^8 km

Sidereal day   23.93447 hr or 23 hr 56'4.09"

Angular velocity of earth 2π rad per sidereal day or 7.29×10^{-5} sec^{-1}